XII International Scientific and Research Conference on Topical Issues in Aeronautics and Astronautics 2016

IOP Conference Series: Materials Science and Engineering
Volume 155

Krasnoyarsk, Russia
11 - 15 April 2016

ISBN: 978-1-5108-3380-7
ISSN: 1757-8981

Printed from e-media with permission by:

Curran Associates, Inc.
57 Morehouse Lane
Red Hook, NY 12571

Some format issues inherent in the e-media version may also appear in this print version.

Printed by Curran Associates, Inc. (2017)

For permission requests, please contact the Institute of Physics
at the address below.

Institute of Physics
Dirac House, Temple Back
Bristol BS1 6BE UK

Phone: 44 1 17 929 7481
Fax: 44 1 17 920 0979

techtracking@iop.org

Additional copies of this publication are available from:

Curran Associates, Inc.
57 Morehouse Lane
Red Hook, NY 12571 USA
Phone: 845-758-0400
Fax: 845-758-2633
Email: curran@proceedings.com
Web: www.proceedings.com

Table of contents

Volume 155

XII International Scientific and Research Conference "Topical Issues in Aeronautics and Astronautics"

11–15 April 2016, Krasnoyarsk, Russia

Accepted papers received: 20 October 2016
Published online: 23 November 2016

Preface

Papers

IOP Conf. Series: Materials Science and Engineering **155** (2016) 011001 doi:10.1088/1757-899X/155/1/011001

INTRODUCTION

International Scientific and Research Conference "Topical Issues in Aeronautics and Astronautics"

The International Scientific and Research Conference "Topical Issues in Aeronautics and Astronautics" is one of the most significant scientific conferences arranged by the Reshetnev Siberian State Aerospace University (SibSAU), the flagship university of the Krasnoyarsk Region (Russia).

The Conference is annually arranged in April and dedicated to the Cosmonautics Day, the anniversary of the first manned flight to space.

This Conference is arranged for undergraduate, graduate and postgraduate students, scientists and lecturers, as well as developers, designers and constructors representing leading companies and enterprises of the aerospace sector to give opportunities to present their projects, research work and results. The Conference is a great chance to connect scientists and highly-qualified and skilled specialists with a new community of future scientists and practitioners in the aerospace sector.

The Conference proceedings include papers presented by creative young specialists closely connected with aviation and space vehicles - design, production, problem-solving in space machine building and aerospace education, macro- and microeconomic development of the field, new approaches to solving philosophical and social problems, - experienced scientists and specialists, and all those who want to dedicate themselves to aeronautics and astronautics.

The selected papers are presented in these proceedings to share University research results, innovations and cutting-edge technologies with the international community to develop aeronautics and astronautics on a global scale.

Peer review statement

All papers published in this volume of *IOP Conference Series: Materials Science and Engineering* have been peer reviewed through processes administered by the proceedings Editors. Reviews were conducted by expert referees to the professional and scientific standards expected of a proceedings journal published by IOP Publishing.

Control of technical object on the basis of the multi-agent system with neuroevolution and student-teacher of-line learning

E A Engel[1], I V Kovalev[2], N E Engel[3]
[1] Associate Professor, First Russian doctoral degree in computer sciences, Katanov State University of Khakassia, Abakan, RF,
[2] Professor, Second Russian doctoral degree in computer sciences, Siberian State Aerospace University, Krasnoyarsk, RF,
[3] Student, Katanov State University of Khakassia, Abakan, RF

E-mail: Angel@khsu.ru

Abstract. This paper presents algorithm for generating neuroevolutionary multi-agent system that allows agents to learn from high-quality activities. Dissimilar traditional learning algorithms proposed algorithm combines student-teacher of-line learning and teaching agents based on sufficient activities producing by any agent in its subculture. The simulation studies demonstrated that the proposed algorithm is effective at rapidly generating near-optimal control agents.

Introduction

This research formalizes problem of technical object's control solving by multi-agent system with neuroevolution and student-teacher of-line learning. Unlike existing single-agent and team-search problems in proposed algorithm the agents collaborate through knowledge of other agents during the dynamic changes in the technical system [1]. Neuroevolution is evolution of neuronets, has arisen as a powerful approach to fulfilling intelligent activities for agents. Evolved networks take a number of simulated or real sensor signals as input, and generate a nonlinear mapping to outputs that define control signal. Recurrent neuronets integrate information over time, and generate output signals robustly under random perturbations and partially observable as opposed to standart value-function based reinforcement learning methods have trouble. Therefore, automatic design methods utilizing intelligent techniques [2-6] such as neuroevolution are effective in fulfilling intelligent agents. NeuroEvolution of Augmenting Topologies (NEAT [1]) is a method for evolving neuronets that adjusts both the topology and the weights as part of the learning process. NEAT has been shown to be successful in several open-ended design domains. Student-teacher techniques can be fulfilled in different ways. We are implemented of-line learning.

Nowadays around 70% of electric power is consumed by electric drives. Therefore, we are considering induction motor as an example of common technical system.

1. Model of the technical object's control on the basis of the multi-agent system with neuroevolution and student-teacher of-line learning

The multi-agent system consists of autonomous control agents evolved through neuroevolution and student-teacher of-line learning. Each control agent represents artificial neuro-controller. Fulfilling an operation of adding and pruning neurons and adjusting weights, NEAT evolves genomes that interpret as networks. The agent's success in control of technical system defines its fitness. Every generation reproduced networks with the highest fitness, while those with the lowest fitness are unlikely to do so. Therefore, NEAT provides genetic diversity through speciation and supports innovation through explicit fitness sharing. Fig.1 presents the proposed algorithm for generating the multi-agent system with neuroevolution and student-teacher of-line learning.

```
Input: generations; maxTimesteps
population ← InitializePopulation()
g ← 0
t ← 0
loop
        while g < generations do
        subcultures ← FormSubcultures(population)
                while t < maxTimesteps do
                        for each subculture in subcultures do
                                for each agent in subculture do
                                        s ← ReceiveAgentInputs()
                                        a ← GetResponse(agent; s)
                                        r ← TakeAction(a)
                                        agent.memory ← UpdateMemory({s; a; r})
                                end for
                                for each agent in subculture do
                                        if Acceptable(agent.memory) then
                                                for all observer in subculture do
                                                        Train(observer, agent.memory)
                                                end for
                                        end if
                                end for
                        end for
                        t ← t + 1
                end while
                population ← SelectAndReproduce(population)
                g ← g + 1
        end while
end loop
```

Figure 1. The proposed algorithm for generating the multi-agent system with neuroevolution and student-teacher of-line learning

During the generating the multi-agent system with neuroevolution and student-teacher of-line learning individuals within the population were ranked depend on the weighted sum of the fitness parameters. The experiment was conducted to measuring the performance of a monocultural algorithm of the multi-neuroagent search in Darwinian and Lamarckian evolutionary paradigms. There are two main categories of the genetic inheritance paradigms in evolution: Darwinian and Lamarckian. During Darwinian evolution, genomes are fixed and any knowledge or qualities shared during their lifetimes are not passed on to their offspring at birth. Unlike, in Lamarckian evolution an agent's genome adjusts as it learns during its life, and these changes are passed on to each of its offspring. In the context of neuroevolution, this corresponds to whether the changes in each agent's neuronet weights, as a result of social learning, are propagated to their genome at the end of the generation. Also experiment was conducted to compare the performance of the proposed algorithm for generating the multi-agent system with neuroevolution and student-teacher of-line learning with Darwinian, Lamarckian and simple neuroevolutionary algorithms.

2. Simulation and results

All the simulations for this study are implemented in MATLAB, Simulink. Figures 2 to 4 illustrate the simulation's results. Figure 2 shows the results of applying a monocultural algorithm of the neuroevolutionary multi-agent search to the induction motor's control in both the Lamarckian and Darwinian algorithms. The aforementioned algorithms quickly converged, but performance Darwinian algorithm better than Lamarckian. For on-line evolution Darwinian algorithm is likely to be preferred to Lamarckian in dynamic environments such as induction motor's control because adaptation is important. The adaptation is necessary for agents to control induction motor during the dynamic changes in the technical system and generate control signal robustly under random perturbations and partially observable.

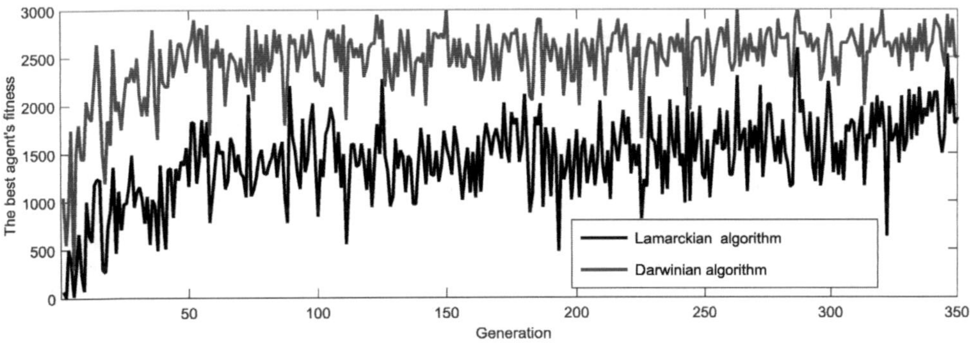

Figure 2. Plot of the performance Darwinian algorithm and Lamarckian algorithm respectively

Figure 3 shows the results of applying the proposed algorithm for generating the multi-agent system with neuroevolution and student-teacher of-line learning to the induction motor's control. Ultimately, in all aforementioned cases performance of the proposed algorithm converges to a higher score than that of simple neuroevolution, Darwinian and Lamarckian algorithms (Figure 3).

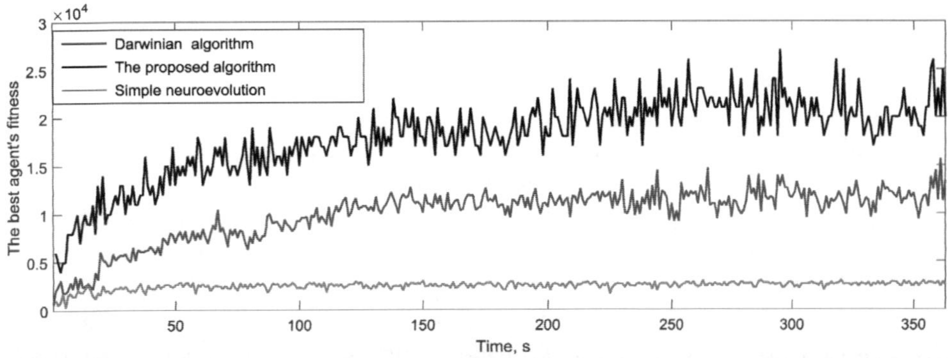

Figure 3. Plot of the performance Darwinian algorithm, the proposed algorithm for generating the multi-agent system with neuroevolution & student-teacher of-line learning and simple neuroevolution algorithm respectively

The proposed algorithm through evolution of recurrent neuronet generates a more effective intellectual model of the induction motor's control, as compared with the standard control model based on PI controller (Figure 4).

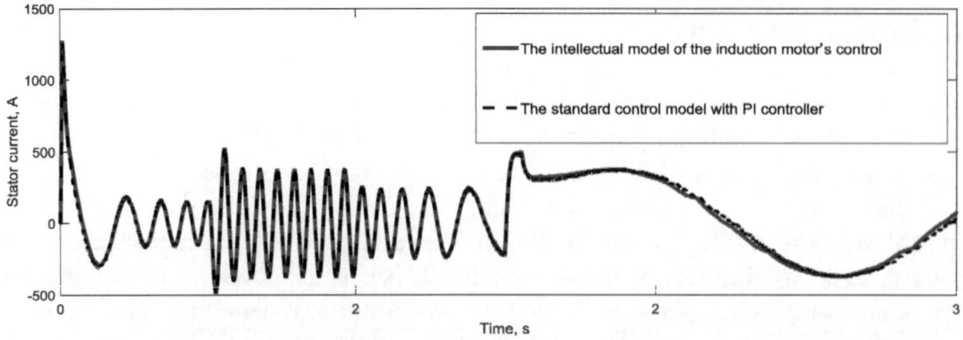

Figure 4. Plot of the control signal provided by the intellectual model of the induction motor's control and the standard control model with PI controller respectively

Extensive simulation studies on Simulink model have been carried out on different initial conditions, different disturbance profiles and variation in induction motor's control system parameters. It shows consistent performance has been achieved for the proposed algorithm on the basis of the recurrent neuronet with good stability and robustness, as compared with Darwinian, Lamarckian and simple neuroevolution algorithms.

3. Conclusions

This paper presented the algorithm for generating the multi-agent system with neuroevolution and student-teacher of-line learning that enables control agents to learn from high-quality activities. Unlike traditional social learning algorithms that follow a student-teacher model, proposed algorithm combines student-teacher of-line learning and teaching agents based on sufficient activities producing by any agent in its subculture. Extensive simulation studies demonstrated that consistent performance has been achieved for the proposed algorithm on the basis of the recurrent neuronet with good stability and robustness, as compared with Darwinian, Lamarckian and simple neuroevolution algorithms. The proposed algorithm generates a more effective intellectual model of the induction motor's control, as compared with the standard control model based on PI controller.

References

[1] Stanley K O and Miikkulainen R 2004 Competitive coevolution through evolutionary complexification *Journal of Artificial Intelligence Research* **21** pp 63-100

[2] Engel E A 2012 Solving control problems, decision making and information processing by fuzzy selective neural network *Information Technology* **2**

[3] Kovalev I, Zelenkov P, Ognerubov S, Bahmareva K, Denisova E 2015 The efficiency analysis of the automated plants *IOP Conference Series: Materials Science and Engineering* **70 (1)**

[4] Engel E, Kovalev I V, Kobezhicov V 2015 Intelligent control of non-linear dynamical system based on the adaptive neurocontroller *IOP Conference Series: Materials Science and Engineering* **94 (1)**

[5] Kovalev I V, Kovalev D I, Zelenkov P V, Voroshilova A A 2015 The pseudo-Boolean

IOP Conf. Series: Materials Science and Engineering **155** (2016) 012001 doi:10.1088/1757-899X/155/1/012001

optimization approach to form the N-version software structure *IOP Conference Series: Materials Science and Engineering* **94 (1)**
[6] Engel E, Kovalev I V, Karandeev D 2015 Energy-saving technology of vector controlled induction motor based on the adaptive neuro-controller *IOP Conference Series: Materials Science and Engineering* **94 (1)**
[7] Kovalev I, Zelenkov P, Ognerubov S 2014 The Efficiency Analysis of Automated Lines of Companies Based on DEA Method *Lecture Notes in Economics and Mathematical Systems* **675** pp 107
[8] Kovalev I V, Zelenkov P V, Ognerubov S The practical approach to the reliability analysis of the software architecture of a complex company control system *IOP Conference Series: Materials Science and Engineering* **94 (1)**

IOP Conf. Series: Materials Science and Engineering **155** (2016) 012002 doi:10.1088/1757-899X/155/1/012002

Model of interaction in Smart Grid on the basis of multi-agent system

E A Engel[1], I V Kovalev[2], N E Engel[3]
[1] Associate Professor, First Russian doctoral degree in computer sciences, Katanov State University of Khakassia, Abakan, RF,
[2] Professor, Second Russian doctoral degree in computer sciences, Siberian State Aerospace University, Krasnoyarsk, RF,
[3] Student, Katanov State University of Khakassia, Abakan, RF

E-mail: Angel@khsu.ru

Abstract. This paper presents model of interaction in Smart Grid on the basis of multi-agent system. The use of travelling waves in the multi-agent system describes the behavior of the Smart Grid from the local point, which is being the complement of the conventional approach. The simulation results show that the absorption of the wave in the distributed multi-agent systems is effectively simulated the interaction in Smart Grid.

Introduction

At the present time, multi-agent systems among others intelligent methods [1-7] fulfilled intelligent energy control and management systems in Smart Grid. Traditional power systems used Supervisory Control and Data Acquisition (SCADA) systems. SCADA systems provide communication architecture capable of controlling and maintaining power system hardware using certain signaling protocols. Many off-the-shelf SCADA are installed with vendor specific control protocols software which limits communication with different DER devices. This results in increased deployment costs. The use of the multi-agent systems in Smart Grid has recently been provided to overcome this disadvantage. Multi-agent systems offer various advantages over the SCADA system for the implementation of Smart Grid. Reactive agents are associated with the subsumption model of intelligence. The core property of reactive agents is that they do not perform reasoning through symbolic manipulation; instead they react to inputs from their environment and signals from other agents [8]. Ease of implementation is an advantage of this approach. It is urgent to consider the influence and time delay when agents issue control signals to control physical devices in Smart Grid during the commands execution. Agents must be able to rejoin to the signals sensed from the external environment quickly enough to control the Smart Grid in an appropriate time. These issues are considered during modeling the interaction in Smart Grid on the basis of the travelling waves in the distributed multi-agent system.

1. Model of the distributed multi-agent system based on wave's notation

The distributed multi-agent system consists of four agents, namely control agent (1), DER agent (2), user agent (3), central agent (4) and database agent (5). In this research we consider absorption of the wave in the distributed multi-agent system where the agents are not identical. The behavior of *n*th agent in a multi-agent system is described as

$$X_n(s) = W^d_{n,n+1}(s) + W^a_{n,n+1}(s) = W^d_{n,n-1}(s) + W^a_{n-1,n}(s),$$

IOP Conf. Series: Materials Science and Engineering **155** (2016) 012002 doi:10.1088/1757-899X/155/1/012002

where agents denoted *(n−1)* and *(n+1)* are neighbours of *n*th agent; $W^{a}_{n,n+1}(s) = G(s)W^{d}_{n,n+1}(s)$. The first lower index indicates the agent where the wave departs, while the second lower index indicates the agent where the wave arrives. The upper index denotes if the wave belongs to the departing ('*d*') or arriving ('*a*') agent. A complex boundary is assumed to be in the multi-agent system. Fig.1 presents the Simulink model of the distributed multi-agent system.

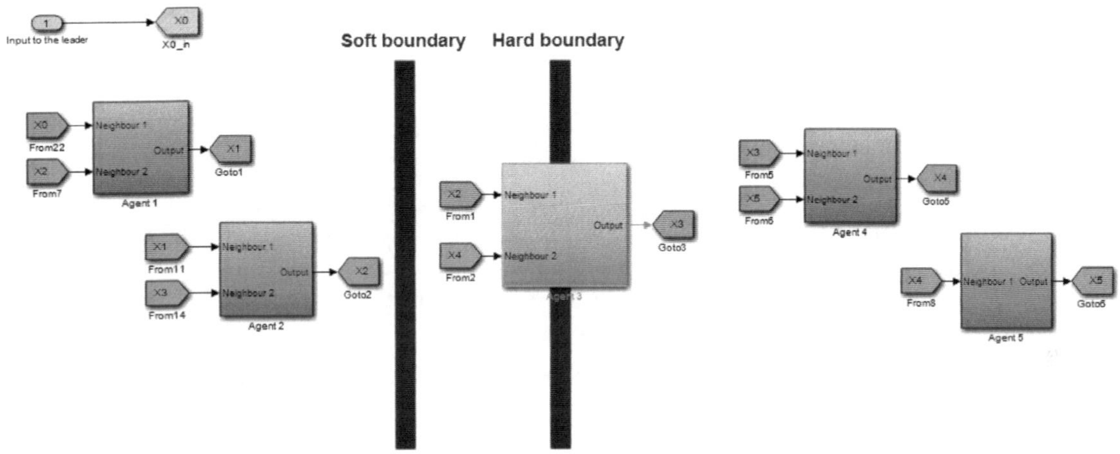

Figure 1. The Simulink model of the distributed multi-agent system

Time step for simulations is 0.005. The terminal simulation time is 50 s. Open loop transfer functions of the individual agents are specified in the symbolic form. We consider the simple model of interaction in Smart Grid on the basis of the travelling waves in the multi-agent system The front and rear node transfer functions of multi-agent system (1) respectively are defined as follows:

ol_tf_front = (5*s+5)/(s^2*(s+5)) and ol_tf_rear = (s+1)/(s^2*(s+3)). (1)

The front and rear node transfer functions of multi-agent system (2) respectively are defined as follows:

ol_tf_front = (5*s+7)/(s^2*(s+5)) and ol_tf_rear = (4*s+5)/(s^2*(s+4)). (2)

2. Simulation and results

All the simulations for this study are implemented in MATLAB, Simulink. The absorption of the wave in the distributed multi-agent systems is demonstrated through a simulation study of the Simulink model (Fig. 1). Figures 2 and 3 show an animation of the travelling waves from above-mentioned multi-agent systems (1) and (2).

IOP Conf. Series: Materials Science and Engineering **155** (2016) 012002 doi:10.1088/1757-899X/155/1/012002

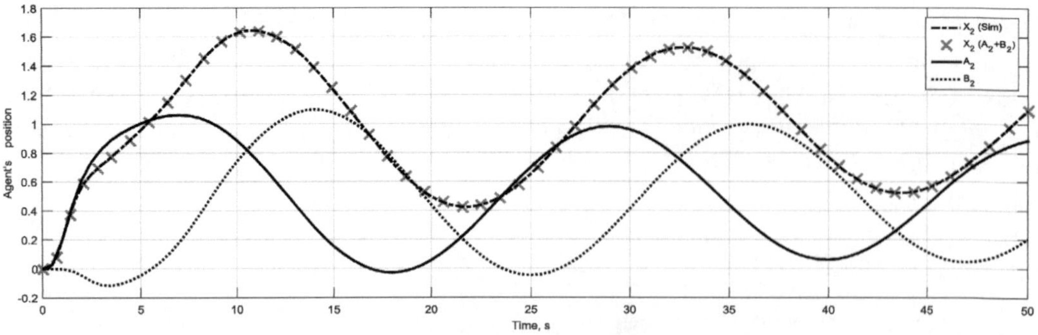

Figure 2. Plot of individual contributions of waves arriving from the neighboring agents of the distributed multi-agent system (1)

The wave travelling from the first agent to the rear-end agent is indicated by 'A'. The wave travelling in the opposite direction is indicated by 'B'. The derivation of the closed form formula for the agent's output necessitates taking into concern all the rejoins in the system, which is, in this case, troublesome.

The output of the multi-agent systems is computed by the following equations:

$$X\_2 = A\_2 + B\_2; \quad (3)$$

$$A\_2 = G^\wedge 2 * X\_ref - B\_2 * G^\wedge 4; \quad (4)$$

$$B\_2 = T\_ab * A\_2 + T\_bb * B_{\_3,L}; \quad (5)$$

$$B_{\_3,L} = bar\_TAB * A_{\_3,L} + bar\_TBB * B_{\_3,R}; \quad (6)$$

$$B_{\_3,R} = G^\wedge 5 * A_{\_3,R}; \quad (7)$$

$$A_{\_3,R} = bar\_TAA * A_{\_3,L} + bar\_TBA * B_{\_3,R}; \quad (8)$$

$$A_{\_3,L} = Taa * A\_2 + T\_ba * B_{\_3,L}; \quad (9)$$

In this study, the absorption of the wave in the distributed multi-agent systems (1) and (2) is compared under the same conditions. Figures 2 to 3 illustrate the simulation's results.

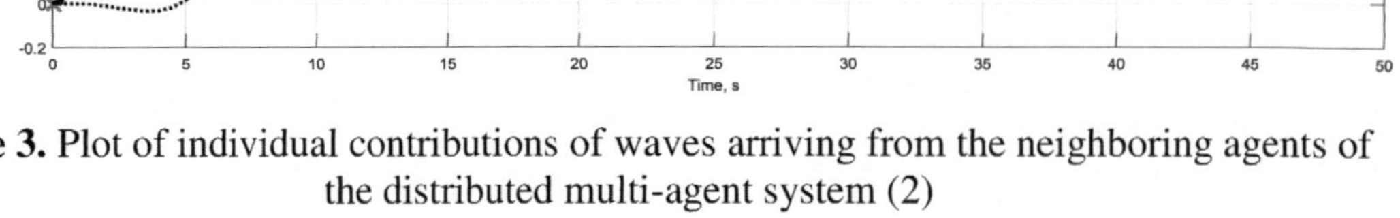

Figure 3. Plot of individual contributions of waves arriving from the neighboring agents of the distributed multi-agent system (2)

Extensive simulation studies on Simulink model have been carried out on different initial conditions, different disturbance profiles. It shows the wave similar behaviour, which is specifically plausible in a large scale Smart Grid.

3. Conclusions

Extensive simulation studies show that the absorption of the wave in the distributed multi-agent systems is effectively simulated the interaction in Smart Grid. Unlike traditional state space description of the multi agent system the important benefit of the Simulink model with travelling wave is that it explicates the comportment of particular agents. By the analysis of the travelling wave and its absorption on the agents, we can measure, how the output of one agent transmits in the multi-agent system.

References

[1] Engel E A 2012 Solving control problems, decision making and information processing by fuzzy selective neural network *Information Technology* **2**

[2] Kovalev I, Zelenkov P, Ognerubov S, Bahmareva K, Denisova E 2015 The efficiency analysis of the automated plants *IOP Conference Series: Materials Science and Engineering* **70 (1)**

[3] Engel E, Kovalev I V, Kobezhicov V 2015 Intelligent control of non-linear dynamical system based on the adaptive neurocontroller *IOP Conference Series: Materials Science and Engineering* **94 (1)**

[4] Kovalev I V, Kovalev D I, Zelenkov P V, Voroshilova A A 2015 The pseudo-Boolean optimization approach to form the N-version software structure *IOP Conference Series: Materials Science and Engineering* **94 (1)**

[5] Engel E, Kovalev I V, Karandeev D 2015 Energy-saving technology of vector controlled induction motor based on the adaptive neuro-controller *IOP Conference Series: Materials Science and Engineering* **94 (1)**

[6] Kovalev I, Zelenkov P, Ognerubov S 2014 The Efficiency Analysis of Automated Lines of Companies Based on DEA Method *Lecture Notes in Economics and Mathematical Systems* **675** pp 107

[7] Kovalev I V, Zelenkov P V, Ognerubov S The practical approach to the reliability analysis of the software architecture of a complex company control system *IOP Conference Series: Materials Science and Engineering* **94 (1)**

[8] Martinec D, Herman I, Hurak Z, Sebek M 2014 Wave-absorbing vehicular platoon controller *European Journal of Control* **20** pp 237-248

IOP Conf. Series: Materials Science and Engineering **155** (2016) 012003 doi:10.1088/1757-899X/155/1/012003

Technology platforms: opportunities and development perspectives

G P Belyakov, N T Avramchikova, G Y Belyakova, M N Chuvashova
Reshetnev Siberian State Aerospace University
31, Krasnoyarsk Rabochy Av., Krasnoyarsk, 660037, Russian Federation

E-mail: nikolesole@yandex.ru

Abstract. The term "technology platform" is used to define thematic scopes in terms of which technological priorities are formed concerning state, business, science and education cooperation that are aimed to solve the problem of state technological independence. The authors have examined the implementation of this tool so that to concentrate essential resources in priority-driven vectors of science and technological advancement of a guiding cluster of innovative technologies in the field of space applications.

The term "technology platform" refers to a number of sufficiently multifunctional opportunities with relation to its application. Particularly its functional qualities are revealed not only in economical science that is special to up-to-date publications on the basis of this field, but primarily in technological. The later traditionally use this term to define the fundamentals of some manufacturing operations, which are accomplished on the basis of exact progression.

Current trends and the processes of knowledge globalization have considerably expanded the scopes of application range of the concept being under study. At present stage, the concept of the technology platform constitutes a set of technologies characterized by an integral field of knowledge.

Having analyzed the main conditions and descriptions of domestic and foreign technology platforms it can be stated that they have a lot in common concerning the ways of formation and development. At the same time there are considerable differences, i.e. European technology platforms have their own background (TP date of birth - 1988) and developed innovative infrastructure. In Russia (date of birth 2010 - 2011) the tool formation of TP is taking place under tough conditions such as low structuredness, lack of valid and well-timed data and backup of R&D. Besides, budgetary funding has great importance. In this case, private business, which is formed on the basis of extractive industry branch revenues, banking and trading activity, maintains its traditional focus and extremely unwilling attitude towards grand and long-term investment projects connected with large-scale reindustrialization of the country. In Europe, project financing is initiated by private business and demands minimum costs from the state. In Russia, innovative activity of enterprises is approximately five times lower in comparison with EU countries.

Basically, the practice analysis of this term implementation in modern publications shows that nowadays the term "technology platforms" is a macroeconomic guideline of innovative management.

Contemporary definitions of the term "technology platforms", which are represented in Chart 1, allow asserting that it is substantially a matter of innovative infrastructure, which makes it possible to provide the integration of state, science and business for the unity and concentration of essential resources in priority areas of science and technological advancement of the country.

Table 1 – Macroeconomic content of "technology platform" definition

Defined by European Council	Defined by the Russian Federation Government
"It is the union of state, business, science and education representatives around the overall opinion of the trend in science and technological advancement and building general approaches for development and industrial assimilation of the corresponding technologies"	"Communicative tool aimed to activate the efforts in creation of commercial proprietary technology, new products (services) for extra fundraising to implement the researches and investigations on the basis of all the related parties contribution (business, science, state, civil society), regulatory and legal framework improvement in the field of science and technological as well as innovative development".

The analysis of the information represented in the chart makes it clear that in Russia the interpretation of the "technology platform" definition does not include such strategic aim as the overall opinion of the trend in science and technological advancement on the basis of general approaches for development and industrial assimilation of the corresponding technologies. This circumstance considerably reduces the opportunities of technology platforms in coordination the activities of all the entities in the long-term outlook.

In Russia, a mixed type of funding is supposed at the stage of decision making of the kind of technology platforms financial backing. Particularly, public contacts will be executed with the award-winning companies of open contests to fill orders in realization of research, development and experience efforts in terms of technology platforms. Besides, the maintenance of a flexible approach in the solution of this issue is possible that is financing of fundamental researches, less attractive for the business, could be mainly effected by the state. While, business community in its turn will execute financial investments into technological projects or R&D, which are attractive for the business from the point of view of their corporatization.

The implementation of this approach, which is considered to be much more effective in terms of existing practice in financing activity of clusters, will beyond question promote to the optimization of the technology platforms financing process that will result in the formation of conditions for their successful establishment, performance and further development.

Technology platforms feature as a sort of mechanism, designed to unite the efforts in formation of consolidated strategy of the trend development taking into account all the parties interests. The efficiency of projects, working on the basis of partnership approach of state, business and expert community, was clearly justified in the countries of the European Union. That is why European Technology Platforms, which have been implemented for decades, are chosen as a model for Russian technology platforms.

Technology platform represents a crucially new and quite complex instrument of national policy; it is the instrument of strategic rather than on-the-spot solution of existing problems, which must be disclosed in mid- and long-term plans.

Successful experience of technology platforms implementation encounters in innovative technologies cluster of CATU (Closed Administrative Territorial Unit) Zheleznogorsk, Krasnoyarsk region where the technology platform called "National information satellite system" (TP NISS), which is a kind of public-private partnership and a way of mobilization of stakeholders opportunities.

The strategic aim of TP "NISS" focuses on the creation of breakthrough technologies combination for:

– Drastic indexes increase in terms of user properties of new generation spacecrafts and availability of personal space packet service;
– Significant world market representation of advanced technology products and space, telecommunication services as well as other non-space branches of economy.

The aim of TP "NISS" is achieve such a level of consumer property in terms of national space communication cluster that would correspond to the interests of state, business and population.

Within a framework of the existing technology platform, a group of technologies will be developed in innovative technologies cluster in CATU Zheleznogorsk according to the following priority areas:

– Satellite construction. Unmanned spacecrafts: technologies of creation new generation aerospace equipment, technologies of information, control and navigating systems; space technologies of orbital attendance; technologies of creation energy efficient engines and drivers for transport systems;
– Microsystem electronics and professional equipment for space hardware: technologies of designing gadgets on the basis of micro- and nano-electromechanical systems; technologies of creation radiation-resistant computer-assisted participating databases of high reliability; technologies of high-performance autonomous energy systems creation including solar dynamics; technologies of micro-and nano-electromechanical systems.
– Advanced materials and technologies: technologies of modern versatile materials creation; nanotechnologies and nanocoating; technologies of creation and usage of composite materials; technologies of creation bulky flexible mechanical system; technologies of production specific optical, thermal, radio-technical blanketing of space vehicles external surfaces.
– Information and telecommunication systems: technologies of designing software and algorithmic software; technologies of math modeling of machine activity; technologies of math modeling of airborne antennas.
– Ground infrastructure improvement. Development of production and testing facilities: technologies of computer-aided testing; technologies of material and blanketing testing for space applicability; technologies of production metal nets out of gold-plated microwire with specific mechanical, radio-technical, optical and thermal characteristics to manufacture bulky flexible antennas; technologies of honeycomb panels production.
– Space services: technologies of broadband access to multimedia services; technologies of monitoring and forecasting of environmental status, prevention and clearance of its contaminations; technologies of prevention emergency situations of

natural and anthropogenic character; technologies of mining geophysics; technologies of monitoring and forecasting of atmosphere, hydrosphere, lithosphere and biosphere condition; technologies of creation GIS (geoinformation systems).

Summing up what has been said the technology platform, which is the instrument of research and technology as well as innovative policy used to promote innovative development and retooling of economy, takes part in the solution of business problems in order to elevate the availability and drastically extend user properties of Russian orbit group.

References

[1] Evaluation Best Practices and Results: The Advanced Technology Program *U.S. Department of Commerce, Technology Administration, NIST, Advanced Technology Program* 2005 URL: http://www.atp.nist.gov/eao/ir05-7174/nistir05-7174.pdf.

[2] Dirk Dohse. Cluster-Based Technology Policy – The German Experience *Industry and Innovation.* Vol. 14. No. 1. February 2007. P. 69–94; European Technology Platforms-2020. Draft Strategy. European Commission. Brussels. November 2012.

[3] Thierry VAUTRIN Innovation and Competitiveness clusters Policy in France. Mexico, March 6th 2009. URL: http://proinno.intrasoft.be/index.cfm?fuseaction=wiw.measures&page=detail&ID=8922.

[4] Avramchikova N T, Belyakova G Y, Proskurnin S D Organization forms of management of innovative development of local administrative-territorial formations (using the example of CATU Zheleznogorsk, Krasnoyarsk region). *Vestnik SibGAU* Vol. 16, No. **4**, pp 997–1006

The microstrip filters based on electromagnetic crystal of resonators 2d disposition*

B A Belyaev[1, 2], S A Khodenkov[1], Ya F Balva[1, 2], S S Aplesnin[1], O N Bandurina[1]
[1] Reshetnev Siberian State Aerospace University, 31 "Krasnoyarskiy Rabochiy" prospect, Krasnoyarsk, 660037, Russia.
[2] Kirensky Institute of Physics
50, stroenie № 38 «Akademgorodok», Krasnoyarsk, 660036, Russia.

E-mail: hsa-sibsau@mail.ru

Abstract. Filters of the 6th order, based on 2D electromagnetic crystals were developed. Application in structures of resonators with convolved and earthed on its base with strip conductors enables the implementation of filters having high frequency-selective properties and demanded in the aerospace equipment.

While developing and researching new structures of frequency-selective microwave devices [1] used in missile-space technology, developers traditionally try to increase their selective properties [2-3], to improve manufacturability, to reduce their size and to lower the cost of finished products.

This paper presents the bandpass filters with perspective microstrip resonators periodically arranged on two spatial coordinates that electromagnetically interact with each other.

It is necessary to note that in such resonator (Fig. 1) with original folded segments of strip conductors *2-1* connected together through general grounding *4* on the basis, due to a large step of wave resistance in line segments, the frequency of higher oscillation modes increase significantly.

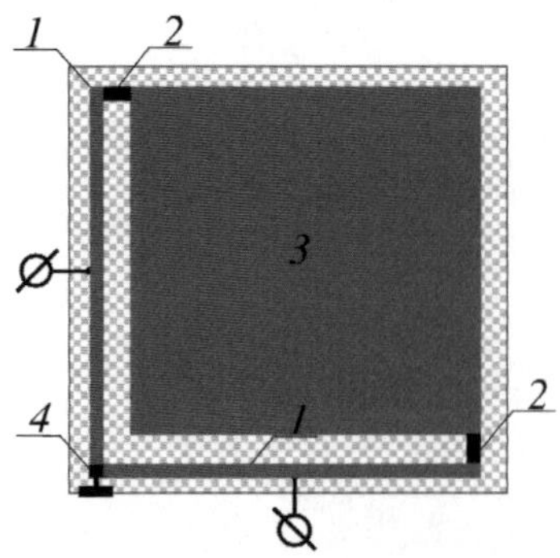

Figure 1. Topology of the strip conductors (black or gray) of microstrip resonator.

*Research was done supported by the Ministry of Education and Science of the Russian Federation, grant MK-9119.2016.8 of the President of the Russian Federation for State Support of Young Russian Scientists.

Parametric filters' synthesis on the basis of electromagnetic crystal was conducted by numerical electrodynamic analysis of 3D models in which ports with impedance of 50 ohms were conductively connected with the strip conductors of input and output resonators. For objective comparison of the characteristics of the devices, the same substrates with dielectric permittivity $\varepsilon = 9.8$ and thickness $h = 1$ mm (material - polykor) were used. Also, the center frequency of the bandwidth (BW) $f_0 \approx 1.0$ GHz and the relative width of the BW - $\Delta f/f_0 \approx 20\%$ were recorded.

The location of six resonators in two filters' structure is chosen in such a way that the inductive bond coupling between them prevails (Fig. 2). Herewith through holes *1* in the substrate are filled with a conductor grounded at one end on the base.

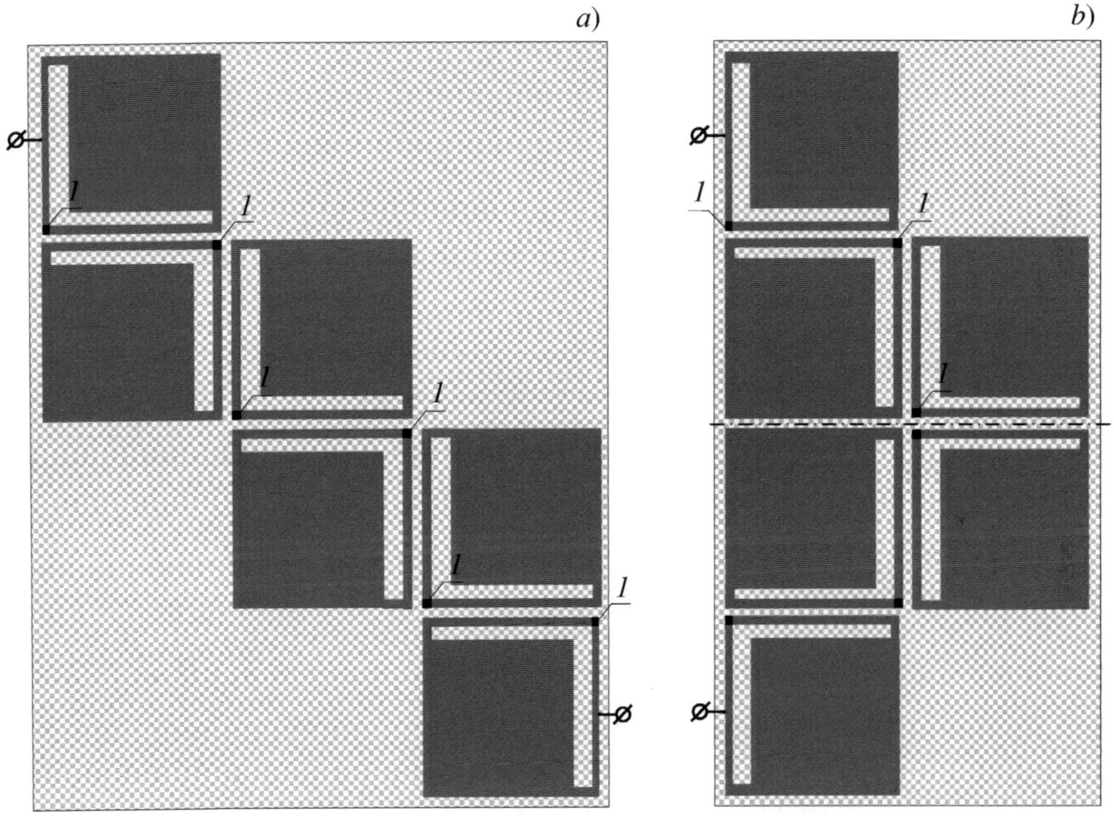

Figure 2. Topology of the 6th order filters strip conductors, based on 2D electromagnetic crystal. *a*) – with vertical displacement of resonators ranks, *b*) – with a defect of two absent outermost resonators.

In amplitude-frequency characteristics (AFCs) of the synthesized filter of 6th order one can be seen (Fig. 3) power attenuation poles, which increase the steepness of the slopes of the bandwidth and strengthen the suppression of power at low-frequency and extended high-frequency stop band.

IOP Conf. Series: Materials Science and Engineering **155** (2016) 012004 doi:10.1088/1757-899X/155/1/012004

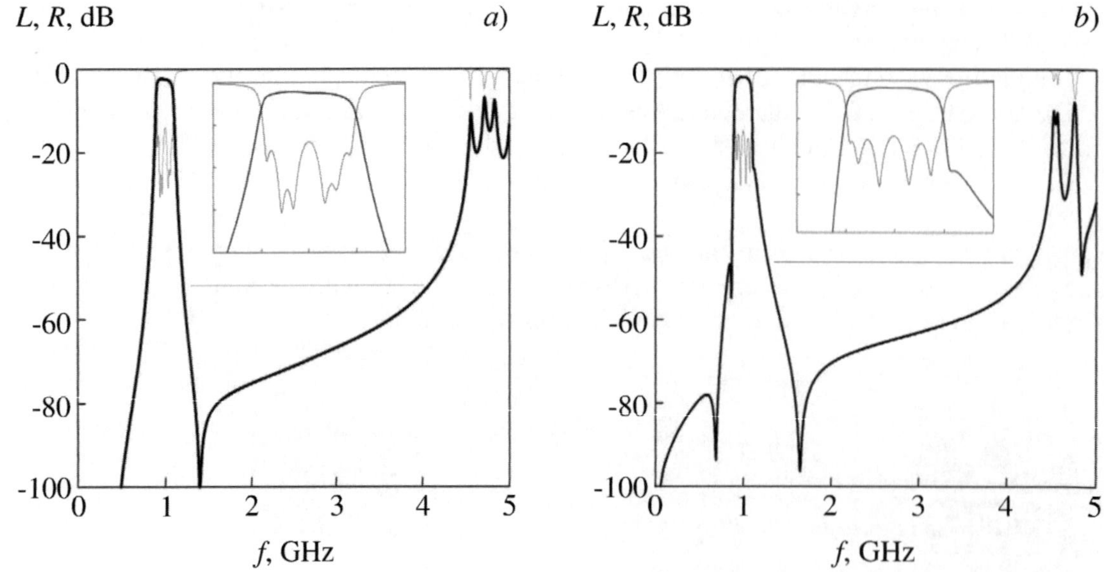

Figure 3. AFCs (in the wide and narrow band) of 6th orders filters. *a*) – used in calculations the topology of conductors presented in Fig. 2*a*, *b*) – in Fig. 2*b*.

In this connection, new structures of bandpass filters based on two-dimensional electromagnetic crystal are proposed. It has been shown that it is necessary to find such spatial configuration of the microstrip resonators with folded and grounded at the base of the strip conductors to make the propagation of electromagnetic waves occur consistently from one resonator to another. This allows realizing 6th order filters with high frequency-selective properties.

References

[1] Borisenkov D V *et al* 2015 *Materials XIX Intern. Scientific. Conf. «Reshetnev reading»* **1** 222.
[2] Belyaev B A *et al* 2013 *Microwave Opt. Technol. Lett.* **9** 2186.
[3] Khodenkov S A *et al* 2015 *Materials XIX Intern. Scientific. Conf. «Reshetnev reading»* **1** 260.

IOP Conf. Series: Materials Science and Engineering **155** (2016) 012005 doi:10.1088/1757-899X/155/1/012005

The microstrip wideband filter*

S A Khodenkov[1], B A Belyaev[1, 2], Ya F Balva[1, 2], S S Aplesnin[1], O N Bandurina[1]

[1] Reshetnev Siberian State Aerospace University, 31 "Krasnoyarskiy Rabochiy" prospect, Krasnoyarsk, 660037, Russia.

[2] Kirensky Institute of Physics, 50, stroenie № 38 «Akademgorodok», Krasnoyarsk, 660036, Russia.

E-mail: hsa-sibsau@mail.ru

Abstract. The filter of high frequency-selective properties is developed. The central six-mode resonator of the design which can be used in the aerospace equipment is electromagnetically connected with six single-mode resonators. The good agreement of the calculated data in comparison with the data received on the experimental model of a design is shown.

Recently microstrip multimode resonators attract the developers of frequency-selective microwave devices [1, 2] that are widely needed in space communication and navigation systems. First of all, it is connected with the possibility of a substantial reduction of structural dimensions by decreasing the number of resonators in them, moreover, without deterioration of frequency-selective properties. In such resonators due to the use of specific form of strip conductors, one can bring natural frequencies of the lowest n-oscillation modes together. As a result multimode resonator filter has N-order, by which, as it is well known [3, 4], its frequency selective properties are determined, exceeding the number of resonators in it in n-times.

In the present work, the structure of developed broadband microstrip filter (Fig. 1) in which central 6-mode resonator *4* is electromagnetically connected with six single-mode resonators *1-3* is given. Parametric filter synthesis was carried out with the help of numerical electrodynamic analysis of 3D models. The substrate with a dielectric constant $\varepsilon = 9.8$ and thickness $h = 1$ mm (material - polykor), while the center frequency of the bandwidth and its relative width were recorded as: $f_0 \approx 3.5$ GHz and $\Delta f/f_0 \approx 140\%$, respectively.

In this case the device adjustment was executed by «manual» parametric synthesis with the selection of the strip conductor portions topology geometrical sizes, i.e. length, width and gaps of regular areas, specified on Fig. 1 by positions numbers.

While 3D modeling, for added tuning of return loss level in the bandwidth a discontinuity of lines resistance $Z_1/Z_2 \approx 0.22$ was used. It is realized by introduction into the structure (perpendicular to the conductor segments *1* from each side) the two sequentially connected segments of strip conductors with wave resistance Z_1 and Z_2 (Fig. 1), respectively. In this case the first segment of this kind becomes a filter port. This makes the coordination of the structure with the microwave path easier.

*Research was done supported by the Ministry of Education and Science of the Russian Federation, grant MK-9119.2016.8 of the President of the Russian Federation for State Support of Young Russian Scientists.

IOP Conf. Series: Materials Science and Engineering **155** (2016) 012005 doi:10.1088/1757-899X/155/1/012005

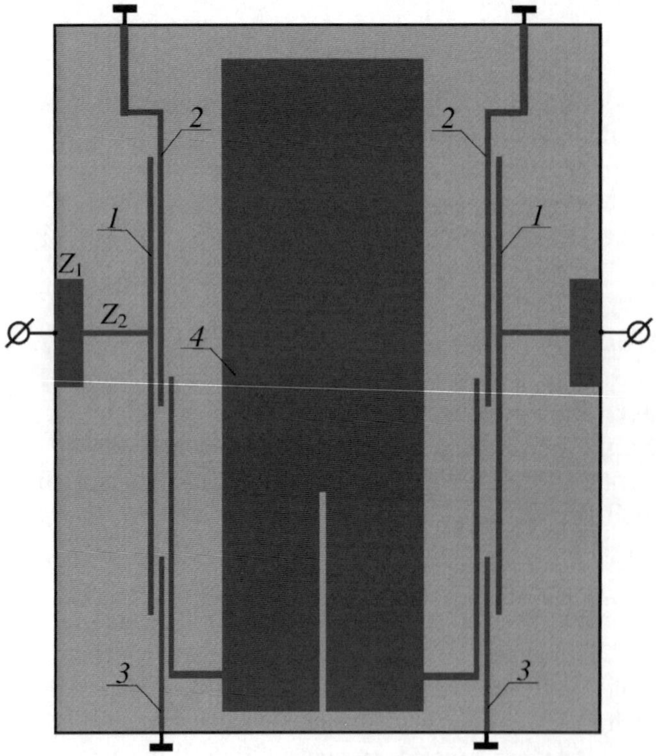

Figure 1. Topology of the strip conductors of microstrip wideband filter.

The geometrical dimensions of the experimental structure (Fig. 2), produced by photolithography, were as follows: 22.6×16.7×1.0 mm^3.

Figure 2. The photograph of fabricated filter.

IOP Conf. Series: Materials Science and Engineering **155** (2016) 012005 doi:10.1088/1757-899X/155/1/012005

In the amplitude-frequency characteristic (AFCs) of high order filter one can monitor power attenuation poles (Fig. 3), which increase the steepness of the slopes of the bandwidth and increase the power suppression at frequencies of stop band. Minimum losses in bandwidth do not exceed the value of -1.1 dB.

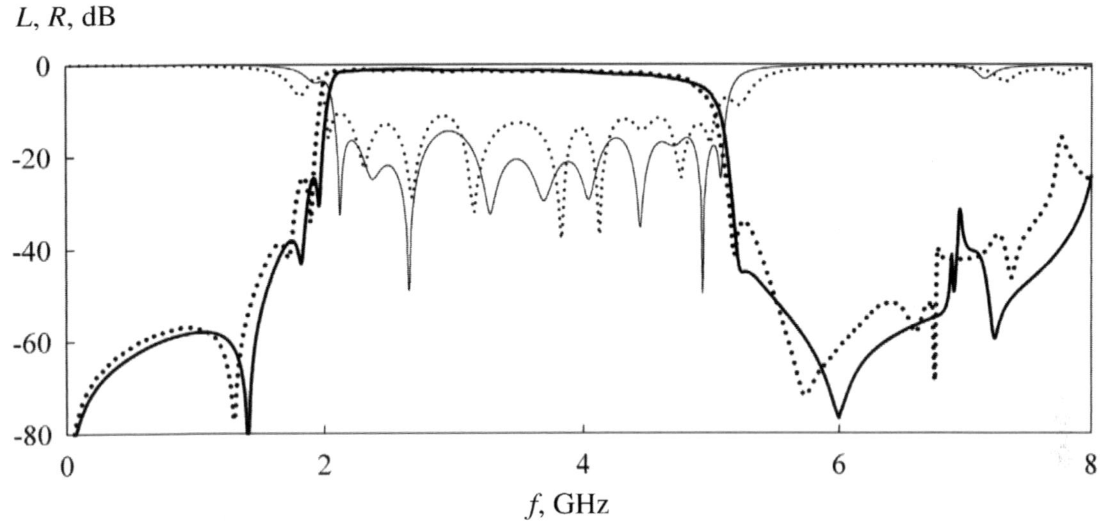

Figure 3. AFCs of microstrip wideband filter.
Lines are the calculation, and points are the measurements.

Thereby, a new structure of microstrip filter with a relative width of bandwidth $\Delta f/f_0 \approx 140\%$ is suggested. Due to the special form of strip conductors of interacting 6-mode resonator and single-mode resonators, one monitors high-frequency selective properties in the microwave device. Measurements of the test-piece of high-order filter characteristics gave good coincidence with the results of numerical electrodynamic analysis.

References

[1] Khodenkov S A *et al* 2014 *Materials All-Russian Scientific.–Practical. Conf. «Actual problems of aviation and cosmonautics»* 165.
[2] Crnojevic-Bengin V *et al* 2013 *Microwave Opt. Technol. Lett.* **55** 1440.
[3] Belyaev B A *et al* 2014 *Materials All-Russian Scientific.–Practical. Conf. «Actual problems of aviation and cosmonautics»* 158.
[4] Belyaev B A *et al* 2013 *Materials XVII Intern. Scientific. Conf. «Reshetnev reading»* **1** 209.

IOP Conf. Series: Materials Science and Engineering **155** (2016) 012006 doi:10.1088/1757-899X/155/1/012006

Conditions for creating perfectly secure systems*

M Styugin

Research fellow, Reshetnev Siberian State Aerospace University, Krasnoyarsk, Russia

E-mail: styugin@gmail.com

Abstract The present paper reviews a method for establishing secure information systems by complicating the possibility to research them for potential adversaries. A formalized model of a researcher and a definition of a research secure system are presented. A theorem for conditions required for creating a system perfectly secured from research. The Shannon's theorem of absolute security of perfect secrecy ciphers in cryptography is an instance of the theorem presented in the paper.

1. Introduction

Complexity of information systems has been constantly increasing lately. Both hardware and software complexity is increasing. This also influences the development of information security technologies. As vulnerabilities in complex systems cannot be detected and eliminated completely at the development stage, therefore information on the system's operating principles and algorithms has to be concealed. Technologies for such "concealment" are based on security methods for protecting information systems from research. In the last few years, there were numerous studies in this area and the number is growing every year.

In 2015 alone, there were over 150 articles on different methods of concealing data/processes in the field of information systems and malicious software. Some rather specific issues can be pointed out there, which are quite frequently reviewed in the researches including those presented at top-tier conferences.

One of the current trends is development of technologies, which complicate data gathering for adversaries by implementing Data Mining [1] and technologies for concealment of user actions. The most widely known research on that problem was conducted by Steven Alpern [2] and Shmuel Gal [3].

The technology of constant system modification to protect the system from research (Moving Target Defence - MTD) [4] and the technology for protection software code and algorithm should also be mentioned. Recently there were over 150 different MTD techniques [5] concerning such spheres as local area network security [6], protection from software code injection [7], protection from XSS attacks [8], protection from DDoS attacks [9], etc.

However, despite the avalanche-like growth of publications in that area there are still no exact mathematical methods for analysis and comparison of research secure systems. The existing mathematical models study only separate system properties to compare their effectiveness as in [10, 11]. Still there is no general model that would define ideal objects such as an absolutely research-secure system for example a definition of an absolutely

* The reported study was funded by RFBR according to the research project No. 16-29-09456 офи_м

indistinguishable data transmission channel in [12] actually does not provide a formal definition of its indistinguishability.

In order to create such theory, perfect secrecy conditions shall be established, which are implemented in cryptography as applied to information systems in general and formulate a general kind of Shannon Secrecy Theorem, for which the Shannon absolute secrecy theorem would be an instance.

2. Formalization of Objective for Protection against Research

The key feature of the present study is formalization of the objective for protection against research represented as aggregate of inputs, outputs and a discrete internal state. That kind of formalization allows introducing ideal objects such as a Perfectly Secure System and demonstrating feasibility of implementing some of the solutions in information security.

Conditions, in which an intruder may operate relatively to the targeted item, have been formalized earlier [13]. Assume that the intruder's targeted item has *f(par)* function, which connects the item's inputs and outputs, where *par* is a set of all possible arguments of function, i.e. $f(p_1, \ldots, p_n)$. Inputs of the item may also be parameters unknown, that is the current set of the item's inputs is selected from a set $<par> \in \{<par>\}$. Curly brackets will be used for indication of all possible values of a variable. In the latter case it denotes a set of all possible combinations of *<par>*, which the intruder can consider as the target's inputs. A set of input parameters may be divided into a $\{par\}^{v}$ set of parameters visible for the intruder and a $\{par\}^{nv}$ set of parameters invisible for the intruder.

$$\{par\} = \{par\}^{v} \cup \{par\}^{nv}, \{par\}^{v} \cap \{par\}^{nv} = \varnothing.$$

The set of the item's possible outputs is similar to the previous set:

$$\{f\} = \{f\}^{v} \cup \{f\}^{nv},\ \{f\}^{v} \cap \{f\}^{nv} = \varnothing.$$

Hence, the set of conditions in which an intruder may operate relatively to the targeted item can be represented with a tuple of thee values:

$$\{<par>' = <par>, par \in \{par\}^{v}, \mathrm{f} \in \{f\}^{v}\}.$$

Where *<par>'* is the number of variables, which the intruder selected as the item's "input", *par* is the current item's input state, *f* is the current item's output state. The above formula is slightly different from the one presented in [13]. Here it is represented in a structure, which is more convenient for conclusions below. The above tuple contains three conditions: an intruder may select the set of parameters for <par>' item's input incorrectly; the true set of parameters may not be a part of its ($\{par\}^{v}$) parameter visibility and the item's output may be outside the set of its functional visibility ($\{f\}^{v}$). The "worst" state of a targeted system for an intruder may be represented by the tuple below:

$$\{<par>' \neq <par>, par \notin \{par\}^{v}, f \notin \{f\}^{v}\}.$$

In that case the intruder will not even be able to set the objective for their activity as shown in [14].

Additional parameters need to be added to the tuple for further formalization:

$$\{f'(par') = f(par), f(par) \in \{f(par)\}^{v}, <par>' = <par>, par \in \{par\}^{v}, <f>' = <f>, f \in \{f\}^{v}\}.$$

Equality of $f'(par') = f(par)$ indicates that an intruder establishes $f'(par')$ function of *f(par)* target item correctly. As well as equality of $<f>' = <f>$ indicates that the intruder established the item's output $<f>'$ correctly. Condition of $f(par) \in \{f(par)\}^{v}$ indicates that the item's function is visible.

In addition it should be noted that from a practical perspective there is normally no point in differentiating between visibility of some of the variables and awareness of their values. In case a variable is visible, then its value is known. Therefore, only the property of being one of the visible elements shall remain in the tuple:

$$\{par \in \{par\}^{v}, f(par) \in \{f(par)\}^{v}, f \in \{f\}^{v}\}.$$

For the sake of notation convenience, compliance to the condition (membership) will be designated as 1 and non-compliance to the condition (not a part of the set) will be designated as 0. Therefore a system which is perfectly suitable for an intruder should be represented as (1, 1, 1). The above notation describes a system complying to conditions $\{par \in \{par\}^{v}$, $f(par) \in \{f(par)\}^{v}$, $f \in \{f\}^{v}\}$. First a definition for "intruder's objective" shall be stipulated. Intruder's objective is to bring a system to the state with values (1, 1, 1). Hence, a secure system is a system that cannot be induced into a state of (1, 1, 1). It should also be noted that in a real-life system when each two of the system parameters are visible it always implies visibility of a third parameter. For example, an input variable and a system's function are known, hence an output value can be found.

Therefore, a secure system can exist only in states with one visible variable. Such states have the structures shown below:

$$(0, 0, 1) - \{par \notin \{par\}^{v}, f(par) \notin \{f(par)\}^{v}, f \in \{f\}^{v}\}$$
$$(0, 1, 0) - \{par \notin \{par\}^{v}, f(par) \in \{f(par)\}^{v}, f \notin \{f\}^{v}\}$$
$$(1, 0, 0) - \{par \in \{par\}^{v}, f(par) \notin \{f(par)\}^{v}, f \notin \{f\}^{v}\}$$

Definition 1. *A secure system* is a system with one known (visible) parameter (input value, output value, input-output dependence function) that does not allow finding the other two system parameters.

Absolutely Secure System

We can represent a Perfectly Secure System with the Shannon Secrecy encryption principle [5] for each of the conditions.

Definition 2.1 *A Perfectly Secure System in state (0, 0, 1)* is a system in which conditions (1) are maintained:

$$\Pr(f(par) = \{f(par)\}) = \Pr(f(par) = \{f(par)\} \mid par = \{par\}),$$
$$\Pr(f(par) = \{f(par)\}) = \Pr(f(par) = \{f(par)\} \mid f = \{f\}).$$

Definition 2.2 *A Perfectly Secure System in state (0, 1, 0)* is a system in which conditions of (2) are maintained:

$$\Pr(f(par) = \{f(par)\}) = \Pr(f(par) = \{f(par)\} \mid par = \{par\}),$$
$$\Pr(f(par) = \{f(par)\}) = \Pr(f(par) = \{f(par)\} \mid f = \{f\}).$$

Definition 2.3 *A Perfectly Secure System in state (1, 0, 0)* is a system in which conditions of (3) are maintained:

$$\Pr(par = \{par\}) = \Pr(par = \{par\} \mid f = \{f\}),$$
$$\Pr(par = \{par\}) = \Pr(par = \{par\} \mid f(par) = \{f(par)\}).$$

Then it shall be defined in which conditions a system can exist in states (1), (2) and (3). They shall be designated as $S(1) = 1$ for fulfillment of conditions for (1) and $S(1) = 0$ for failure to fulfill conditions. $S(2)$ and $S(3)$ shall be designated in the same manner.

Theorem 1.1 $S(1) = 1 \Rightarrow |\{f(par)\}| \geq \max(|\{f\}|, |\{par\}|)$

Here the system can be in state $S(1)$ when a set of different functions between system input and output are not less than the maximum between sizes of input and output value sets.

Two conditions shall be reviewed for theorem proof.

1. When $|\{par\}| < |\{f\}|$ and let us review non-fulfillment of the first part of the condition:

$$|\{f(par)\}| < |\{f\}|$$

Set $\{f_{\widetilde{par}}\}$ shall be defined so that it would include all values of f from a specific $\widetilde{par}$ parameter by enumeration of all possible functions of $\{f(par)\}$. The following is obtained

$$\{f_{\widetilde{par}}\} \stackrel{\text{def}}{=} \{\tilde{f} | \tilde{f} = f(\widetilde{par}) \text{for some } f(par) \in \{f(par)\}\}.$$

It is evident that $|\{f_{\widetilde{par}}\}| \leq |\{f(par)\}|$. Hence, it may be concluded that there is at least one $f' \in \{f\}$, which is such that $f' \notin \{f_{\widetilde{par}}\}$. Thus,

$$\Pr(\{f\} = f' | \{par\} = \widetilde{par}) = 0 \neq \Pr(\{f\} = f').$$

The obtained expression contradicts S(1) condition, which stipulates that

$$\Pr(f = \{f\}) = \Pr(f = \{f\} \mid par = \{par\}).$$

2. Here is a case when $|\{par\}| > |\{f\}|$ and the first part of the condition is not observed:

$$|\{f(par)\}| < |\{par\}|$$

The set of $\{par_{\tilde{f}}\}$ shall be defined so that it would include all the values of *par,* which can be used for obtaining the result of $\tilde{f}$ by enumerating all possible functions of $\{f(par)\}$. As a result the following is obtained:

$$\{par_{\tilde{f}}\} \stackrel{\text{def}}{=} \{\widetilde{par} | f(\widetilde{par}) = \tilde{f} \text{for some } f(par) \in \{f(par)\}\}.$$

It is evident that $|\{par_{\tilde{f}}\}| \leq |\{f(par)\}|$. Hence, there is at least one $par' \in \{par\}$, which is so that $par' \notin \{par_{\tilde{f}}\}$. Thus,

$$\Pr(\{par\} = par' | \{f\} = \tilde{f}) = 0 \neq \Pr(\{par\} = par').$$

The obtained expression contradicts S(1) condition, which stipulates that

$$\Pr(f = \{f\}) = \Pr(f = \{f\} \mid par = \{par\}),$$

which is equivalent to

$$\Pr(par = \{par\}) = \Pr(par = \{par\} \mid f = \{f\}).$$

Theorem 1.2 $S(2) = 1 \Rightarrow |\{par\}| \geq |\{f\}|$.

This theorem assumes that a system can exist in state *S*(2) when size of the set with potential system's inputs is greater or equal to the size of the system's input set.

Proof is similar to Theorem 1.1

Theorem 1.3 $S(3) = 1 \Rightarrow |\{f(par)\}| \geq \max(|\{f\}|, |\{par\}|)$

This theorem assumes that the system in state *S*(3) shall have a size of possible outputs set that is not less than the maximum value of size of potential system's input and output values.

Proof similar to Theorem 1.1

All the above conditions are necessary, but they are not enough for establishing a Perfectly Secure System.

According to the commonly used terminology a term "uniform" will be used for denoting an absolutely random function or a bit chain where each subsequent bit can be predicted with a probability not higher than ½ regardless of computational asymptotic complexity.

Here is a definition for the uniform-function

$f(par) -$ uniform $\Leftrightarrow \forall par\colon \Pr(f) = 1/|\{f\}|$.

That means that a function is uniform when it is established in such a way that any output function value for an intruder is equally probable. That is the function itself is not probabilistic, but a determinate one. However, it should be randomly selected hence it does not provide any information to an intruder.

Theorem 2.1

1. When $f(par)$ is uniform and $|\{f\}| \geq |\{par\}|$, then $S(1) = 1$ then and only then when for any $par \in \{par\}$ there are $f_1 \dots f_n$, where for each $f_i, i = 1 \dots n$ there are $f_i, i = 1 \dots n$, in

which each $f_i, i = 1 \dots n$ there are c, where $m = \frac{|\{f(par)\}|}{|\{f\}|}$ so that $f_i = f_k^i(par), i = 1 \dots n, k = 1 \dots m$.

2. When $f(par)$ is uniform and $|\{f\}| \leq |\{par\}|$ then $S(1) = 1$, then and only then when for any $f \in \{f\}$ there are $par_1 \dots par_n$, where for each $par_i, i = 1 \dots n$ there are $f_1^i(par_i), \dots, f_m^i(par_i)$, where $m = \frac{|\{f(par)\}|}{|\{par\}|}$ so that $f = f_k^i(par_i), i = 1 \dots n, k = 1 \dots m$.

The first part of the theorem asserts that for each *par* there are *n* function results and for each parameter and function result there are *m* functions, that can be matched with them. The second part of the theorem asserts that for each *f* output value there are *n* different parameters, which can be used for obtaining it by *m* different functions.

Formal proof of the above theorem is quite lengthy, so only the general principle of its development is presented here. Let's review the first part of the theorem, when $|\{f\}| \geq |\{par\}|$. In this case there shall be as many functions for transforming *par* into *f* to have the opportunity to obtain each value of *f* from each *par*. Otherwise, it will result in non-fulfillment of *S*(1) condition. When the number of functions is sufficient to obtain each *f* from each *par* then they also shall be evenly distributed throughout all the pairs to make sure that probability of a pair's occurrence is not higher than the occurrence probability for the other pairs, which also causes violation of *S*(1) condition. Hence, the number of functions *f(par)* shall be divisible by the number of output results designated by *f*.

The second part of the theorem has a similar proof.

Theorem 2.2

When *par* is uniform then *S*(2) = 1, then and only then, when for each *par* and for each *f* there is a function $f_1(par), \dots, f_m(par)$, where $m = \frac{|\{f(par)\}|}{|\{par\}|}$ which are such that $f = f_k(par),\ k = 1 \dots m$.

Proof is similar to Theorem 2.1.

Theorem 2.3

1. When *par* is uniform and $|\{f\}| \geq |\{par\}|$ then *S*(3) = 1, then and only then, when for each $par \in \{par\}$ there are $f_1 \dots f_n$, where for each $f_i, i = 1 \dots n$ there are $f_1^i(par), \dots, f_m^i(par)$, where $m = \frac{|\{f(par)\}|}{|\{f\}|}$ are such that $f_i = f_k^i(par), i = 1 \dots n, k = 1 \dots m$.

2. When *par* is uniform and $|\{f\}| \leq |\{par\}|$ then $S(3) = 1$ then and only then, when for each $f \in \{f\}$ there are $par_1 \dots par_n$, where for each $par_i, i = 1 \dots n$ there are $f_1^i(par_i), \dots, f_m^i(par_i)$, where $m = \frac{|\{f(par)\}|}{|\{par\}|}$ are such that $f = f_k^i(par_i), i = 1 \dots n, k = 1 \dots m$.

Proof is similar to Theorem 2.1.

Conclusion

The paper presented a formalization of an information system researcher including its limitations. A research secure system and an absolutely research secure system can be defined based on the researcher's information limitations. A generalized theorem formulating the conditions required and sufficient for an absolutely research secure system was presented. The Shannon absolute secrecy theorem in cryptography is an instance of the presented theorem.

The created model confirms the result provided in [12]. The obtained theoretic results can be used for designing indistinguishable information systems when it is required to complicate research for an external agent.

References

[1] Chengy P, Lee I, Pan J-S and Lin C.-W., Roddick J F 2015 Hide association rules with fewer side effects. *IEICE Transactions on Information and Systems.* Volume E98D, Issue **10**, Pages 1788-1798
[2] S. Alpern and Gal S 2002 Searching for an agent who may or may not want to be found. *Operations Research*, 50(2):311–323
[3] Gal S 2001 On the optimality of a simple strategy for searching graphs. *International Journal of Game Theory*, **29(4)** :533–542
[4] Jajodia S 2011 Moving Target Defense. Creating Asymmetric Uncertainty for Cyber Threats. Series: Advances in Information Securit . - 184 p
[5] Okhravi H, Hobson T, Bigelow D and Streilein W 2014 Finding Focus in the Blur of Moving-Target Techniques. *IEEE Security and Privacy*. vol. **12** pp 16-26
[6] Carvalho M. and Ford R. 2014. Moving-target defenses for computer networks. *IEEE Security and Privacy*. Vol. **12(2)** pp 73-76
[7] Larsen P, Brunthaler S and Franz M 2015 Automatic Software Diversity. *IEEE Security and Privacy*. Vol. **13 (2)** pp 30-37
[8] Portner J, Kerr J and Chu B 2015 Moving target defense against cross-site scripting attacks. *Lecture Notes in Computer Science (including subseries Lecture Notes in Artificial Intelligence and Lecture Notes in Bioinformatics)*. Vol. **8930** pp 85-91
[9] Ma D, Xu Z and Lin D 2015 Defending blind DDoS attack on SDN based on moving target defense. *Lecture Notes of the Institute for Computer Sciences, Social-Informatics and Telecommunications Engineering, LNICST*. Vol. **152** pp 463-480
[10] Jin B Hong and Dong Seong Kim 2014 Scalable Security Models for Assessing Effectiveness of Moving Target Defenses. *Dependable Systems and Networks (DSN), 44th Annual IEEE/IFIP International Conference*. pp 515 - 526
[11] Jin Hong and Dong-Seong Kim 2012 HARMs: Hierarchical Attack Representation Models for Network Security Analysis. *Originally published in the Proceedings of the 10th Australian Information Security Management Conference*, Novotel Langley Hotel, Perth, Western Australia, 3rd-5th December, 2012
[12] Mikhail Styugin. 2015. Absolutely Indiscernible Data Transfer Channel. *Proceedings of The 14th European Conference on Cyber Warfare and Security (ECCWS-2015)*. pp 286-293
[13] Styugin M 2014 Protection against System Research. *Cybernetics and Systems: An International Journal*. Vol. **45 (4)** pp 362-372
[14] Katz J and Lindell Y 2014 Introduction to Modern Cryptography, Second Edition. N.Y.: Chapman and Hall/CRC. 603 p

IOP Conf. Series: Materials Science and Engineering **155** (2016) 012007 doi:10.1088/1757-899X/155/1/012007

The automated system for technological process of spacecraft's waveguide paths soldering

V S Tynchenko, A V Murygin, O A Emilova, A N Bocharov, V D Laptenok

Reshetnev Siberian State Aerospace University 31, Krasnoyarsky Rabochy Av., Krasnoyarsk, 660037, Russian Federation

E-mail: vadimond@mail.ru

Abstract. The paper solves the problem of automated process control of space vehicles waveguide paths soldering by means of induction heating. The peculiarities of the induction soldering process are analyzed and necessity of information-control system automation is identified. The developed automated system makes the control of the product heating process, by varying the power supplied to the inductor, on the basis of information about the soldering zone temperature, and stabilizing the temperature in a narrow range above the melting point of the solder but below the melting point of the waveguide. This allows the soldering process automating to improve the quality of the waveguides and eliminate burn-troughs. The article shows a block diagram of a software system consisting of five modules, and describes the main algorithm of its work. Also there is a description of the waveguide paths automated soldering system operation, for explaining the basic functions and limitations of the system. The developed software allows setting of the measurement equipment, setting and changing parameters of the soldering process, as well as view graphs of temperatures recorded by the system. There is shown the results of experimental studies that prove high quality of soldering process control and the system applicability to the tasks of automation.

1. Introduction

Waveguide systems are widely used in spacecraft, ensuring apparatus and its on-Board equipment. The thin-walled tubes made of aluminum alloy with rectangular cross section are used as a waveguide system in JSC "Information satellite systems" named after academician M. F. Reshetnev (JSC "ISS") of the city Zheleznogorsk, Krasnoyarsk Krai, Russian Federation. The induction soldering is used for production of waveguide systems in JSC "ISS" [1, 2]. However, induction soldering of thin-walled waveguides is a complex process with a set of features [3].

Firstly, the melting temperature of silumin solder is close to the melting temperature of the base material. Secondly, the components of the waveguide systems such as a waveguide pipe and flange are independent systems until solder between them will not be melted [4]. Thus, there is practically no heat between the elements, and their heating occurs independently from each other. Thirdly, it was found that during induction heating of the waveguide paths zones of the maximum heat generation in the sections of elements (maximum density of induced eddy currents) do not coincide with the zones of the soldering [5].

IOP Conf. Series: Materials Science and Engineering **155** (2016) 012007 doi:10.1088/1757-899X/155/1/012007

Besides, a large range of brazed products with different mass-dimensional parameters leads to the difference by mass of the same name products to 25%. Also the elements of the waveguide paths do not change their color during heated, which eliminates visual inspection, of brazed products local overheating [6].

Listed factors significantly complicate the reproduction of the waveguide path induction soldering technological process. There are requirements to positioning accuracy, to the inductor form, to the generator equipment parameters and to the highly skilled personnel [7, 8].

The analysis of the above listed features shows the need for technological process automation and led to the development of a hardware-software complex, which allows accurately reproduce the process. Developed automated system produces the control of the product heating process, adjusting the power, which is delivered to the inductor. The control in system is based on the information about the temperature from the soldering zone [9].

2. Software structure and algorithms

The software product of induction soldering process automation, is a Windows application, which can be used in operation systems such as Windows XP/7/8/8.1/10.

Software product developed in C++ programming language using an object-oriented approach, which allows to make a modern and scalable applications [10].

The system works with COM ports, which is realized by WinApi [11, 12, 13], work with the PCI1710 card by implementation of freely distributed library bdaqctrl.h [14].

Structural scheme of software product is shown in figure 1.

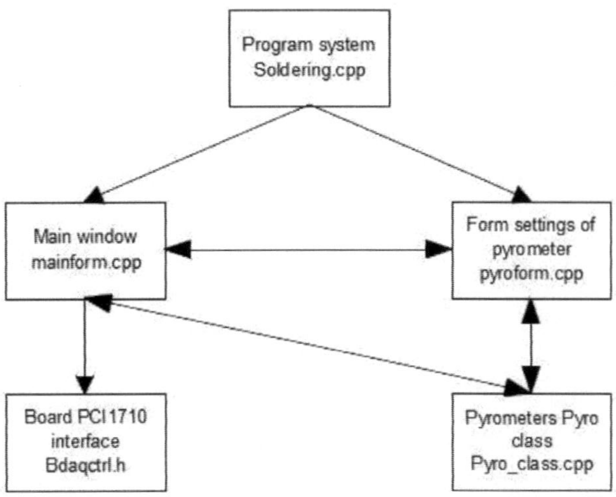

Figure 1. Scheme of the software product

The software system project consists of five source code files:

1. soldering.cpp – contains the main program function WinMain(), which starts the program system;
2. mainform.cpp – contains algorithms of the application main window;
3. pyroform.cpp – contains pyrometers information forms;

4. bdaqctrl.h – contains the PCI1710 card interface, which is used for thermocouple connection and generator control;

5. Pyro_class.cpp contains the description of the pyrometers class, algorithms of interaction with them, as well as algorithms for the pyrometers adjustment.

A single custom class Pyro is implemented to store all information about pyrometers, their connection and parameters.

By using this class, it is possible to produce:

1) the switching on of the pyrometer laser;
2) the switching off of the pyrometer laser;
3) query of the pyrometer laser state (0 - disabled, 1 - enabled);
4) request of the thermometer serial number;
5) request of the minimum temperature received by the pyrometer;
6) request of the maximum temperature received by the pyrometer;
7) request of the current temperature measured;
8) the request of the pyrometer emissivity parameter;
9) the setting of the pyrometer emissivity parameter;
10) the automatic calculation of emissivity parameter.

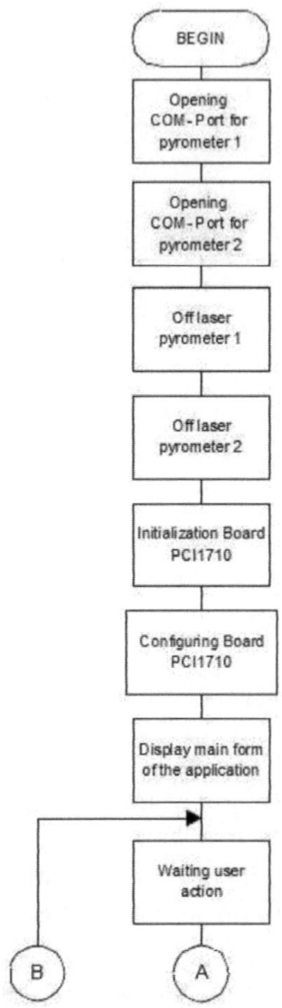

Figure 2. The block diagram of the software system.

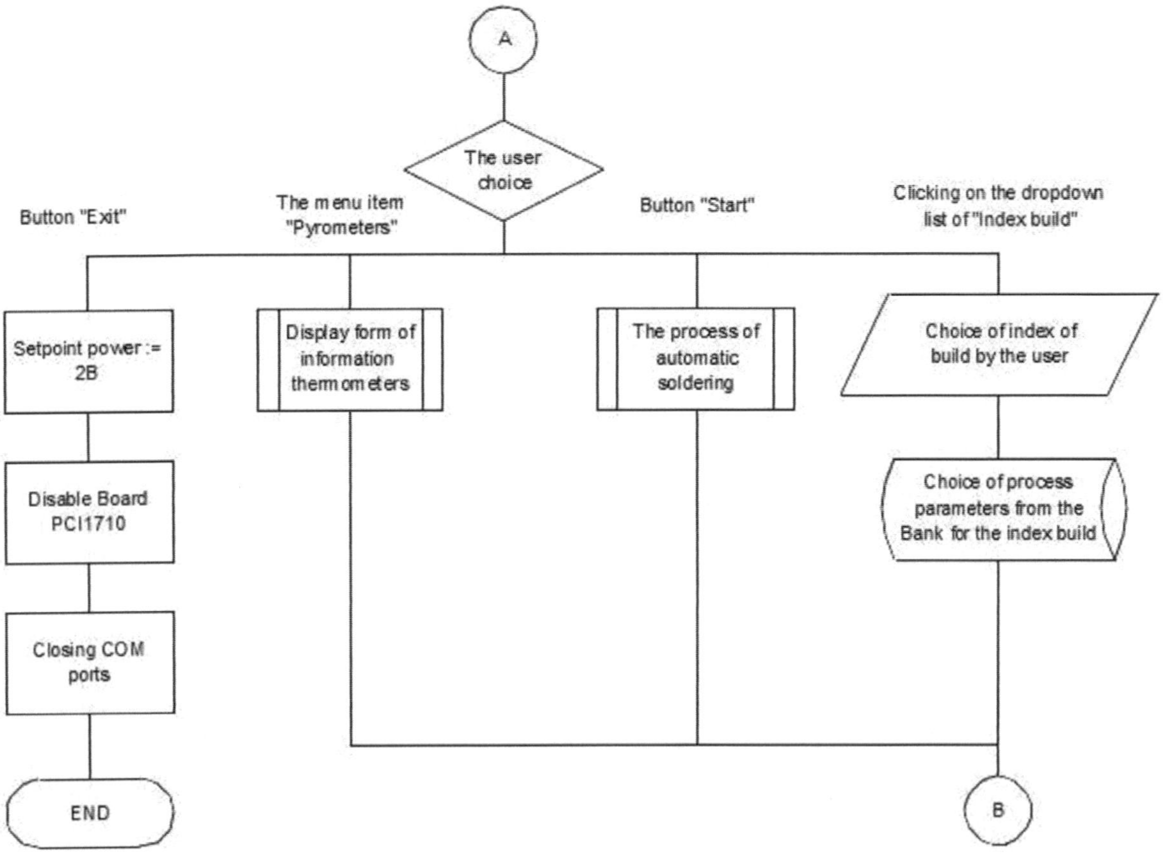

Figure 2 (cont.). General block diagram of the of the application operation

From the block scheme it is obvious that initialization of all data structures, ports and interfaces configurations are made on the first phase. Also, before the application main window occurs, the disconnection of the pyrometers lasers is made, because laser significantly affects the measurement accuracy.

During the application closing all ports will be closed, the PCI1710 card will be deactivated and the inductor will be switched off.

At the initial stage before the form displaying on the screen sequential survey of the pyrometers is conducted and the following information from them is receiving:

- serial numbers;
- the minimum temperature measured pyrometer;
- maximum temperature measured pyrometer;
- coefficient of the material emissivity that is installed on the thermometer.

Further, all information is displayed on the appropriate fields of the form.

When you click on the "START" button the automatic soldering process starts, during which product heating control based on the proportional controller is carried out [15, 16] with aim to maintain the required heating rate [17].

Figure 3 shows a block diagram of the soldering process control.

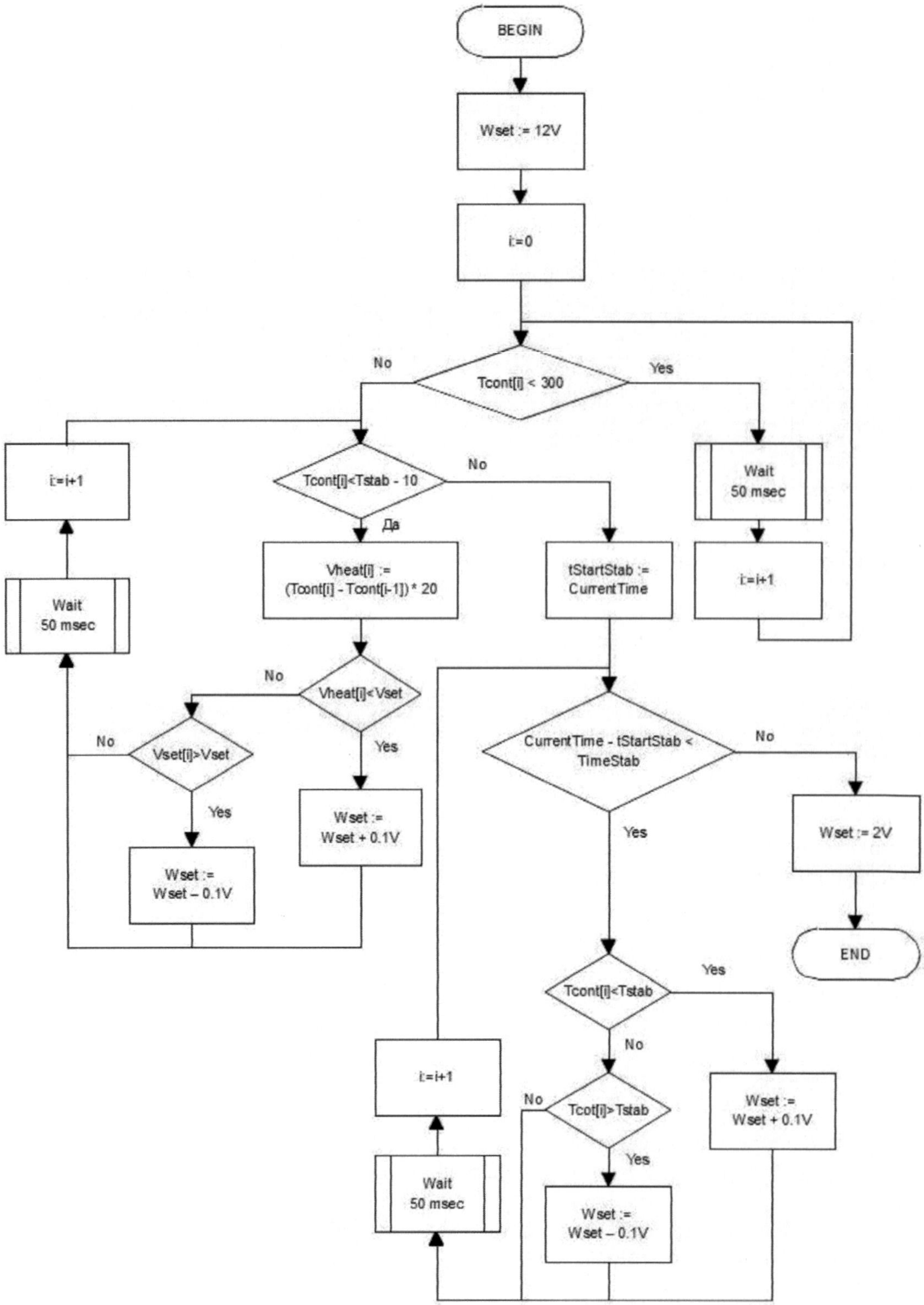

Figure 3. The Block diagram of the soldering process control algorithm

Here:

– Wset – setting power of the generator in volts. Changes from 2V to 12V, where 2V is 0% power generator, 12V is 100% capacity;

– Tcont – temperature value obtained from the pyrometer, introduced in the control loop;

– Tstab – stabilization temperature value, which is required for the solder melting and specified as technological parameter of the process;

– Vheat – heating rate of the product;

– Vset – required heating rate, which is necessary to maintain. Is specified as technological parameter of the process;

– tStartStab – the time stamp when the system begin the stabilization process;

– TimeStab – the time during which is necessary to sustain the product when T_{stab}. Is specified as a parameter of the process.

This management scheme implemented on the base of the proportional regulator, allows effective control for 20 times per second. In a series of experiments, it was found that this controlling actions discretization is sufficient for effective management of the induction soldering process.

3. The automated system description

The system provides two access levels:

- operator access which allows to control the equipment and to view a database of the waveguides brazing technological parameters;
- technologist access which allows to view and edit a database of the waveguides soldering technological parameters and to control equipment complex.

The main window of the automated system in the operator regime is shown in figure 4.

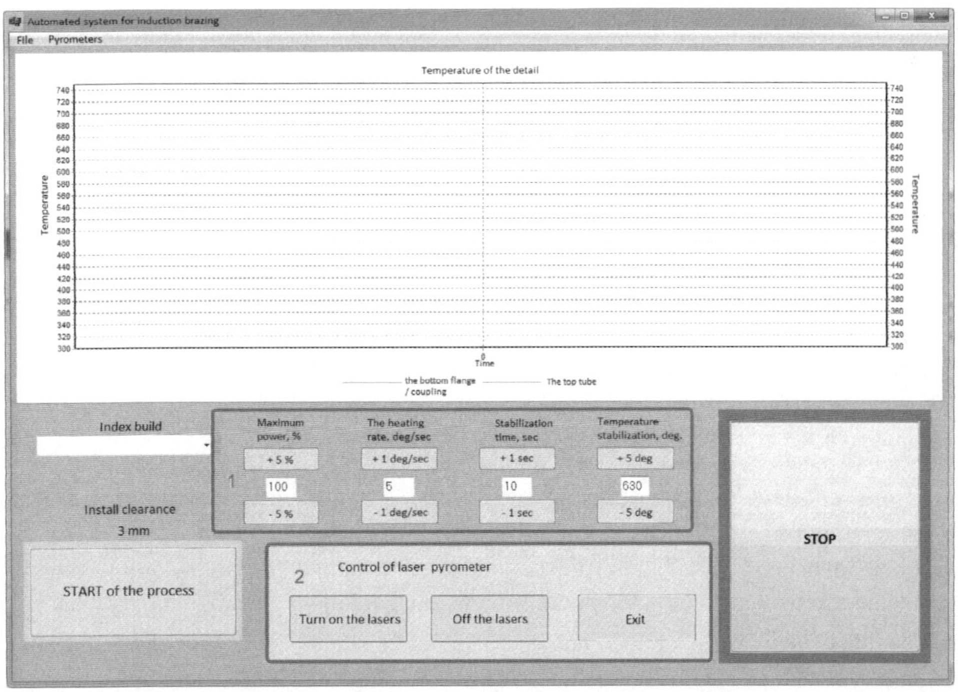

Figure 4. The Main window of the software system (operator)

In this window the user has several possibilities.

First, it is the view of the heating process graph.

Also, there are a number of control elements grouped by functionality. They are highlighted in the image frames and marked with numbers 1 and 2. Here, the first control elements group allows to set soldering parameters, the second group – the control buttons for the pyrometers laser.

In the emergency heat mode it is possible to shutdown the system by pressing the button "STOP".

Also, operator has to choose the technological process parameters from the database before soldering.

The technologist additionally has the opportunity to manage database of the soldering technological parameters (figure 5). For this purpose, there is the button "Edit database", where the technologist can see in tabular form the data of process parameters and manage them: edit, add new or remove the outdated.

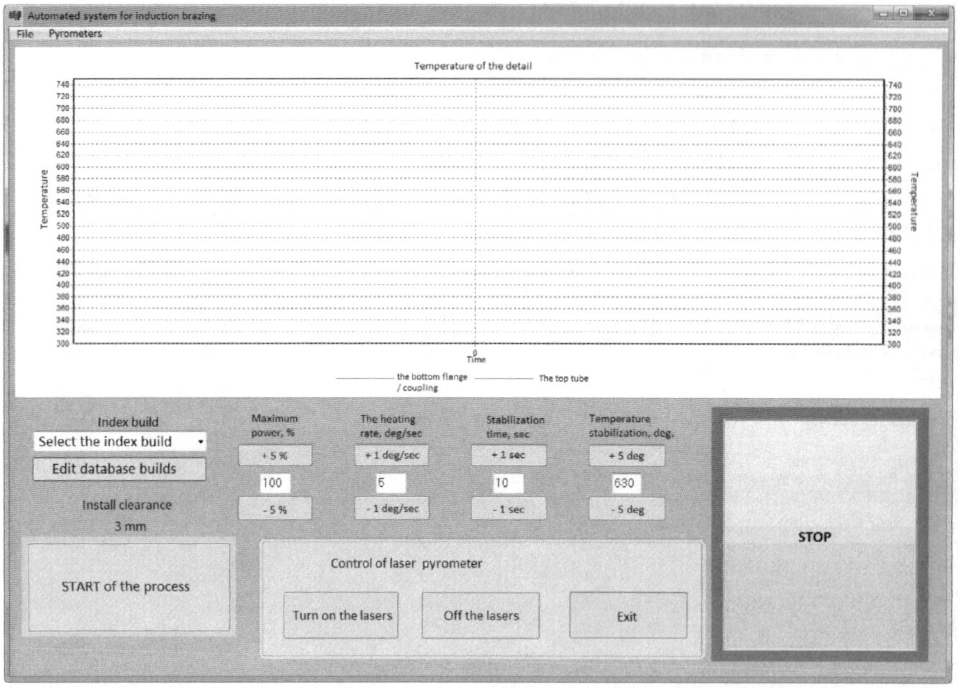

Figure 5. The Main window of the software system (operator)

For providing analysis of the soldering results there is the option of logging processes in graphical form (images of temperatures graphs) and in the text.

If it is necessary to check the connection of the measuring devices and to adjust their parameters, user can look at the pyrometers information form (figure 6).

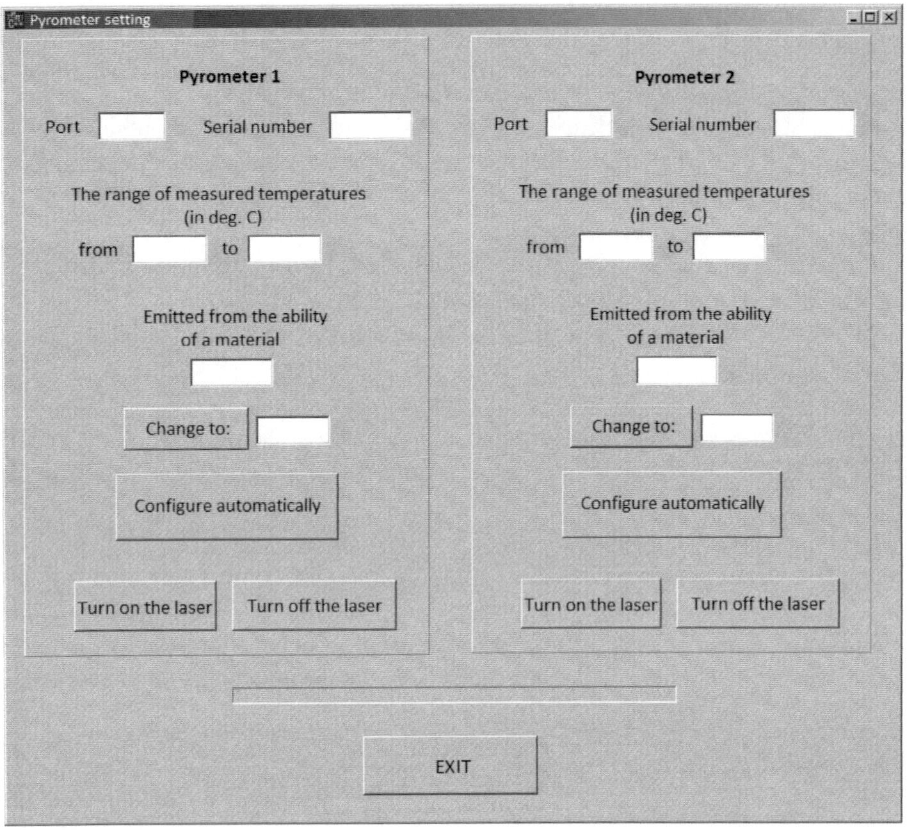

Figure 6. The Window of information about the thermometers

The are two tools to adjust the value of the material emissivity:

1) manual correction – when you click on the button "Change to", value of the material emissivity in the pyrometer will be replaced by the included in the box to the right of the button. This value is expressed in the range from 0 to 1;

2) automatic configuration – when you click on the button "Configure automatically" the selecting process of the best emissivity value will start. This process requires a connection of at least one thermocouple to the system. Average time of the automatic setting varies from 40 to 50 seconds.

4. Experimental study of the automated system effectiveness

To investigate the effectiveness of the developed system a number of laboratory experiments with the experimental equipment with various sizes of pipes and flanges/couplings of waveguides was conducted.

While conducting the experiments we used the information from two pyrometers directed to the pipe and flange/coupling.

The purpose of the first set of experiments was information about the thermal process, which must be entered in the control loop, selections.

Three options were considered:

- using temperature data from the pyrometer directed to the flange/coupling;

- using information about temperatures from the pyrometers directed to the pipe and flange/coupler;
- using temperature data from the pyrometer directed to the pipe.

Disadvantage of the flange/coupling temperature control was solders insufficient flow into the full depth of the joints due to the natural cooling pipe in the aging process. Using of the weighted-average temperature from both thermometers directed to the pipe and flange/coupling has often led to the pipe burn-through.

The pipe temperature process control showed the best results, thus the melting of the parts was pretended and the complete flow of the solder was ensured.

The aim of the second set of experiments was to determine the soldering technological regimes for the following sizes of waveguides: 28.5x12.6 mm; 35x15 mm; 58x25 mm and 19x9.5 mm.

For these experiments the single heating rate 5 deg/sec was implemented. The increasing of the heating rate led to increasing of the temperature difference between the upper and lower surfaces of the flange. The reducing of the speed led to the deterioration of the flux and, consequently, to the solder flow.

Through the experiments the necessary clearances and temperature stabilization were established. The choice of an appropriate gap between inductor and the flange allows to achieve the minimum temperature difference between the joined elements - pipe and flange.

Bank of the waveguide paths soldering technological process was made as the results of experiments.

Examples of the obtained processes graphs are shown in figures 7, 8.

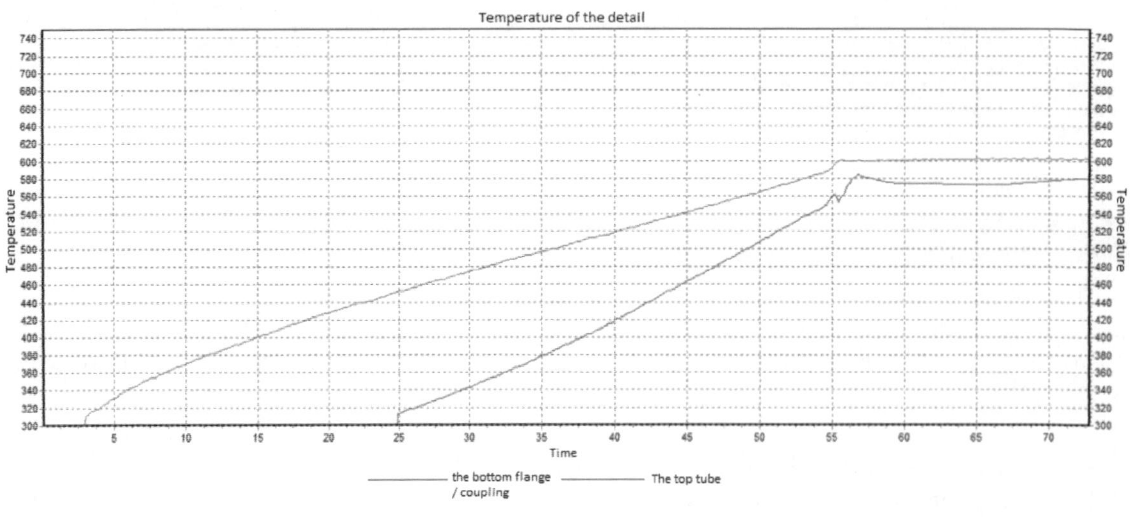

Figure 7.The Graph of the soldering process pipe flange 19x9.5 mm, gap 3 mm, the stabilization temperature of 600 degrees

IOP Conf. Series: Materials Science and Engineering **155** (2016) 012007 doi:10.1088/1757-899X/155/1/012007

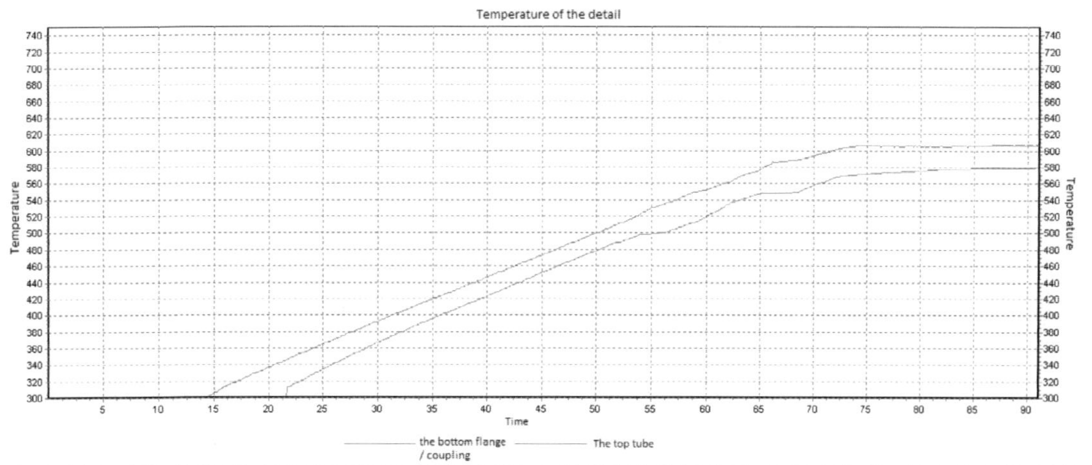

Figure 8. Chart of the soldering process pipe flange 58x25 mm, gap 6 mm, the stabilization temperature of 600 degrees

Conclusion

In presented study the automated system is developed, which allows to solder waveguide paths of different structures and to reduce the human factor.

The system has the following basic functions:

1) automated control of the induction soldering process;
2) visualization of the temperature value in the soldering area;
3) setting the parameters of the soldering process;
4) calibration of the measuring devices – pyrometers;
5) maintenance of technological processes database;
6) access rights to the system.

Developed software helps the operator to choose the induction soldering parameters and to provide the reproducible heating modes.

Thus, the application of the proposed automated system will improve the quality of the waveguide paths, reduce human factor and improve the staff working conditions.

References

[1] Zlobin S K, Mihnev M M, Laptenok V D, Zaitsev R V 2012 Features of soldering waveguide distribution paths of aluminum alloys using induction heating source *Reshetnev reading : proceedings of the XVI Intern. scientific. conf.* in 2 p. Krasnoyarsk : SibSAU pp 192-193

[2] Silchenko P N, Mihnev M, Ankudinov A V, Kudryavtsev I V 2012 Ensuring the strength and accuracy of large-sized waveguide-distributive systems of space communication devices *Problems of mechanical engineering and reliability of machines* **1** pp 112-117

[3] Vologdin V V, Kushch E V, Assam V V 1989 *Induction brazing.* – L. : «Mechanical engineering»,. –72 p

[4] Kudryavtsev I V, Barykin E S, Gotselyuk O B 2013 Mathematical model of heating of the waveguide during transmission of high signal power *Young scientist* **9** pp 52-57

[5] Laptenok V D, Zlobin S K, Mihnev M M, Bocharov A N 2013 Features of the production of the waveguide-junction of paths of antenna-feeder devices of spacecraft *Vestnik SibGAU* **6(52)** pp 196-201
[6] Laptenok V D, Dolgopolov B B, Bocharov A N *The Technique of debugging of technological process of soldering the waveguide paths* Krasnoyarsk, SibSAU. 2009. 112 p
[7] Slugocki A E, Ryskin S E *Inductors for induction heating* Leningrad : "Energy", 1974. – 418 p
[8] Slugocki A. E. Inductors – L.: *"Engineering"*, 1989. – 462 p
[9] Zlobin S K, Mihnev M M, Laptenok V D, Seregin Yu N, Bocharov A N, Tynchenko V S, Dubtsov Yu P, Dolgopolov B B 2014 Automated equipment and technology for soldering waveguide paths of spacecraft *Vestnik Vestnik SibGAU* **№4(56)** pp 219-229
[10] Laforêt R *Object-oriented programming in C++: textbook* – SPb.: Peter, 2011. 928 p
[11] Arkhangelsky A I *Programming Techniques in C++ Builder 6 and 2006*. Mechanisms Windows, networks: the textbook. Moscow: Binom-Press, 2010. 992 p
[12] Work with COM ports using threads. Mode of access: http://piclist.ru/S-COM-THREAD-RUS/COM_port_potoki_TThread_WINAPI.pdf. Date: 22.03.2016
[13] Titov O Work with the communications ports (COM and LPT) in programs for Win32. Mode of access: http://www.realcoding.net/article/view/2416. Date: 12.04.2016
[14] Tynchenko V S, Bocharov A N, Seregin Yu N, Laptenok V D Interaction Module with the hardware of the ACS "Pike": *a certificate of registration of computer programs //* Certificate of state registration of computer programs №2015611846. 2015
[15] The types of controllers and control laws. Mode of access: http://automation-system.ru/main/15-regulyator/type-of-control.html. Date: 22.03.2016
[16] Automatic regulators and their types. Mode of access: http://mashmex.ru/metallurgi/104-osnovi-avtomatizacii.html?start=6. Date: 20.03.2016.
[17] Porro V, Sedyakin A Automation of processes. Mode of access: http://opiobjektid.tptlive.ee/Automatiseerimine. Date: 25.03.2016

IOP Conf. Series: Materials Science and Engineering **155** (2016) 012008 doi:10.1088/1757-899X/155/1/012008

Analysis of complexity metrics of a software code for obfuscating transformations of an executable code*

M A Kuznetsov, V O Surkov
Reshetnev Siberian State Aerospace University
31 "KrasnoyarskiyRabochiy" prospect, Krasnoyarsk, 660014, Russia.

E-mail: programist1024@yandex.ru

Abstract. The complexity metrics of the software code applicable for the analysis of binary files, as well as the methods used at overlaying the obfuscating transformations of these files are considered in the article. The results of the research of the metrics dependence on the performed by special software products obfuscating transformations of a vulnerable binary file are presented. The research confirms the applicability of the technique of the automated vulnerabilities search in binary files, whose analysis is difficult due to obfuscating transformations overlay.

Modern software tends to more effective problem solving in all spheres of human activity, including the aim of increasing the demand for the product. Together with this desire the product internal design is changing, as a rule towards more complex mechanisms and logic. Such changes lead to an increase in the number of lines of the program source code which in its turn produces more errors in the program. According to the research conducted by Coverity Company with increasing the code base the average number of errors / vulnerabilities grows in 741 products with an open source code (Table 1). The only exceptions are projects with a number of lines of more than 1 million, which may be due both to a statistical error and increasing the code quality requirements for large projects (more stringent requirements for code coverage, use of static and dynamic tools for source code analysis). [1]

Table 1. The statistics of a number of errors/ vulnerabilities for programs in C/C++.

Size of the code base	Average number of errors/ vulnerabilities per 1000 lines of the code
Less than 100 000	0,35
From 100 000 to 500 000	0,5
From 500 000 to 1 000 000	0,7
More than 1 000 000	0,65
Average in all projects (252 010 232 lines of the code)	0,59

The presence of errors in a software product allows causing significant damage by such actions as the unauthorized copying of programs, their illegal distribution and use, thereby reducing its quality, until the complete cessation of its operation, which ultimately causes

* This work supports by the president's grant for young Russian scientists. Contract No. 14.Z56.15.6012.

significant material damage to the manufacturers of software. This fact gives rise to the need to use different methods of analysis and evaluation of the program code in order to detect errors in the source code.

To date, one of the most popular tools used to assess the program code are metrics. They are a set of software assessments that enable developers and project managers to get an idea of a developing or existing product: to understand which methods need re-working or additional testing; to see the potential risks; to understand the current state of the project, to track the progress of work in the software development process. [2]

A typical scenario of using metrics in the development and modification of the software is as follows:

1. A set of measures that characterize the metrics complexity, the laboriousness of its design and analysis, style, reliability, etc. in accordance with the standards that are adopted in a company or an organization is calculated for an existing program in accordance with metrics accepted for its evaluation. It is assumed that the existing code corresponds to the reference values of the used set of measures.
2. After making changes to a program the evaluation of its metrics performance is re-executed. New values of the measures are compared with both the reference values and the characteristics of the program before its modification. Thus, not only checking of a new code for compliance with the developer standards is performed, but also the effects of changes on the previously achieved performance (complexity, reliability, clarity, etc.) of a software product are evaluated.
3. The decision on the possibility and feasibility of introducing a new code into the project is made. The effort and personal contribution of the developers to making changes in the program are evaluated.

During the development of programs and evaluation of developers' efforts some metrics (especially quantitative ones) may serve rather as testimonials as all sorts of tricks on the part of not quite bona fide developers are possible with the aim of both to minimize and maximize some used measures (for example, one and the same code structure can be written in different number of lines and steps). At the same time, little quantitative evaluation can indicate not a lack of effort, but the complexity of the performed work – both the development or modification and the analysis of programs. For example, searching for an error in the code snippet may take a long time, and correction of the detected errors may be reduced to a single line of code. The latter problem can be largely solved by the use of the quantitative metrics instead of complexity metrics to assess the complexity of the problem to be solved by the developer. [4]

Application analysis task in the absence of source codes sets slightly different requirements to the choice of code metrics than the development and modification of applications as a number of quantitative metrics mainly those used for the evaluation of project labor costs is based on the characteristics of the source code.

In [4-12] a number of metrics used to evaluate the complexity of the binary code is considered.

Quantitative metrics:

- File size [4];
- the number of lines of a source code [4];
- the average number of lines for functions [4];
- total number of functions [4];

- Jilb metrics [5];
- cyclomatic complexity [6];
- Halstead metrics [7];
- ABC-metrics [4];

Metrics of complexity of the control flow:

- McCabe metrics [5];
- Pivovarsky metrics [5];
- Harrison and Magel metrics [8];
- boundary values metrics [5].

Metrics of data stream complexity evaluation:

- span metrics [9];
- Henry and Kafura metrics [9];
- Card and Glass metrics [10];
- Oviedo metrics [11];
- Chapin metrics [12];
- metrics of appeal to global variables [4].

Hybrid metrics:

- Cocol metrics [5];

It should be noted that the evaluation complexity metrics are not limited to the above mentioned ones. There is a large set of object-oriented metrics, the metrics that take into account previous changes or observations in the code that are not applicable in the framework of a binary code analysis. [4-12]

Besides the absence of the source codes, the software application analysis can be complicated by special obfuscating transformations the use of which is due to the protection of intellectual property of developers, who are usually not interested in publishing or transmitting their designs such as the authoring algorithm and the key data used in the program to third parties. This fact contributes to the appearance of random errors and vulnerabilities, as well as software bugs deliberately included in the program code. Such state of affairs explains the urgency of the problem of finding vulnerabilities in binary files

For reasons stipulated by the influence of various types of protection on the efficiency of code execution, these protective mechanisms are used heterogeneously. Critical in terms of security areas whose implementation time is relatively small are protected with the most powerful means; fragments that are important, but critical in terms of counting time are protected by less costly transformations. A considerable part of the application which does not contain an important or unique code and is of no interest for analysis may be generally free from any protection. It seems expedient to choose a set of metrics applicable to both software application units (functions and modules) and arbitrary application fragments of different lengths (sequences of listing instructions, performance routes and subroutes of selected algorithms). These metrics should quite clearly characterize the code fragments according to the presence of obfuscation and other security mechanisms in them, and accurately locate the differently protected areas within the analyzed fragments. The applicability of metrics to sufficiently short code or route fragments should allow building profiles of complexity and security of the analyzed code, allowing only localizing protective equipment in the application code. [4]

Such techniques as data obfuscation and control obfuscation are used for binary files obfuscation.

Work [4] examines these techniques in terms of their impact on the previously listed code metrics. We will briefly mention the effect of different methods of binary files obfuscation on the software code metrics.

Data obfuscation:

– conversion of data placement and encoding (Chapin metric, span metric, Kafura metrics, Cocol metrics, McCabe metrics);
– conversion of data aggregation (data structure complexity metrics);
– conversion of data sorting (control complexity metrics).
– Control flow obfuscation:
– aggregation conversion (all complexity measures of the control flow based on the assessment of cyclomatic complexity, ABC metrics, Jilb metrics);
– calculations conversion (control flow metrics);
– opaque predicates (McCabe metrics, Myers metrics, Hansen metrics, Harrison and Magel metrics, Pivovarsky metrics, Jilb metrics);
– tabular interpretation (control flow complexity metrics).

For automatic analysis and searching for vulnerabilities in binary files which contain confusing transformations it is necessary to study the behavior of code metrics during the above mentioned transformations. To solve this problem we conducted the experiment in which with the help of software package (idametrics plugin for disassembler IDA Pro) [13-15] metrics have been received for a sample of 10 binary files containing a vulnerability (an open database of vulnerable applications exploit-db [16] was used as a data source) prior to and after obfuscating transformations.

To perform transformations, the analysis of existing means of binary file obfuscating was carried out in which the following software tools have been chosen (as they are free software or have a trial period of use): VMProtectUltimate 3.0.6 Demo, TheEnigmaProtector 4.40, yodaProtector 1.03.

The presentation of the results for each of the metrics is not possible in a single article. However, the behavior of 18 of the 20 metrics is similar; therefore we restrict ourselves to only some of the results of metrics. We consider the behavior of the metrics: the file size, Halstead B and ABC (Tables 2-4 and Figures 1-3).

The data in Table 2 and Figure 1 provides insights about the original size and the size received after the transformation of the binary files sample. This result does not allow drawing conclusions about the behavior of given metrics.

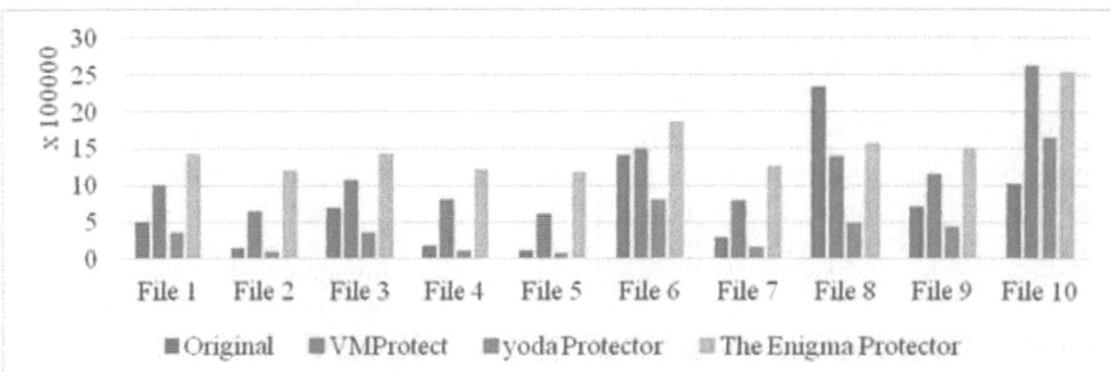

Figure 1. Metrics: Size of a file

Table 2. Metrics: Size of a file.

File Name	Original	VMProtect	yodaProtector	TheEnigmaProtector
File 1	507904	999424	356352	1421312
File 2	139264	638976	98304	1200128
File 3	688128	1069056	352256	1421312
File 4	176128	802816	118784	1216512
File 5	110592	614400	77824	1179648
File 6	1406280	1495040	815104	1856840
File 7	282624	790528	163840	1265664
File 8	2334720	1384448	483328	1568768
File 9	704512	1142784	438272	1503232
File 10	1024000	2625536	1646592	2547712

The data in Table 3 and Figure 2 provides insights about the source ABC-metrics value and the received one after the binary files sample transformations. This result leads to the conclusion about the behavior of the metrics during obfuscating transformations of a binary file: the value of the metrics is reduced by orders of magnitude.

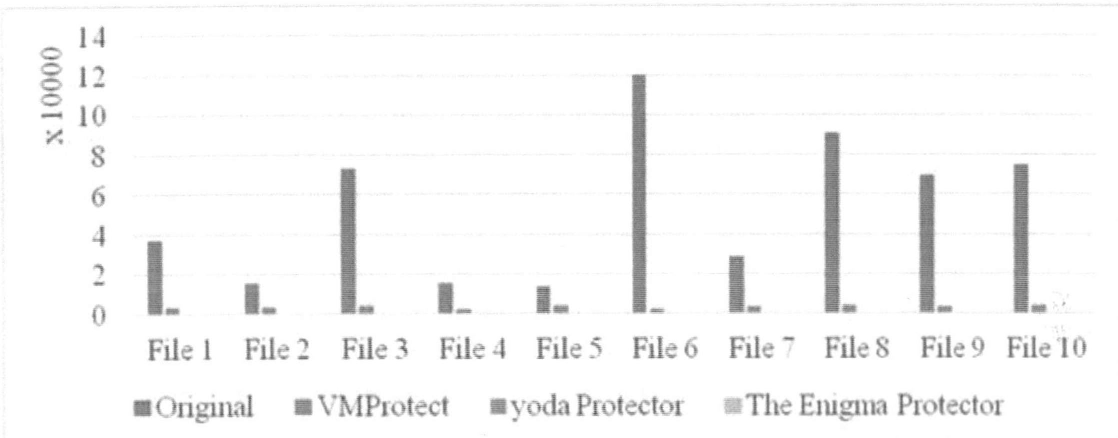

Figure 2. Metrics: ABC-Metrics.

Table 3. Metrics: ABC-metrics.

File name	Original	VMProtect	yodaProtector	TheEnigmaProtector
File 1	36758,57	3639,49	28,72	45,88
File 2	15115,50	3278,71	28,72	26,79
File 3	73331,94	4123,88	28,72	77,93
File 4	15187,43	2340,86	28,72	88,26
File 5	13818,99	4039,84	28,72	16,41
File 6	119812,77	2469,51	34,72	106,31
File 7	28792,80	3159,90	28,72	85,56
File 8	90420,16	3968,47	28,72	38,09
File 9	69389,09	3639,13	28,72	14,87

In addition, let us consider Halstead metrics (data in Table 4 and Figure 3). This metrics has a behavior similar to the results obtained for the ABC-metrics.

As a result of the pairwise comparison of metrics values for a binary file and metrics for the binary file with the presence of obfuscating changes, the following conclusion was drawn: 18 of 20 metrics of a binary file with the presence of obfuscating transformations have a clear decrease in value. As for the other two metrics: the file size, the average number of lines of the function code – the behavior is not clear.

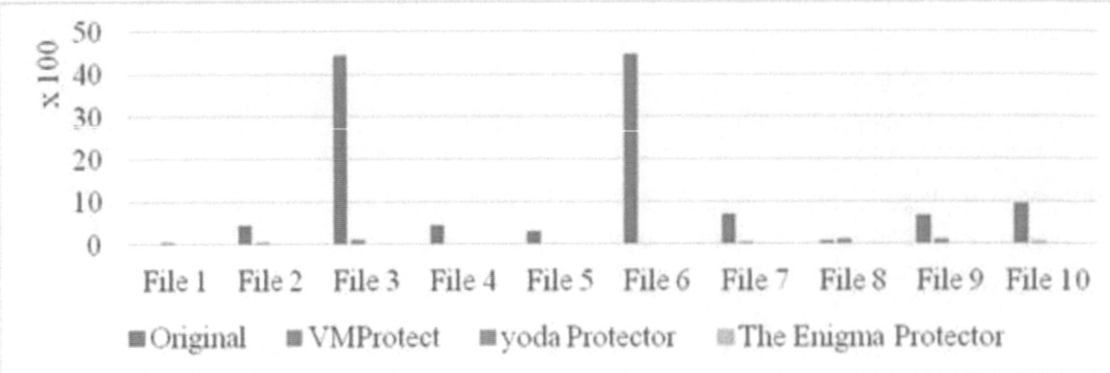

Figure 3. Metrics: Halstead metrics

Table 4. Metrics: Halstead metrics.

File name	Original	VMProtect	yodaProtector	TheEnigmaProtector
File 1	10,22	55,66	0,18	0,40
File 2	434,28	37,35	0,18	0,18
File 3	4425,78	93,88	0,18	0,67
File 4	454,93	7,43	0,18	0,75
File 5	294,46	6,62	0,18	0,08
File 6	4466,61	6,02	0,24	0,20
File 7	704,62	59,43	0,18	0,61
File 8	79,53	97,59	0,18	0,23
File 9	672,28	96,23	0,18	0,06
File 10	945,90	42,80	0,18	0,23

To obtain more information on binary metrics changes in the process of obfuscating transformations let us conduct another study in which we compare the values of code metrics for each function of the binary file before and after transformations.

Let us consider the study results illustrated by the example of the vulnerable function of a binary file (Table 5 as well as Figure 4 show the change of metrics after file transformation).

On the basis of research results data (Table 5) it can be seen that the change in the metrics values: H.B, ABC, H.V, Assign, Cocol, LOC, H.D, R is within 13% (in particular H.B metrics – 5.82% and ABC – 12.53%).

In [13] code metrics are used in order to prioritize the search for vulnerabilities in one or another function of a binary file. The effectiveness of each metrics in the analysis of the file is determined by examining the effectiveness of each metric in searching for vulnerabilities. The results of the study are shown in Figure 6.

Table 5. Analysis of changes in metrics of a vulnerable function of a binary file at overlaying the obfuscating transformations.

Metrics name	Obfuscating transformations		
	Before	After	% of change
H.B	0,287049	0,270337	-5,82%
ABC	35,49648	31,04835	-12,53%
H.N*	240,7106	315,5177	31,08%
H.V	935,9502	962,3421	2,82%
Assign	34	30	-11,76%
Cocol	77,28705	74,27034	-3,90%
LOC	68	69	1,47%
Harr	401	140	-65,09%
H.D.	27	24	-11,11%
H&C	45	6	-86,67%
Ovideo	40	17	-57,50%
Span	25	9	-64,00%
Bound	130	15	-88,46%
Pi	131	15	-88,55%
BBLS	18	10	-44,44%
CC	9	5	-44,44%
Chapin	21	12	-42,86%
C&S	4	0	-100,00%
R	1	1	0,00%
Jilb	0,176471	0,115942	-34,30%

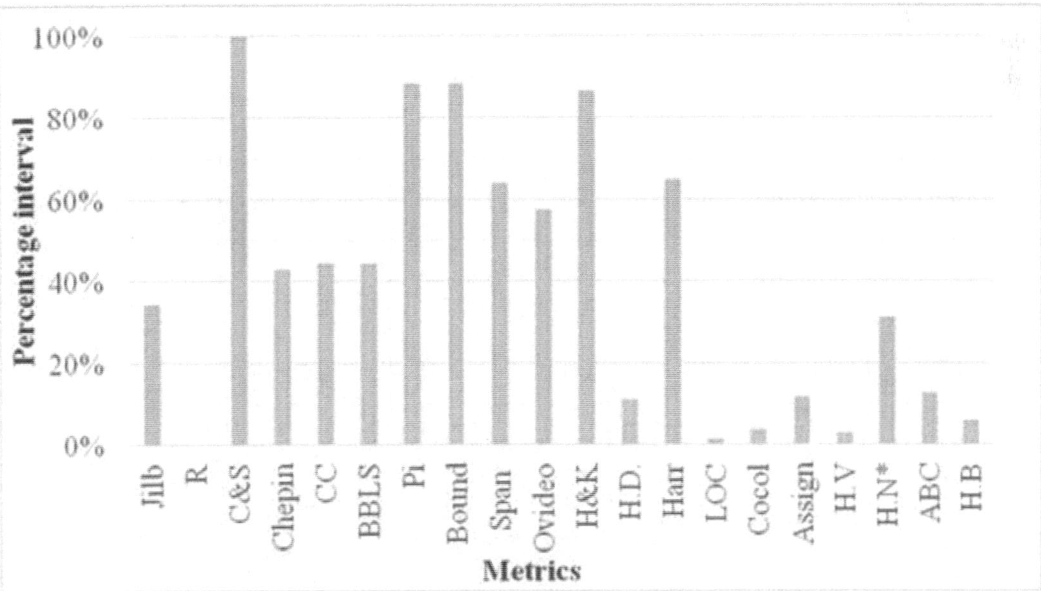

Figure 4. Graph of absolute metrics values change when applying obfuscating transformations.

Table 5 and Figure 4 use the following symbols for the metrics: H.B, H.D, H.V, H.N* – Halstead metrics B, D, V and N; ABC – ABC metrics; Assign – metrics of the number of assignments of values to variables; Cocol – Cocol metrics; LOC – metrics of the number of code lines; Harr – Harrison metrics; H & C – Henry and Kafura metrics; Oviedo – Oviedo metrics; Span – Span metrics; Bound – metrics of boundary values; Pi – Pivovarsky metrics; BBLS – metrics of the number of basic blocks; CC – cyclomatic complexity; Chapin – Chapin metrics; C & S – Card and Glass metrics; R – metrics of ratio of the graph arcs number to the number of base blocks; Jilb – Jilb metrics.

Metrics marked according to the results of the analysis in Table 5 (H.B, ABC, H.V, Assign, Cocol, LOC, H.D., except for R metrics) are included in the 10 most effective ones (on the basis of the data in Table 5 and Figure 5).

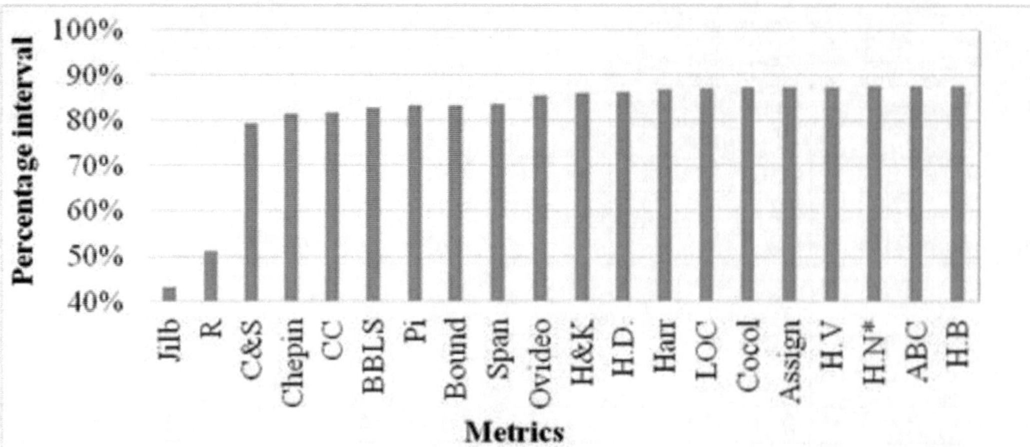

Figure 5. Effectiveness of code metrics in searching for a vulnerable function of a binary file.

It is worth noting that in order to increase the reliability of the study results it is required to conduct research on a bigger experimental sample. However, the received results allow us to draw a preliminary conclusion about insignificant dependence of complexity metrics of a software code of a vulnerable function: changes in the values of the metrics most effective in searching for vulnerabilities (H.B, ABC, H.V, Assign, Cocol, LOC, H.D.) are within 13%.

Such a result (with some adjustment for the change of metrics – 13%) allows considering the possibility of using the method of automated searching for vulnerabilities in the executable code, proposed in [13], for automated searching for vulnerabilities in binary files, which are difficult to analyze due to overlay of obfuscating transformations.

References

[1] CoverityScan: 2013 OpenSourceReport. http://softwareintegrity.coverity.com.
[2] Code Metrics Values https://msdn.microsoft.com/en-us/library/bb385914.aspx.
[3] Software code metrics http://www.viva64.com/en/a/0045.
[4] Software complexity metrics. Technical reporthttp://www.ispras.ru/preprints/docs/prep_25_2013.pdf.
[5] Abran A 2010 Software Metrics and Software Metrology Hoboken, NJ: Wiley-IEEE Computer Society

[6] McCabe T J 1976 A complexity measure. IEEE Transactions on Software Engineering, (4), 308-320
[7] Halstead M H 1977 Elements of Software Science. Amsterdam: Elsevier North-Holland Inc 127 pp
[8] Harrison W A, Magel K I 1981 A complexity measure based on nesting level. ACM Sigplan Notices, 16(3), 63-74
[9] Henry S, Kafura D 1981 Software structure metrics based on information flow. IEEE Transactions on Software Engineering, (5), 510-518
[10] Card D, Glass R 1990 Measuring Software Design Quality. Prentice Hall
[11] Oviedo E I 1993 Control flow, data flow and program complexity. In Software engineering metrics I (pp. 52-65). McGraw-Hill, Inc
[12] Chapin N 1989 An entropy metric for software maintainability. In System Sciences. Vol. II: Software Track, Proceedings of the Twenty-Second Annual Hawaii International Conference on (Vol. 2, pp. 522-523). IEEE
[13] Shudrak M, Zolotarev V 2016 Improving fuzzing using software complexity metrics *Lecture Notes in Computer Science* Volume **9558** pp. 246-261
[14] Shudrak M, Zolotarev V 2012 The new technique of decompilation and its application in information security *UKSim 6th European Symposium on Computer Modeling and Simulation*. Valletta, Malta. pp. 115 – 120
[15] Shudrak M, Zolotarev V 2013 The technique of dynamic binary analysis and its application in the information security sphere *In Proceedings of IEEE EuroCon*, Zagreb, Croatia. pp. 112-118
[16] Vulnerable applications and exploits database: http://www.exploit-db.com/.

IOP Conf. Series: Materials Science and Engineering **155** (2016) 012009 doi:10.1088/1757-899X/155/1/012009

Automatic control of positioning along the joint during EBW in conditions of action of magnetic fields

A A Druzhinina[1], V D Laptenok[2], A V Murygin[3] and P V Laptenok[4]
[1]Cand. Sc., docent, Siberian Federal University, 79, Svobodniy Av., Krasnoyarsk, 660041, Russian Federation
[2]Dr. Sc., professor, Reshetnev Siberian State Aerospace University, 31, Krasnoyarsky Rabochy Av., Krasnoyarsk, 660037, Russian Federation
[3]Dr. Sc., professor, Reshetnev Siberian State Aerospace University, 31, Krasnoyarsky Rabochy Av., Krasnoyarsk, 660037, Russian Federation
[4]Postgraduate student, Reshetnev Siberian State Aerospace University, 31, Krasnoyarsky Rabochy Av., Krasnoyarsk, 660037, Russian Federation

E-mail: avm514@mail.ru

Abstract. Positioning along the joint during the electron beam welding is a difficult scientific and technical problem to achieve the high quality of welds. The final solution of this problem is not found. This is caused by weak interference protection of sensors of the joint position directly in the welding process. Frequently during the electron beam welding magnetic fields deflect the electron beam from the optical axis of the electron beam gun. The collimated X-ray sensor is used to monitor the beam deflection caused by the action of magnetic fields. Signal of X-ray sensor is processed by the method of synchronous detection. Analysis of spectral characteristics of the X-ray sensor showed that the displacement of the joint from the optical axis of the gun affects on the output signal of sensor. The authors propose dual-circuit system for automatic positioning of the electron beam on the joint during the electron beam welding in conditions of action of magnetic interference. This system includes a contour of joint tracking and contour of compensation of magnetic fields. The proposed system is stable. Calculation of dynamic error of system showed that error of positioning does not exceed permissible deviation of the electron beam from the joint plane.

Introduction

Electron beam welding (EBW) is widely used in aerospace engineering, nuclear power, automotive, shipbuilding to create permanent connections, when other types of welding cannot provide the required weld quality.

Often the process of electron beam welding is accompanied by disturbances in the form of magnetic fields caused by the residual magnetization of equipment and parts to be welded, the work of various electromagnetic devices, and the thermoelectric currents flowing in the details during welding of dissimilar materials. As a rule, the effect of magnetic fields on the electron beam cannot be eliminated completely. These fields deflect the electron beam. This leads to formation lack of fusion. In addition, magnetic fields are nonstationary. They can change during the welding process. Therefore, the possibility of using of positioning systems with a preliminary record a trajectory is very limited.

The method of control and compensation of displacement of a scanning electron beam is designed to eliminate the effect of magnetic fields on electron beam during the welding process. The measuring device in the system is a collimated X-ray detector. The sensor is placed on the electron gun. The collimator slit of sensor is aimed at the optical axis of the gun. X-ray sensor signal is processed by the synchronous detection method, with frequency separation of first and second harmonics of frequency of scanning the electron beam across the joint [1].

Elimination the effect of magnetic fields on the electron beam deflection from the optical axis of electron beam gun is realized by introducing the compensating magnetic fields in the zone of action of magnetic interference. Special electromagnetic devices produce the compensating magnetic field directed towards the magnetic interference [1-2].

Analysis of the spectral characteristics of the X-ray sensor

The X-ray sensor includes X-ray detector and collimator. The collimator is made as slotted hood which limits the field of view of the sensor. The collimator is made of material which absorbs X-ray radiation. Slit of collimator must coincide with the optical axis of the gun and the direction of welding (Fig. 1). When the electron beam scans across the joint in the output of the X-ray sensor harmonic components appear. Harmonic components have frequencies that are multiples of the frequency of scanning ω.

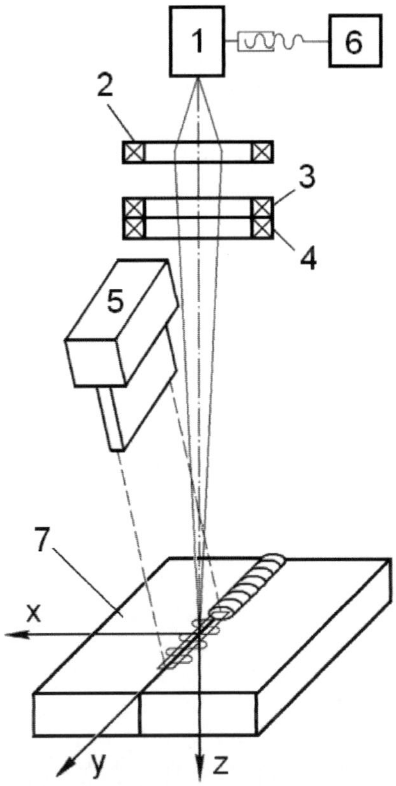

Figure 1. Scheme of control of the effect of magnetic interference on the deflection of the electron beam: 1 – electron beam gun; 2 – focusing system; 3, 4 – deflection system; 5 – collimated X-ray sensor; 6 – electric drive of the electron beam gun; 7 – welded product.

The output signal of the X-ray sensor can be presented as a Fourier series [3]. If ω=1, the output signal can be written as

$$I_d(t) = \frac{a_0}{2} + \sum_{n=1}^{\infty}\left(a_n \cos nt + b_n \sin nt\right), \tag{1}$$

where coefficients of the series are given by the expressions:

$$a_0 = \frac{1}{\pi}\int_{-\pi}^{\pi} I_d(t)dt, \tag{2}$$

$$a_n = \frac{1}{\pi}\int_{-\pi}^{\pi} I_d(t)\cos ntdt, \tag{3}$$

$$b_n = \frac{1}{\pi}\int_{-\pi}^{\pi} I_d(t)\sin ntdt. \tag{4}$$

Here, the intensity of X-ray radiation from the surface of the work piece is determined by expression

$$I_d = kk_1U_0^2ZI_b\frac{1}{\sigma_x\sqrt{2\pi}}\int_{-\infty}^{\infty} f_d(x)\exp\left(-\frac{(x-\varepsilon_0-\varepsilon_m\sin\omega t)^2}{2\sigma_x^2}\right)dx, \tag{5}$$

where $f_d(x)$ – function of the view area of the collimator of X-ray sensor; k=1,5·10^{-9} B^{-1} – proportionality coefficient; k_1 – coefficient considering the orientation of the sensor in the space; I_b – electron beam current; Z – atomic number of the material of welded product; U_0 – accelerating voltage; σ_x – root mean square deflections of electrons from the axis of the beam along the x axis; ε_0 – deflection of the beam from the optical axis of the electron gun; ε_m – amplitude of the beam scanning across the joint.

Function of the view area of the collimator of X-ray sensor has the form

$$f_d(x) = \begin{cases} 0, & x < -\frac{l_2}{l_1}h - \frac{h}{2}, \\ \frac{l_1}{l_2h}x + \left(1 + \frac{l_1}{2l_2}\right), & -\frac{h}{2} - \frac{l_2}{l_1}h \le x < -\frac{h}{2}, \\ 1, & -\frac{h}{2} \le x < -\frac{\Delta}{2}, \\ 0, & -\frac{\Delta}{2} \le x \le \frac{\Delta}{2}, \\ 1, & \frac{\Delta}{2} < x \le \frac{h}{2}, \\ -\frac{l_1}{l_2h}x + \left(1 + \frac{l_1}{2l_2}\right), & \frac{h}{2} < x \le \frac{h}{2} + \frac{l_2}{l_1}h, \\ 0, & x > \frac{h}{2} + \frac{l_2}{l_1}h, \end{cases} \tag{6}$$

where h – width of the slit of the collimator of X-ray sensor; l_1 – length of the collimation channel; l_2 – distance from the collimator to the workpiece surface; Δ – width of gap at the joint.

The first harmonic of the scanning frequency allows determining the deflection of the electron beam from the optical axis of the electron gun caused by the action of magnetic interference.

Fig. 2 shows how the amplitude of the first harmonic as a function of displacement of the electron beam varies with different width of gap at the joint. The calculations were performed by the formulas (3) and (4) taking into account (5) and (6) for l_1 = 10 mm, l_2 = 15 mm, h = 0.3 mm.

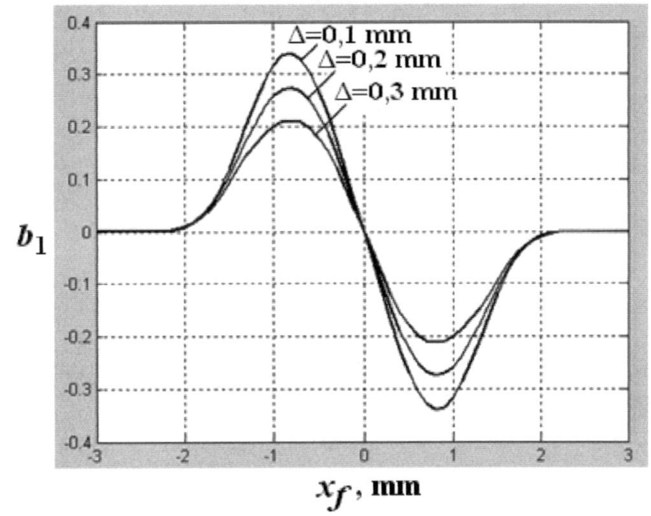

Figure 2. Dependence $b_1(\varepsilon_0)$ at ε_m=1.

Spectral characteristics indicate that an increase of width of gap at the joint leads to a decrease of the amplitude of the first harmonic. EBW is used for the production of important parts that have passed thorough preliminary processing before welding. Therefore, width of gap is constant during the welding process and does not affect the readings of the X-ray sensor.

If joint of welded details is displaced relative to the optical axis of the electron beam gun (Fig. 3), the beam deflection from the joint consists of joint displacement from the optical axis of the gun x_j and the beam deflection from the optical axis of the gun x_f:

$$\varepsilon_j = x_j + x_f.$$

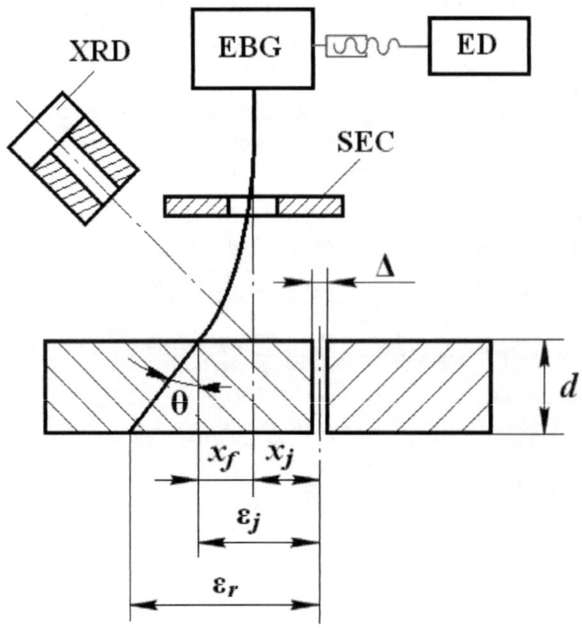

Figure 3. To analysis of the spectral characteristics of the X-ray sensor:
EBG – electron beam gun; ED – electric drive for gun; XRD – X-ray detector;
SEC – collector of secondary electrons; Δ – width of gap at the joint.

Dependencies of amplitude of the first harmonic from the beam deflection from the axis of the gun for different values of the joint displacement are shown in Fig.4.

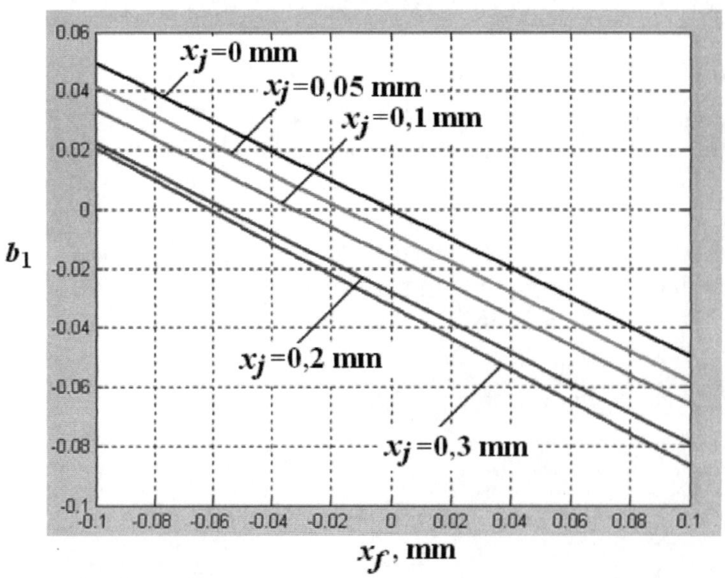

Figure 4. Dependence $b_1(\varepsilon_0)$ at ε_m=1, Δ=0.2 mm.

Analysis of presented characteristics shows that the displacement of joint from the axis of collimation hole of sensor leads to displacement of the spectral characteristics of the sensor respect to the origin. This displacement should be taken into account in the automatic system of tracking along the joint in conditions of action of magnetic fields.

Automatic system of tracking along the joint during EBW in conditions of action of magnetic fields

Structural diagram of dual-circuit system of tracking along the joint during EBW in conditions of action of magnetic fields is presented in Fig. 5. Due to the effect of displacement of joint on the signal of the collimated X-ray sensor of beam position relative to the optical axis of the gun the dual-circuit system becomes an interconnected.

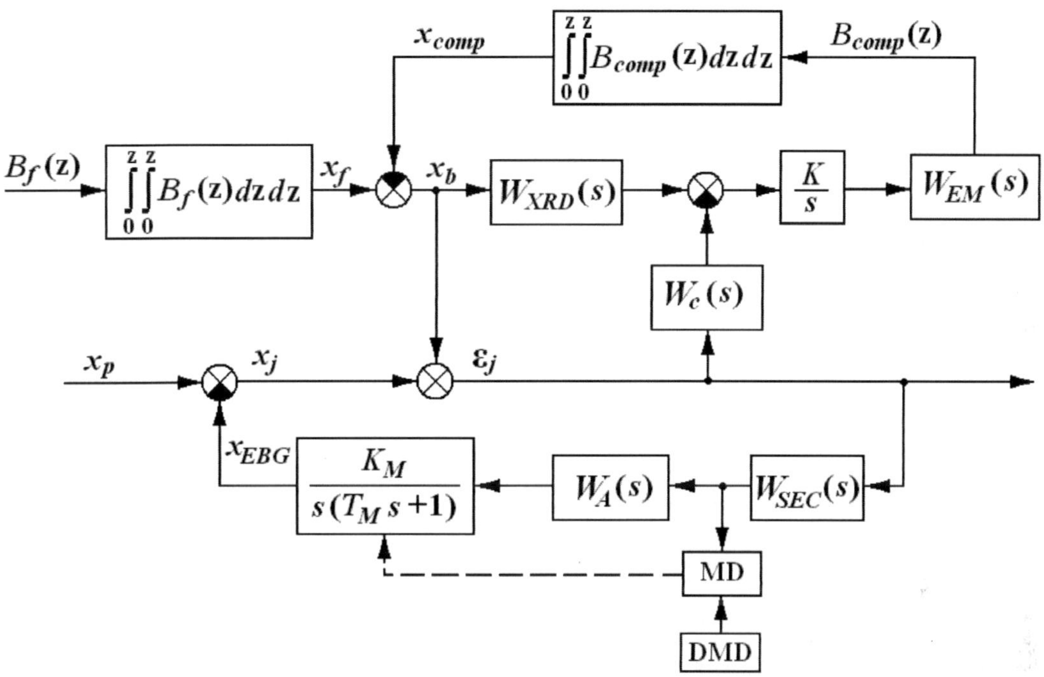

Figure 5. Structural diagram of dual-circuit system of tracking along the joint.

The following notation is used in the diagram: x_p – coordinate of weldment; x_{EBG} – coordinate of gun; $B_f(z)$ – magnetic induction of the field of interference; $B_{comp}(z)$ – magnetic induction of compensating field; x_f – displacement of electron beam from the action of the field of interference; x_{comp} – displacement of electron beam from the action of compensating field; x_b – displacement of the electron beam from the optical axis of the gun; x_j – displacement of the joint from the optical axis of the gun; $\varepsilon_j = x_j + x_b$ – electron beam deflection from the joint; $W_{SEC}(s)$ – transfer function of measurement device the position joint; $W_{XRD}(s)$ – transfer function of the measuring device of the beam deflection; $W_c(s) = K_c \cdot W_{XRD}(s)$ – transfer function of communication, where $K_c < 1$ – coupling coefficient; $W_{EM}(s)$ – transfer function of the electromagnet; $W_A(s)$ – the transfer function of the amplifier; K – transfer coefficient of the integrator; K_M – transfer coefficient of the motor; T_M – time constant of the motor; s – Laplace operator; MD – memory device; DMD – motion sensor of product.

Secondary emission sensor (collector) is used as a measuring device of the joint position. Secondary emission sensor (collector) collects electrons which are reflected from the surface of the product during scanning the joint by an electron beam [4].

SimuLink MatLAB was used to analysis of stability and the quality of the tracking system along the joint. Functional diagram of dual-circuit tracking system for calculations in SimuLink is presented in Fig. 6.

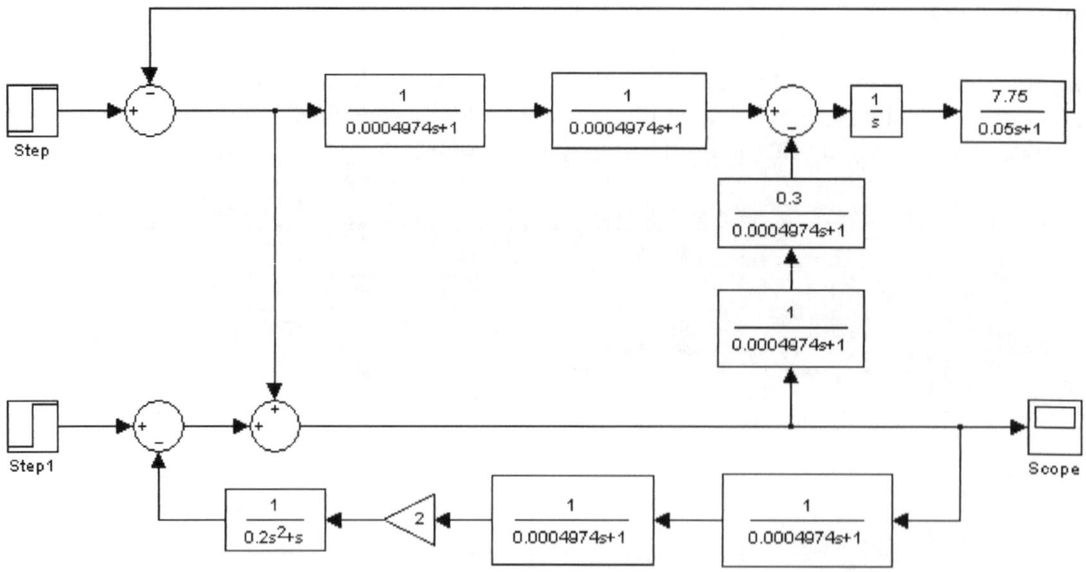

Figure 6. Model of dual-circuit tracking system in SimuLink.

Fig. 7 shows the transient processes in the system. Analysis of transient processes shows that the system is stable and the required performance of quality of control process can be achieved.

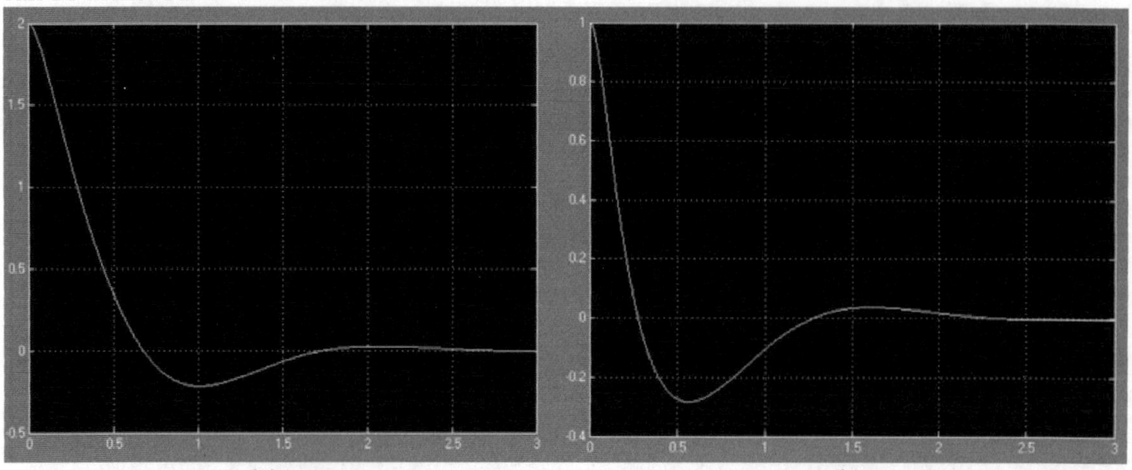

(a) (b)

Figure 7. Transient processes in the system (a) at a deviation the joint and in the presence magnetic interference, (b) in the presence of magnetic interference and absence of deviation the joint.

The error of alignment the electron beam with the joint in the root of weld will be

$$\varepsilon_r = \varepsilon_j + \theta \cdot d.$$

Dynamic error of the system is determined by the formula

$$\varepsilon_v = \frac{v_{max}}{K_v},$$

where v_{max} – maximum rate of change of the input signal; K_v – transfer coefficient of the system.

Preparation of joints connection for electron beam welding is performed with the use of highly accurate processing equipment. Therefore change of joint position is usually associated with an error of the installation of parts in welding manipulator. When performing circumferential welds error of joint positioning is described by a sinusoidal function $x(t)=A_{max}\cdot\sin\omega t$, where A_{max} – maximum amplitude of the input signal, ω – frequency of the input signal.

If the signal $x(t)=A_{max}\cdot\sin\omega t$ arrives at the input of the system then the rate of change of the input signal is

$$v = \frac{dx(t)}{dt} = A_{max}\,\omega\cos\omega t,$$

and maximum rate of change of the input signal is

$$v_{max} = A_{max}\,\omega.$$

For the contour of tracking along the joint at the welding speed 30 m/h and the diameter of the product 1 m (length of weld is 3.14 m) frequency of the input signal is ω = 0.0167 radian/second. At the maximum deviation of joint from the optical axis of the gun A_{max} = 2 mm dynamic error of contour of tracking along the joint is

$$\varepsilon_j = \frac{A_{max}\,\omega}{K_v} = \frac{2\cdot 0.0167}{2} = 0.0167 \text{ mm}.$$

For the contour of compensation of magnetic fields at the welding speed 30 m/h and the length of weld 500 mm the frequency of magnetic field is ω = 0.1 radian/second. Magnetic fields that accompany welding process can reach values which deflect the electron beam up to 5 mm [5]. In this case the dynamic error in the contour of compensation of magnetic interference is

$$\varepsilon_v = \frac{A_{max}\,\omega}{K_v} = \frac{0.1\cdot 5}{7.75} = 0.0645 \text{ mm}.$$

This error occurs at the full compensation of disturbing magnetic fields. This is performed in the center of the penetration channel (at a depth of $d/2$). Given the shape and dimensions of the channel penetration, in the worst case, full compensation is possible at depth of $d/3$. Angle of deflection of electron beam from the optical axis of the gun leading to such displacement is

$$\theta = \varepsilon_v \frac{3}{d} = \frac{0.0645\cdot 3}{d} = \frac{0.194}{d}.$$

At the root of the weld such an angle will cause an error

$$\theta\cdot d = 0.194 \text{ mm}.$$

Thus, error of alignment the electron beam with the joint in the root of weld would be

$$\varepsilon_r = \varepsilon_j + \theta\cdot d = 0.0167 + 0.194 = 0.2107 \text{ mm}.$$

Error of the method equals $0.2107 \le 0.3$ mm and does not exceed permissible deviation of the electron beam from joint plane.

The contour of compensation of the magnetic field allows reducing by two orders the deflection of the electron beam from the optical axis of the gun as well as angle of inclination

IOP Conf. Series: Materials Science and Engineering **155** (2016) 012009 doi:10.1088/1757-899X/155/1/012009

of trajectory. This is especially important when welding products of large thickness. In systems with preliminary record of joint trajectory the contour of compensation of the magnetic field should operate in the recording mode as well as in playback mode of joint trajectory during the welding. This allows increasing accuracy positioning along a joint in the case of changing magnetic fields. High reliability of the system is preserved. In the welding mode the joint is melted. Therefore, the deviation of joint does not change the signal of the measuring device of the electron beam deflection from the axis of the gun. In this case, the coupling coefficient is equal to zero.

Conclusions

1. Automatic compensation of the magnetic field of interference allows significantly improving the accuracy and reliability of positioning along the joint during electron beam welding.
2. Due to the contour of compensation of magnetic field the system of positioning along the joint becomes a dual-circuit and an interconnected.

References

[1] Laptenok V D, Druzhinina A A, Murygin A V and Seregin Yu N 2016 Method to eliminate the impact of magnetic fields on the position of the electron beam during EBW *IOP Conf. Series: Materials Science and Engineering* Volume **122**, p. 012021

[2] Druzhinina, A A, Laptenok, V D, Murygin, A V 2014 Improvement of the positioning accuracy along the joint connection of details with the residual magnetization in electron beam welding *Vestnik SibGAU* No. **5** (57) pp. 168-173

[3] Druzhinina, A A, Laptenok V D, Murygin A V, Seregin Yu N 2012Method for monitoring of impact of magnetic fields in the process of electron-beam welding by X-ray radiation from treatment area. *Vestnik SibGAU* No. **5** (45) pp. 158-163

[4] Laptenok V D, Murygin A V, Seregin Yu N, Braverman V Ya Control of electron beam welding. Krasnoyarsk, 2000, 234 p.

[5] Wei P S, Chih-Wei Wen 2002 Missed joint induced by thermoelectric magnetic field in electron-beam welding dissimilar metals *Experiment and scale analysis. Metallurgical and Materials Transactions B* Volume **33** Issue 5 pp 765-773

IOP Conf. Series: Materials Science and Engineering **155** (2016) 012010 doi:10.1088/1757-899X/155/1/012010

Implementation of a combined algorithm designed to increase the reliability of information systems: simulation modeling[1]

A Popov, V Zolotarev, S Bychkov
Reshetnev Siberian State Aerospace University
31 "KrasnoyarskiyRabochiy" prospect, Krasnoyarsk, 660037, Russia.

E-mail: bychkovss_va@mail.ru

Abstract. This paper examines the results of experimental studies of a previously submitted combined algorithm designed to increase the reliability of information systems. The data that illustrates the organization and conduct of the studies is provided. Within the framework of a comparison of As a part of the study conducted, the comparison of the experimental data of simulation modeling and the data of the functioning of the real information system was made. The hypothesis of the homogeneity of the logical structure of the information systems was formulated, thus enabling to reconfigure the algorithm presented, – more specifically, to transform it into the model for the analysis and prediction of arbitrary information systems. The results presented can be used for further research in this direction. The data of the opportunity to predict the functioning of the information systems can be used for strategic and economic planning. The algorithm can be used as a means for providing information security.

Introduction

Strategic and economic planning are some of the most important directions of the company's activities, because they determine the development trends of the company.

The more parameters/indicators are taken into account, the higher accuracy of the planning is achieved. From an economic and strategic perspective, uninterrupted functioning of the information system (IS) an important variable that indicates the functioning of the company (no spendings on IS restoration and fixing of IS errors, smooth performance of business-related tasks)[1].

The authors of the current study have previously presented the paper in which they deal with the application options for the combined algorithm designed to increase the level reliability of information systems. During the stage that involved designing the combined methodology to analyze the reliability of information systems[2], the following mathematical methods have been selected:

– methods of calculating the reliability of the information system index, in particular, the dynamic method (dynamic methods work on the condition that an information system is being operated, thus indicating effectively the level of reliability of the information system);

1 This work supports by the president's grant for young Russian scientists. Contract No. 14.Z56.15.6012.

– methods of mathematical modeling of information systems (Mathematical modeling is used to predict the level of reliability of an information system at a particular/specific moment/period of time);

– methods of decreasing MTBF and decreasing the failure rate of information system (aimed at tackling the problems that have already been identified)[3];

As a part of the experimental mathematical modeling of IS, a detailed analysis of the existing IS (their functioning and structure) has been done. Based on its results, the conclusion has been formulated: in terms of mathematical modeling, these IS can be presented as a polyphase multichannel open-loop queuing system that is defined by the following characteristics[4]:

– input data stream;

– processing and combining data modules;

– pending for data verification;

– multichannel service devices (control modules, processing module, data transmission modules);

– output data streams.

The primary task of the experiment was to obtain data on the location and time estimated for the occurrence of error in the IS model that has been put to the test. For the organization of the experiment the aim of which was to determine the reliability of the IS the decision to use the simulation[5] was made. The decision to prepare a simulation model was made in view of the availability of the data on the functioning of this IS system over the period of 5 years. During this period of time, all the errors that have occurred in the IS have been recorded, and the measures to eliminate them have been taken as well. FTo construct the simulation, functioning software, AnyLogic produced by The AnyLogic Company, was used.

The necessity to apply the simulation in the presented algorithm is determined by the following factors:

– the possibility of altering the key parameters of the IS without exposing the real IS to crash risk;

– the possibility to select and combine the parameters of the simulation model that will allow to obtain accurate data based on the collected statistics (congruence of simulation results and actual statistical surveys);

– the possibility of making predictions about the occurrence of failures in the IS based on the current simulation model;

– the possibility to analyze the dependence of some parameters of the simulation model on the others, and to come to the adequate conclusions and recommendations concerning the real IS.

Therefore, the construction of a simulation model provides us with great opportunities for carrying out a detailed analysis of the IS performance, including an exceptional one, that is, to alter key parameters of the IS without causing damage to the system and for completing subsequent analysis of the obtained data in order to make conclusions and suggestions on how to fix the bugs in the IS[6].

The description of the studied IS: (In order to maintain confidentiality, the names of IS are changed; no unique names of the modules as well as no unique features are included in the description; the IS structure is simplified to the level where it still allows to maintain the accuracy of data during the experiment, but doesn't allow to discredit the specific IS. The IS func-

IOP Conf. Series: Materials Science and Engineering **155** (2016) 012010 doi:10.1088/1757-899X/155/1/012010

tion modules, which performs the task of receiving information, its primary inspection and making of its registry record, receive the same type of information from different sources. This information is directed at the processing modules of the same type. This IS is a prime example that indicates the possibility of reducing the simulation to a linear structure.

Table 1. Input data for further research

No	IS	Parameter	Description	Unit	Quantity
1	IS 1	IS module	The IS function module, which performs the task of receiving information, its primary inspection and making of its registry record	PC.	6
		Data processing module	The IS function module, which performs certain operations on the data alteration (these are the types of modules known: processing (conversion of information that involves minor changes), changes (getting totally different information comparing to initially received), storage and generation (getting new data without preliminary processing of received data, where the former is related to the later))	PC.	6
		Data type	The number of types of incoming and outcoming data streams that cannot be merged and that are processed in the separate IS modules	PC.	8
		Time of functioning	The time during which the observation of the real IS has been conducted	days	2000
2	IS 2	IS module	The IS function module that performs the task of the receiving information, its primary inspection and making of its registry record	PC.	25
		Data processing module	The IS function module, which performs certain operations on the data alteration (these are the types of modules known: changes (getting totally different information comparing to initially received), storage)	PC.	27
		Data type	The number of types of incoming and outcoming data streams that cannot be merged and that are processed in the separate IS modules	PC.	25

IOP Conf. Series: Materials Science and Engineering **155** (2016) 012010 doi:10.1088/1757-899X/155/1/012010

		Time of functioning	The time during which the observation of the real IS has been conducted	days	2000
3	IS 3	IS module	The IS function module that performs the task of the receiving information, its primary inspection and making of its registry record	PC.	4
		Data processing module	The IS function module, which performs certain operations on the data alteration (these are the types of modules known: changes (getting totally different information comparing to initially received), generation (getting new data without preliminary processing of received data, where the former is related to the latter))	PC.	6
		Data type	The number of types of incoming and outcoming data streams that cannot be merged and that are processed in the separate IS modules	PC.	5
		Time of functioning	The time during which the observation of the real IS has been conducted	days	2000

Formal description

According to the data presented, the IS models should consist of the following elements (simulations of):

– data (of various types) entry;
– operation of IS modules;
– data processing modules;
– output data verification modules functioning;
– data output.

Logic diagrams for the IS models 1, 2 and 3 are presented in Fig.1., Fig. 2, and Fig. 3.

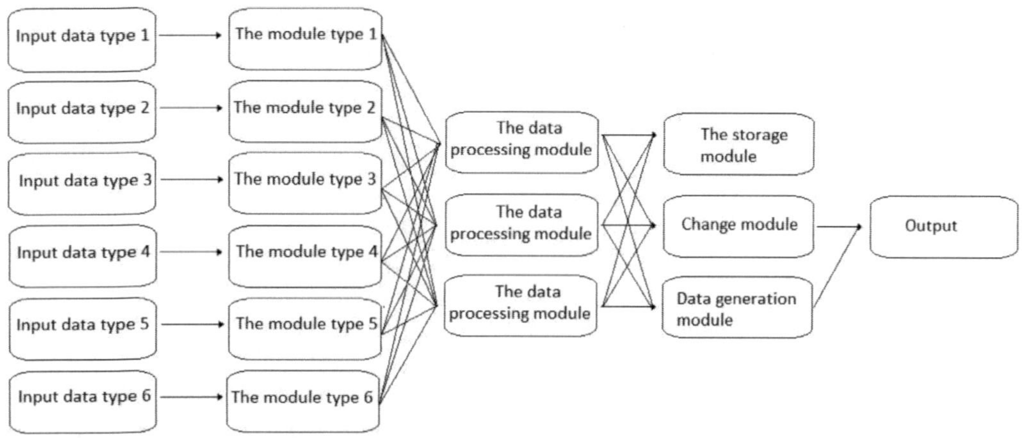

Figure 1. Logic scheme of the simulation model for the IS 1.

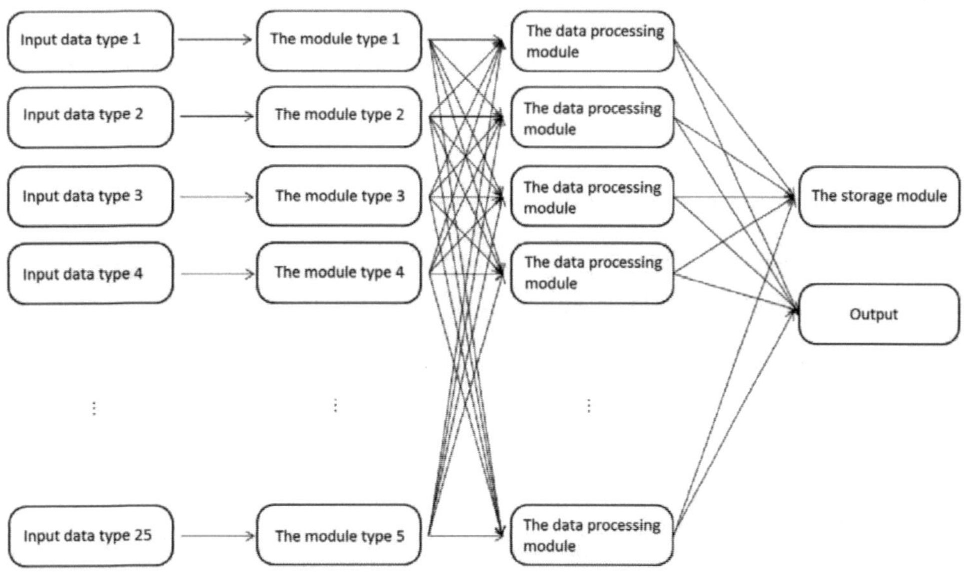

Figure 2. Logic scheme of the simulation model for the IS 2.

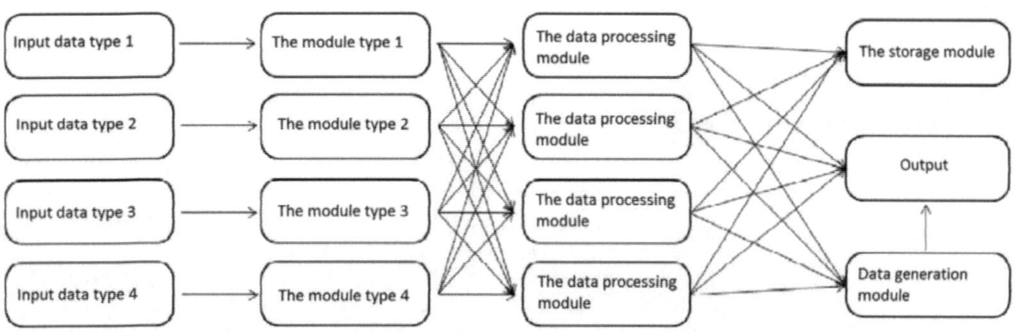

Figure 3. Logic scheme of the simulation model for the IS 3.

The following experiment was conducted on the IS models in order to analyze the operation of the IS.

IOP Conf. Series: Materials Science and Engineering **155** (2016) 012010 doi:10.1088/1757-899X/155/1/012010

There is a number of different data types fed through the IS input. The intervals between the feedthroughs of the data of the same type are random. Each type of the data enters the corresponding IS module and then it is exposed to the processing, the result of which is arbitrary (the data can be accepted, modified, and fed through the output IS, or it can lead to a generation of new data in the IS), and it cannot take more than a predetermined period time. After processing, the data is exposed to the procedure of verification in one of the emission modules, where the time taken to perform this procedure is arbitrary. After verification, a part of the data may be returned for revision, while the rest of the data comes out of the IS.

As a result of the IS operation, there can occur errors in modules. The number of errors depends on the level of workload of the modules, and the error in an arbitrary module leads to an IS downtime. The IS downtime is dependent on the IS restoration process and it is fixed and unchangeable.

As it can be noted from the presented circuit of the simulation IS model, possible errors in the IS operation will occur in the modules that perform the tasks of reception, processing, storage, generation of new data and data alteration, which is confirmed by statistical data gathered from the observations of the IS operation.

In this regard, artificially programmed errors that caused haltings (failures) in the IS operation were installed in the IS models. Programmed failures are probabilistic per se, which is related to the frequency of functioning of the particular IS module. Each halting (failure) is registered, thus leading to the decreased probability of the next one. Therefore, the simulation of failures caused by haltings in hardware and software parts of the IS is performed, which completely eliminates human factor effect. The decision to not consider the errors caused by human factor was based, first of all, on the absence of objective ways to avoid these errors. The statistical compilation of the data gathered from the observations of the current IS operation did not include the data related to this type of errors either.

Analysis of the results

As a result of the experiment, the following results have been obtained:

– both the modules of the IS that are prone to fail and the possible timing of its occurrence have been defined;

– a stage of the IS operation that can be compared to the real data has been simulated.

– The following table (2) presents selected results of the simulation IS model operation as well as the data gathered from the observations of the IS operation. It is important to emphasize that the dates presented are almost or completely identical to each other.

Table 2. Research results.

№	Information system	Fact	Experimental data (date)	Observational data (date)
1	IS 1	Module error 1	07/03/2012	15/03/2012
2		Module error 1	21/07/2013	17/07/2013
3		Module error 2	28/04/2013	24/04/2013
4		Module error 2	27/11/2013	02/12/2013

5		Module error 2	11/06/2014	06/06/2014
6		Module error 3	04/08/2014	08/08/2014
7		Module error 3	19/11/2015	17/11/2015
8		Module error 4	14/06/2015	14/06/2015
9		Module error 5	25/02/2015	28/02/2015
10		Module error 6	20/06/2014	18/06/2014
11		Module error 6	15/04/2015	14/04/2015
12	IS 2	Module error 1	11/02/2012	08/02/2012
13		Module error 1	10/08/2014	14/08/2014
14		Module error 1	23/07/2015	23/07/2015
15		Module error 2	9/03/2013	12/03/2013
16		Module error 2	21/09/2013	23/09/2013
17		Module error 2	12/04/2014	14/04/2014
18		Module error 3	16/07/2012	16/07/2012
19		Module error 3	27/11/2015	24/11/2015
20		Module error 4	23/04/2012	26/04/2012
21		Module error 4	19/02/2013	18/02/2013
22		Module error 5	27/01/2013	31/01/2013
23		Module error 5	24/08/2015	24/08/2015
24		Module error 6	05/11/2013	01/11/2013
25		Module error 6	25/03/2015	30/03/2015
26	IS 3	Module error 1	11/04/2012	11/04/2012
27		Module error 1	10/03/2014	14/03/2014
28		Module error 2	13/04/2012	09/04/2012
29		Module error 2	11/04/2013	11/04/2013
30		Module error 2	24/04/2015	27/04/2015
31		Module error 3	12/06/2012	08/06/2012

32		Module error 3	09/01/2014	09/01/2014
33		Module error 4	27/05/2012	30/05/2012
34		Module error 5	20/05/2014	15/05/2014
35		Module error 6	28/08/2013	29/08/2013
36		Module error 6	26/05/2015	25/05/2015

Also, a series of tests aimed at identifying interdependence between changes in the parameters of failures and errors occurrence in some modules and the occurrence of failures and errors in other modules was completed. The sample of the results of these tests is presented in the table below (3).

An important parameter in this case is a factor increasing the frequency of errors. Example: the given level of error in the test module = 0.1, depending on the module that logged n errors, the higher the level of errors is in the test module and in the identification of errors – the less regular flow of information will be transmitted to the next stage (as a result of the decrease in the input stream related to the dependent module, where the module detects the reduction of errors completed by 6%).

Table 3.Dependence of error occurrence on changes in the parameters (Impact of change in the coefficient on the grand total is chosen according to the initial standard levels of coefficient values).

№	Module name	Parameter	Dependent Modules (name)	Trend (correctly amount of processed data)	Rate of increase in the incidence of errors	Impact of change in the coefficient on the grand total
1	The module type 1	0,3	The data processing module	decline	10%	5%
2	The module type 2	0,5	The data processing module	decline	23%	12%
3	The module type 3	0,15	The data processing module	raise	1%	1%
4	The module type 4	0,01	The data processing module	raise	2%	1%
5	The data processing module	0,2	The storage module	raise	5%	4%
6	The data processing module	0,18	The storage module	decline	4%	4%

7	The data processing module	0,7	The storage module	decline	21%	18%
8	The data processing module	0,35	The storage module	decline	4%	4%
9	Data generation module	0	Output	raise	3%	3%
10	Data generation module	0,17	Output	decline	6%	6%

With reference to these data and according to the IS values of error coefficients, it can be concluded that the optimum values for the construction of three IS that correlate with the real statistical data gathered from observations are as follows:

Table 4. Optimal coefficients optimal for the analysis of the IS 3.

№	Module name	Parameter
	The module type 1	0,12
	The module type 2	0,08
	The module type 3	0,17
	The module type 4	0,04
	The data processing module	0,15
	The data processing module	0,03
	The data processing module	0,09
	The data processing module	0,21
	Data generation module	0,05

Conclusion

As for the results of the data analysis, some discrepancies in the results obtained and the actual IS model can be indicated. However, these results are based on the operation of such a IS model that does not take into account all the peculiarities of the IS operation and its architecture. The future studies will be concerned with the fixation of this error.

In spite of few flaws in the results, the authors of this study can list the following advantages and the positive results of the experiment:

– Authors propose the hypothesis that all the IS (the IS that complete within themselves a real business process that involves human participation) can be represented as queuing systems. In case this hypothesis is proved, the expenses on the construction of the improved IS model to obtain reliable results can be reduced. Thus, the opportunity to create a generalized model for the analysis of any IS becomes available.

– Adjusted for the error in calculations, the results can be used for economic and strategic planning. As seen from the data, the degree of readiness of the analyzed IS is high.

– Algorithm can also be applicable for dealing with the issues of the IS information security. The analysis that demonstrates the potential problems of the IS [7], as well as their actual occurrence, will prove its usefulness in the design of information protection systems. It is also possible to model the information security system itself in order to reveal its vulnerabilities and subsequently adjust them.

References

[1] Anderson R 2013 Measuring the Cost of *CybercrimeThe Economics of Information Security and Privacy*, Part IV, Springer Berlin Heidelberg Pp. 265-300.

[2] Baskerville R, Trevor Wood-Harper A A 2013 Critical Perspective on Action Research as a Method for Information Systems Research. *Enacting Research Methods in Information Systems* Vol. **2** Pp. 169-190.

[3] Allen S 2014 Tam Performance Monitoring with Application of Reliability Growth Analysis. *Engineering Asset Management – Systems, Professional Practices and Certification, Part of the Series Lecture Notes in Mechanical Engineering* Pp 1281-1289.

[4] Kim H, Choi B 2016 The Effects of Three Customer-to-customer Interaction Quality Types on Customer Experience Quality and Citizenship Behavior in Mass Service Settings *Journal of Services Marketing* Vol. **30.** Issue. 4.

[5] Macal C, North M 2014 Introductory Tutorial: Agent-based Modeling and Simulation. *Proceedings of the 2014 Winter Simulation Conference, IEEE Press* Pp. 6-20.

[6] Sargent R 2013 Verification and Validation of Simulation Models *Journal of Simulation* Vol. **7**, Iss. 1,. Pp 12-24.

[7] Shudrak M, Zolotarev V 2016 Improving Fuzzing Using Software Complexity Metrics *Lecture Notes in Computer Science* Vol. **9558** Pp. 246-261.

IOP Conf. Series: Materials Science and Engineering **155** (2016) 012011 doi:10.1088/1757-899X/155/1/012011

Analog signal functional converters for solar array simulators

E A Mizrah, R V Balakirev, N V Shtabel, D N Poymanov
Reshetnev Siberian State Aerospace University
31, Krasnoyarskiy rabochiy ave., Krasnoyarsk, 660037, Russia.

E-mail: shtabnik@gmail.com

Abstract. In the article authors describe different types of functional converters for solar array simulators. Functional converter used to get nonlinear current to voltage characteristic on solar array simulator output. Described and studied two types of digital functional converters and compared with analog functional converter.

1. Introduction

At the present time solar array simulators (SAS) are widely used in terrestrial tests of spacecraft power system with solar arrays as primary power sources. Solar array simulators are special power sources, which are reproducing main output characteristics of solar array (SA) as I-V curve and admittance.

Solar array simulators developed and produced by the authors use functional feedback (FF) to obtain required characteristics. Solar array simulators usually built as current stabilizer with load voltage functional feedback presented as functional converter (FC) to reproduce required

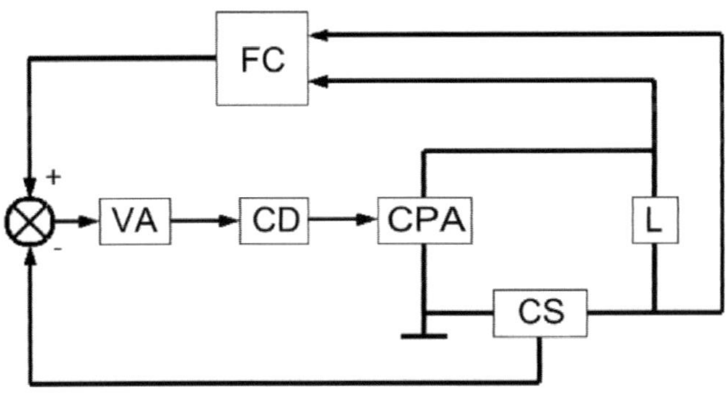

Figure 1. Structure of SAS with functional feedback by load voltage. FC – functional converter, L – load, CS – current sensor, VA – voltage amplifier, CD – correction device, CPA – continuous power amplifier

I-V curve and amplifiers to set I-V curve working range. Structure of SAS shown on Figure 1.

In [1,2] it is shown that FF properties has significant influence on accuracy and dynamical characteristics of SAS, generally: I-V curve reproduction accuracy, SAS bandwidth and output admittance. At the same time FF properties are mostly depends on functional converter realization and characteristics. Therefore, a point of interest is a development of functional converter providing required static and dynamical characteristics of SAS.

2. Functional converters classification

Functional converters classification contain three groups by realization method:

- Reference FC in which nonlinear characteristic obtained by real solar array (or cell) placed in given irradiation, light intensity, temperature, etc. This method was widely used in USA solar array simulators [3].
- Diode type FC. Based on diode I-V curve, which are exponential and roughly can represent I-V curve of solar array [4].
- Circuit FC. Nonlinear I-V curve obtained by scheme on discrete elements. Circuit FC can be of analog type (in which nonlinear circuit based on continuous elements) and digital type in which input signal converted to digital form before processing.

Reference FC considered difficult to realize, because of large number of special hardware such as special lamps, heat chamber, etc. to reproduce orbital environment for reference solar cell or solar array. Due to given difficulties this type of FC has limited usage.

FC of diode type easily realized in comparison with reference FC and circuit FC, but it provides less customizable I-V curve and often does not reproduce real solar array I-V curve. Furthermore, diode also has dependence of I-V curve from environment temperature, which should be considered during solar array simulator development.

Analog FC of circuit type based on transistor scheme [2] can reproduce I-V curve by number of segments, depending on number of transistors. Use of this FC together with scaling amplifiers on FC input and output provides broad range of solar array working modes such as shadowing, temperature change, emergency cases and so on. Main disadvantages of this scheme are fixed I-V curve shape, setup complexity and relatively low reproduction accuracy.

Latest developments in given field are digital functional converters (DFC) which are based on analog-to-digital conversion (ADC) followed by digital signal processing by microcontrollers, DSP, FPGA, logic, etc. This kind of FC provides much more flexible configuration of I-V curve, ability to change I-V curve during test, simplicity of setup and development, higher noise immunity.system should provide continuous long-time testing of accumulators including current and voltage protections and emergency stop in case of energy breakdown.

3. Digital functional converters

Digital FC has three main realizations:

- Calculating FC, which uses microcontroller to calculate output for every input value [5].
- Lookup table FC, based on memory with predefined output values for every input signal value.
- Logic FC based on FPGA containing complex scheme to realize some logic function to convert input values into output.

Regardless of digital signal processor (DSP) digital FC always should make digital-to-analog signal conversion on output.

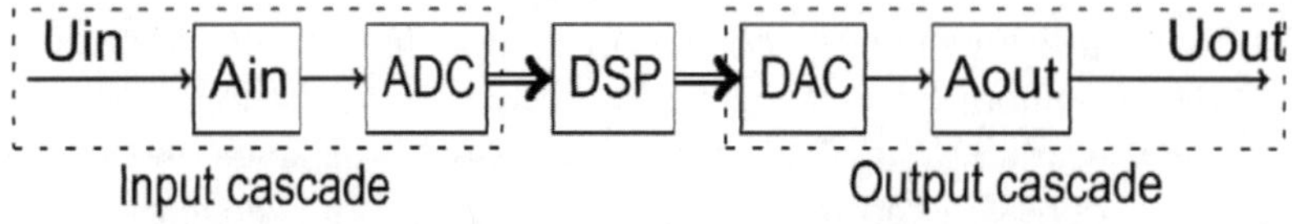

Figure 2. Typical structure of digital functional converter. ADC – analog-to-digital converter, DSP – digital signal processor, DAC – digital-to-analog conveter, Ain, Aout – input and output scaling amplifiers.

Typical structure of digital FC shown on Figure 2.

Digital FC can also be used with scaling amplifiers on input (Ain) and output (Aout) as in case of analog FC to improve accuracy in cases of limited I-V curve region used. For example if maximum current is set to 160 Amperes, but reproduced I-V curve uses only region from 0 to 80 Amperes, effective usage of current range is 0.5. With ADC or DAC resolution of 12-bits it means that in fact only 11 bits will be used. Scaling amplifier can change effective input and output range of digital FC to match used region and FC will operate in full resolution.

One of main drawbacks of digital FC is worse dynamical characteristics which lead to lower reproduction accuracy of SA admittance. This drawback comes from limited analog-to-digital and digital-to-analog conversion speed. In scheme functional converted can be represented as two elements: nonlinear function and signal delay caused by conversion. As functional converter placed in feedback loop of SAS, it leads to effect of parasitic capacitance on SAS output. From SAS admittance equation [2]:

$$Y_{SAS}(s,U) = \frac{Y_{DSA}(U) + Y_{Csas}(s,U)}{1 + R_{PSA}\left(1/R_{DSA}(U) + C_{SAS}(s,U)\right)}$$

We can see that increase of CSAS has significant influence on SAS admittance.

4. Results

The authors studied characteristics and compared three types of functional converters: analog FC with scaling amplifiers, digital FC on lookup table and digital calculating FC. Study was done on two experiments: dynamical characteristics of entire FC and SAS dynamical characteristics with each type of FC.

Experiment on FC dynamical characteristics give us bode diagram for FC, which contain magnitude and phase. On Figure 3 shown phase plot for both digital FC types, from which we can see that lookup table FC has much lower phase shift and therefore has less effect on output SAS capacitance.

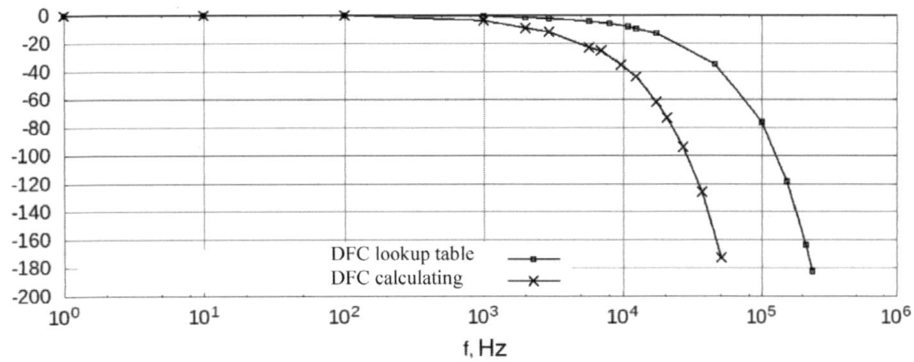

Figure 3. Phase plot of lookup table digital FC and calculating digital FC.

To study effect on SAS output capacitance measured admittance of SAS with analog FC (used as reference) and digital FC with lookup table. Results shown on Figure 4.

Comparison of SAS admittance for digital and analog FC cases shows that characteristics are just similar and on practice analog FC can be replaced by digital FC in SAS, this will allow to get much more functions of SAS like exact change of I-V curve shape.

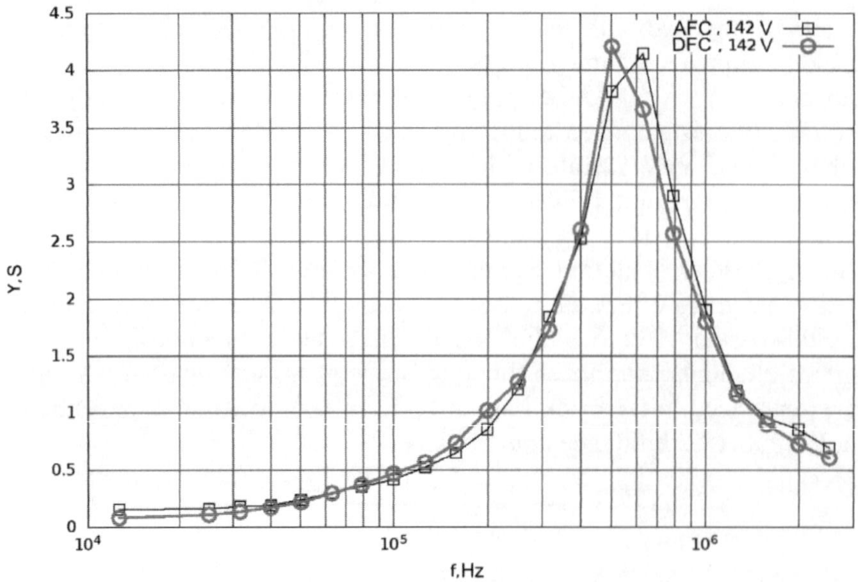

Figure 4. Phase plot of lookup table digital FC and calculating digital FC.

Acknowledgments
This study was supported by the Ministry of Education and Science of the Russian Federation (Government Contract 14.577.21.0082, unique identifier RFMEFI57714X0082).

References

[1] Mizrakh E A, Sidorov A S, Balakirev R V 2006 O sinteze admitansnykh chastotnykh kharakteristik imitatora solnechnoy batarei *Vestnik SibGAU* **2(9)**

[2] Sidorov A S, Mizrakh E A 2007 Issledovanie imitatora solnechnykh batarey s parallel'nym nepreryvnym usilitelem moshchnosti *Vestnik SibGAU* **4(17)**

[3] Martín-Segura G, López-Mestre J, Teixidó-Casas M, Sudrià-Andreu A 2007 Development of a photovoltaic array emulator system based on a full-bridged structure. *Electrical Power Quality and Utilisation (EPQU) 9th International Conference* T. **4.** Vyp. **2** P. 1–6

[4] Kremzukov Yu A 2008 Issledovanie dinamicheskikh kharakteristik imitatora solnechnoy batarei IBS-300/25. *Izvestiya Tomskogo politekhnicheskogo universiteta. Zhurnal Izvestiya Tomskogo politekhnicheskogo universiteta* T. **312.** № **4** Energetika P. 131–135

[5] Shtabel' N V, Balakirev R V 2010 Issledovanie dinamicheskikh kharakteristik tsifrovogo funktsional'nogo preobrazovatelya (tezisy) *Aktual'nye problemy aviatsii i kosmonavtiki : materialy Vseros. nauch.-prakt. konf. studentov, aspirantov i molodykh spetsialistov (12–16 aprelya 2010, g. Krasnoyarsk)* v 2 t. T. **1**.

Nonlinear Compensation of Solar Array Simulators with Dual Power Regulation

E A Mizrah, S B Tkachev, D N Poymanov, A S Fedchenko
Reshetnev Siberian State Aerospace University
31, Krasnoyarskiy rabochiy ave., Krasnoyarsk, 660037, Russia

E-mail: alek.fedchenko@gmail.com

Abstract. During the tests of the spacecraft electrical systems there is a need for simulators of individual parts of the spacecrafts, in particular, solar array simulators (SAS). One of the topologies of medium and high power SAS simulators has dual control of consumed power and contains series or parallel connected linear and switching regulators. This topology allows to provide wide bandwidth and high efficiency, but the range of the resistance change of periodically switched SAS load is limited to the value of the stabilized attribute. Nonlinear compensator (NC) allows to reduce the average feedback voltage of the switching regulator in case of periodic load switching, which, in turn, allows to increase the average value of the stabilized attribute. The describing function method provides a mathematical description of the NC electrical circuit, which allows to select parameters of NC that eliminate the excitation of self-oscillation based on the SAS load switching frequency range and to study the switching regulator stability.

1. Introduction

Validation of electrical systems characteristics of models and experimental prototypes of the spacecraft's requires performing the special electrical tests. During these tests, as power supply sources is used devices that reproduce (with a specified error) the desired static and dynamic characteristics of the real primary power sources, in particular, solar arrays. Such devices are calling solar array simulators (SAS) [1-5]

The quality of ground-based tests is determined by the properties of used simulators, namely:

- accuracy of reproduction of static (current-voltage characteristic) and dynamic (internal impedance) characteristics of the actual solar array;
- performance properties of simulator, such as the possibility of uninterrupted operation and the possibility of durable operation in maximum loading conditions.

The most promising topology of medium and high power SAS at the moment is topology with dual control of consumed power which contains series or parallel connected linear and switching regulators [6].

A linear regulator (LR) in the SAS may be connected in series (Fig. 1a) or in parallel (Fig. 1b) with the load. The LR provides the desired characteristics of a real solar cell (current-voltage characteristics and impedance) and a switching regulator (SR) limits the power that dissipate the LR by stabilizing the LR voltage (series connection) or by stabilizing the LR current (parallel connection).

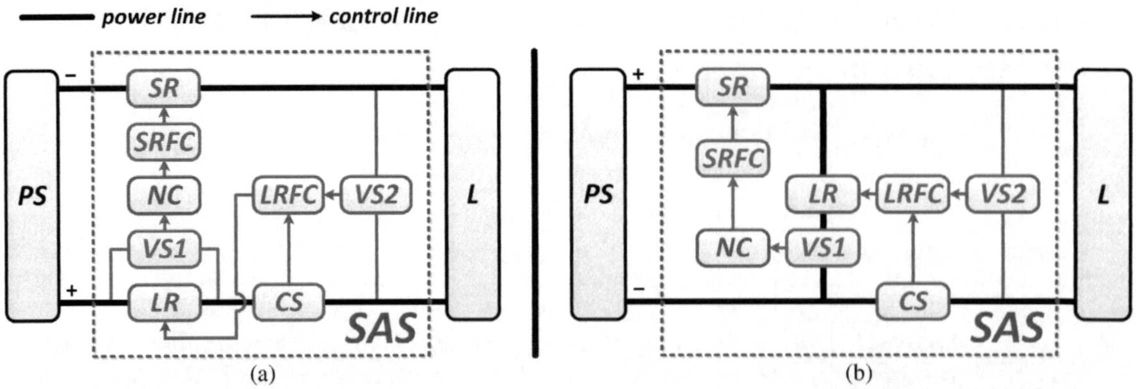

Figure 1. Block diagram of the SAS with dual control of consumed power: (a) series connection, (b) parallel connection. PS – power supply, SR – switching regulator, SRFC – switching regulator feedback circuit, NC – nonlinear compensator, LR – linear regulator, LRFC – linear regulator feedback circuit, VS – voltage sensor, CS – current sensor, L – load.

This topology combines the advantages of simulators containing only one regulator. Simulators based on individual linear regulator allow to achieve high performance rates and to ensure the accuracy of desired dynamic characteristics. Simulators based on individual switching regulator allow to achieve high values of efficiency and good weight/size ratio.

However, simulators with dual power control has problems with a broadband load switching. Since performance of the SR is considerably slower than performance of the LR, during the process of load switching from open-curcuit mode to short-curcuit mode there is a possibility that the LR will stop operate in the active region, which will lead to deterioration of the quality of electricity at a load.

In [7] was proposed the method of improving the quality of SAS operation by introducing a nonlinear compensator (NC) in the SR feedback loop. The NC belongs to the class of dynamic compensators, since the effect of its introducing becomes significant only during transients. During steady state (constant load resistance) stabilized LR attribute remains unchanged.

The purpose of the NC introduction in the feedback loop of SR - reducing the average SR feedback voltage in the case of a periodic load switching, which, in turn, will lead to an increase in value of the stabilized attribute:

- voltage across the LR – in series type simulators;
- current through the LR – in parallel type simulators.

Increasing the value of the stabilized attribute allows the LR to remain in active mode for a wider range of periodically switched resistance of SAS.

The NC may be made as a RC-circuit whose resistance R is shunted by diode VD (figure 2).

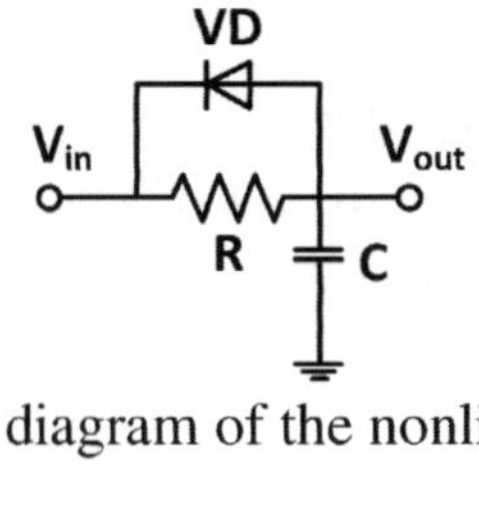

Figure 2. Circuit diagram of the nonlinear compensator.

A capacitor C charges with time constant $T_c = RC$ and discharges through a diode VD with time constant $T_d = R_eC$, where $R_e = \dfrac{RR_{VD}}{R + R_{VD}}$, R_{VD} – differential resistance of open diode.

In this circuit $R_{VD} \square R$, so the discharge time constant T_d is much smaller than the capacitor charge time constant T_c.Consequently, the average voltage across the capacitor decreases with a periodic load switching, which increases the average value of the LR stabilized attribute and increase switching load range, wherein the LR remains is in active mode. Thus, the introduction of the NC improves performance properties of SAS.

2. Mathematical description of solar array simulator with the nonlinear compensator

One of the common methods for the study of nonlinear systems is the describing function method [8-10]. This method allows to study the stability of the nonlinear system described by differential equations of any order.

The nonlinear system represented as a feedback connected linear part of whole nonlinear system and a nonlinearity. Assume that the signal at the input of a nonlinear system $v_{in}(\varphi)$ is sinusoidal with constant component V_0:

$$\phi_k(\vec{r}) = (2\pi)^{2/3} \exp\left(i\vec{k}\cdot\vec{r}\right), \tag{1}$$

where $\varphi = \omega \cdot t$ – phase angle, ω – frequency, t – time, A – the amplitude of the harmonic component of the input voltage.

Without the diode VD, output $V_{oRC}(s)$ related to input $V_{in}(s)$ voltage of the NC as the first-order transfer function of with a time constant $T_c = RC$:

$$G_{NC}(s) = \frac{V_{oRC}(s)}{V_{in}(s)} = \frac{1}{T_c s + 1}. \tag{2}$$

The transfer function (2) corresponds to the absolute value $|G_{NC}(jd)| = R(d)$, and the argument $G_{NC}(d) = \varphi_0(d)$:

$$R(d) = |G_{NC}(jd)| = \frac{1}{\sqrt{1+d^2}}, \tag{3}$$

$$\varphi_0(d) = \arg G(d) = -\tan^{-1}(d), \tag{4}$$

where $d = \omega \cdot T_c$ – normalized frequency.

Consider the mathematical description of the NC under asymmetric fluctuations of the input voltage $V_{in}(\varphi)$, assuming that diode is an ideal switch. The input voltage source has the following parameters: $A = 8V$, $U_0 = 20V$. The input voltage V_{in}, the output voltage of the nonlinear compensator V_{out} and the output voltage of RC circuit V_{oRC} are shown in figure 3.

The figure shows that:

- the output voltage of the NC - non-sinusoidal periodic function with period of 2π;
- voltage v_{out} in the intervals $0 < \varphi < \frac{\pi}{2}$ and $2\pi - \varphi_1 < \varphi < 2\pi$ is describing by input signal expression (1), where $\varphi_1(d)$ – angle of switching.

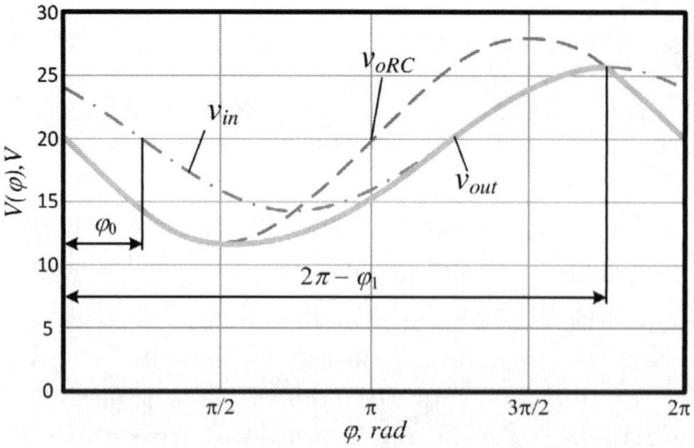

Figure 3. The nonlinear compensator voltages.

An analytical expression that describes the output voltage $v_{out}(\varphi)$ in the interval $\frac{\pi}{2} < \varphi < 2\pi - \varphi_1$ can be found from the differential equation of the NC for nonzero initial conditions and input voltage (1):

$$V_{out}(\varphi) = \begin{cases} V_0 - A \cdot \sin(\varphi), \quad if\ 2\pi n \le \varphi \le \dfrac{\pi}{2} + 2\pi n \\ V_0 - A \cdot R(d) \cdot \sin(\varphi - \varphi_0) - A \cdot (1 - R(d) \cdot \cos(\varphi_0)) \cdot e^{-\frac{\varphi - \left(\frac{\pi}{2} + 2\pi n\right)}{d}} \\ \qquad if\ \dfrac{\pi}{2} + 2\pi n \le \varphi \le 2\pi(n+1) - \varphi_1 \\ V_0 - A \cdot \sin(\varphi), \quad if\ 2\pi(n+1) - \varphi_1 \le \varphi \le 2\pi(n+1) \end{cases} . \tag{5}$$

Switching angle φ_1 (figure 4) can be found numerically from the equation:

$$\frac{\sin(\varphi_1) + e^{-\frac{\frac{3\pi}{2} - \varphi_1}{d}}}{\cos(\varphi_1)} = \frac{R(d) \cdot \sin(\varphi_0)}{1 - R(d) \cdot \cos(\varphi_0)} . \tag{6}$$

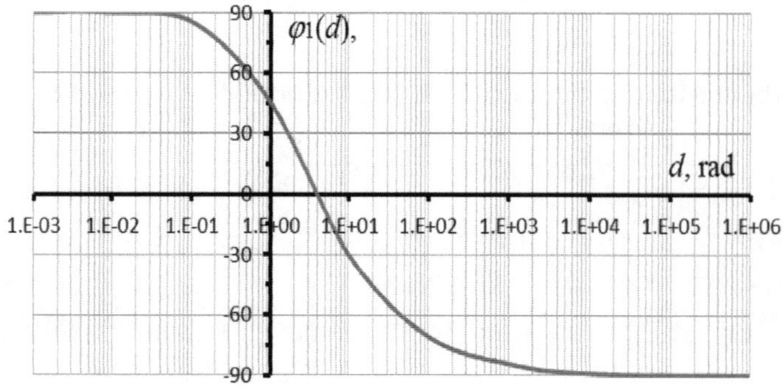

Figure 4. Normalized phase response of $\varphi_1(d)$.

Constant term of the output voltage $V_0^*(V_0, A, d)$ and describing function $H(V_0, A, d)$ of the NC can be found according to the describing function method:

$$V_0^*(V_0, A, d) = \frac{1}{\pi A}\int_0^{2\pi} V_{out}(\varphi)d\varphi, \tag{7}$$

$$\operatorname{Re} H(V_0, A, d) = \frac{1}{\pi A}\int_0^{2\pi} V_{out}(\varphi)\cdot\sin(\varphi)d\varphi, \tag{8}$$

$$\operatorname{Im} H(V_0, A, d) = \frac{1}{\pi A}\int_0^{2\pi} V_{out}(\varphi)\cdot\cos(\varphi)d\varphi, \tag{9}$$

From the expressions (5) and (7-9) can be found expression for constant term of output voltage $V_0^*(A,\omega,V_0)$ (10) and for the describing functions $H(A,\omega,V_0)$ (real (11) and imaginary (12) parts) in closed form:

$$V_0^*(V_0, A, d) = V_0 - \frac{A\cdot R(d)}{2\pi}\sin(\varphi_0) - \frac{A}{2\pi}\cos(\varphi_1) - \frac{A\cdot R(d)}{2\pi}\cos(\varphi_0-\varphi_1) + \frac{A\cdot d}{2\pi}\cdot(1-R(d)\cos(\varphi_0))\cdot\left(e^{-\frac{\frac{3\pi}{2}-\varphi_1}{d}}-1\right), \tag{10}$$

$$\begin{aligned}\operatorname{Re} H(V_0, A, d) = &-\frac{1}{4}-\frac{1}{2\pi}\varphi_1-\frac{R(d)}{2\pi}\cos(\varphi_0)\left(\frac{3\pi}{2}-\varphi_1\right)-\frac{R(d)}{4\pi}\sin(\varphi_0+2\varphi_1)\\ &-\frac{R(d)}{4\pi}\sin(\varphi_0)+\frac{1}{4\pi}\sin(2\varphi_1)-\frac{R^2(d)d}{\pi}(1-R(d)\cos(\varphi_0))\\ &+\frac{R^2(d)d^2}{\pi}(1-R(d)\cos(\varphi_0))\left(\cos(\varphi_1)-\frac{1}{d}\sin(\varphi_1)\right)\cdot e^{-\frac{\frac{3\pi}{2}-\varphi_1}{d}}\end{aligned}, \tag{11}$$

$$\begin{aligned}\operatorname{Im} H(V_0, A, d) = &-\frac{1}{4\pi}+\frac{R(d)}{2\pi}\sin(\varphi_0)\left(\frac{3\pi}{2}-\varphi_1\right)+\frac{R(d)}{4\pi}\cos(\varphi_0+2\varphi_1)\\ &+\frac{R(d)}{4\pi}\sin(\varphi_0)-\frac{1}{4\pi}\sin(2\varphi_1)+\frac{R^2(d)d^2}{\pi}(1-R(d)\cos(\varphi_0))\\ &+\frac{R^2(d)d^2}{\pi}(1-R(d)\cos(\varphi_0))\left(\sin(\varphi_1)+\frac{1}{d}\cos(\varphi_1)\right)\cdot e^{-\frac{\frac{3\pi}{2}-\varphi_1}{d}}\end{aligned}. \tag{12}$$

The expression (10) shows that the constant component V_0^* depends on the constant component V_0, input signal amplitude A and the normalized frequency d. The expressions (11) and (12) show that the describing function H are independent of V_0 and depend only on the variable d ,that is, the charge time constant T_c and frequency ω, therefore denote it $H(d)$.

To find the parameters of the NC circuit (figure 2), the value of the normalized frequency must be determined $d = \omega T_c$. The normalized frequency d is determined as follows:

- assign the following parameters on the assumption of the required power dissipation of the LR and the maximum value of the amplitude fluctuations of the stabilized attribute: constant input voltage V_0, input signal amplitude A and voltage reduction ratio $k_{VR} = \frac{V_R}{V_0}$, where V_R – the desired voltage drop across the switching regulator;

- plot the output voltage $V_0^*(A,d)$ by the expression (10);
- find the value of the normalized frequency d based on the value of the reduced voltage V_R and the amplitude A.

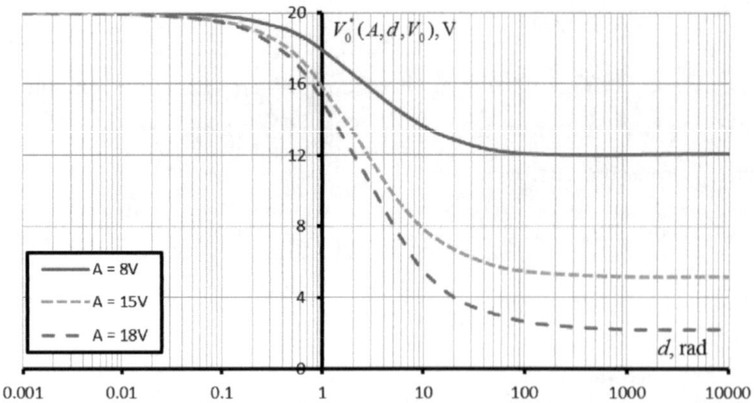

Figure 5. Constant term of the input signal at $V_0 = 20\text{V}$.

For example, for $V_R = 5.2\text{V}$ and $A = 15\text{V}$ from figure 5 can be obtained $d = 100\text{rad}$. Then, assigning the load switching frequency, the time constant can be found as $T_c \approx \frac{d}{\omega} = RC$. Parameters R and C of the NC are selected taking into account the impedance matching of the input circuits of subsequent devices.

After finding the time constant T_c it necessary to check the stability of the SR control system for self-oscillations. Since the bandwidth of the SR is much smaller than the pulse modulation frequency, on the basis of Nyquist-Shannon sampling theorem we will consider the SR as a continuous system.

According to the describing function method [8] characteristic equation is written as negative reciprocal of a linear system transfer function $G(j\omega)$:

$$G(j\omega) = -\frac{1}{H(d)}, \tag{13}$$

For example, we use the following form of the SR linear system transfer function consisting of a second-order filter and compensator:

$$G(j\omega) = \frac{(T_2 s+1)}{(T_1 s+1)} \cdot \frac{K}{(T_3^2 s^2 + 2\xi T_3 s + 1)}, \tag{14}$$

where $T_1 = 0.34\text{s}$, $T_2 = 7.88 \cdot 10^{-5}\text{s}$, $T_2 = 7.74 \cdot 10^{-5}\text{s}$, $\xi = 0.133$, $K = 220$.

Since the curves of $G(j\omega)$ and $-\frac{1}{H(d)}$ (figure 6) do not overlap, there is no real value of the oscillation frequency, and, therefore, no self-oscillation. However, with the increase of the SR bandwidth due to changes in the parameters of $G(j\omega)$, it is possible that loci can intersect and self-oscillation will occur.

3. Simulation results

For the topology of SAS with the LR voltage stabilization (figure 1 (a)), using circuit simulation package Micro-cap, the transients of SAS without the NC (figure 6) and with the NC (figure 7) were plotted for the following conditions: a load resistance R_L changed rapidly from 0.1 Ohm to 1 Ohm (current region of current-voltage characteristics), the LR stabilized voltage $v_{LR} = 10\text{V}$.

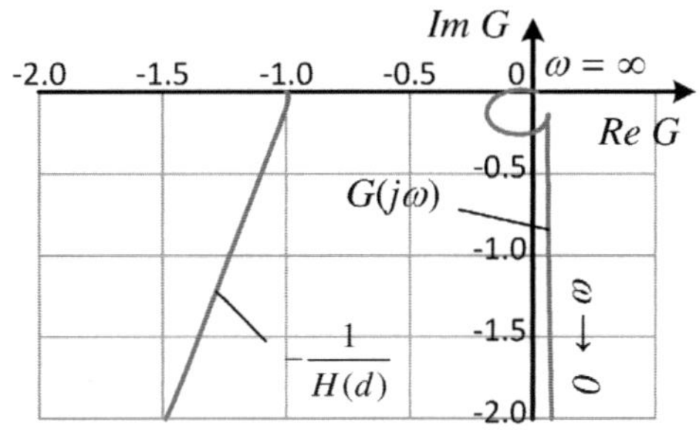

Figure 6. Constant term of the input signal at $V_0 = 20\text{V}$.

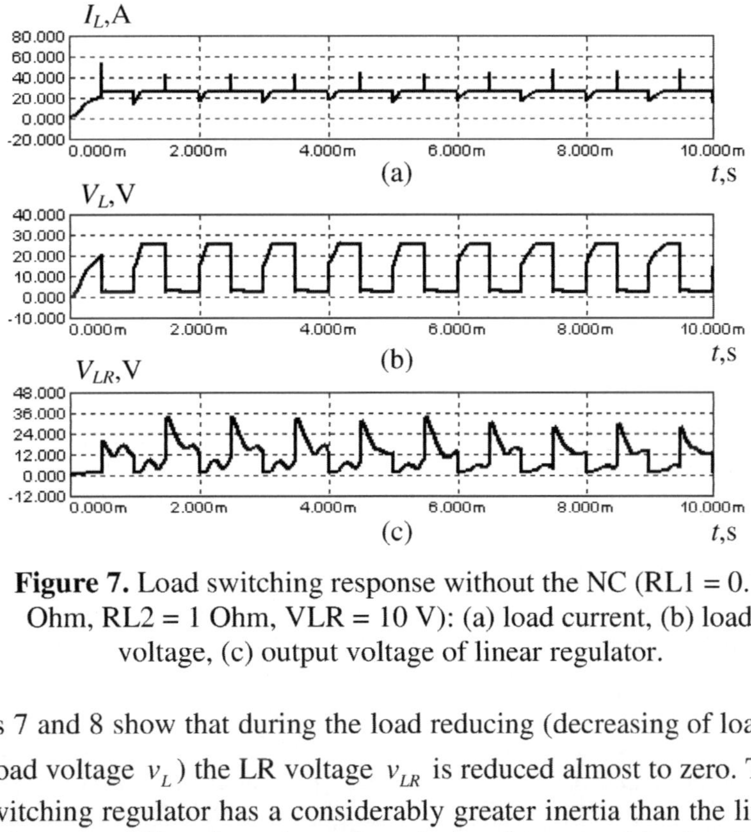

Figure 7. Load switching response without the NC (RL1 = 0.1 Ohm, RL2 = 1 Ohm, VLR = 10 V): (a) load current, (b) load voltage, (c) output voltage of linear regulator.

The figures 7 and 8 show that during the load reducing (decreasing of load current i_L and increasing of load voltage v_L) the LR voltage v_{LR} is reduced almost to zero. This is due to the fact that the switching regulator has a considerably greater inertia than the linear regulator, it does not have time to stabilize the voltage drop during the stepped load resistance switching. Therefore, when the voltage pulse amplitude of the load exceeds the value of the stabilized voltage to the LR, a further increase in the voltage v_L is determined by a switching regulator, and the shape of the voltage v_L is significantly different from the desired square.

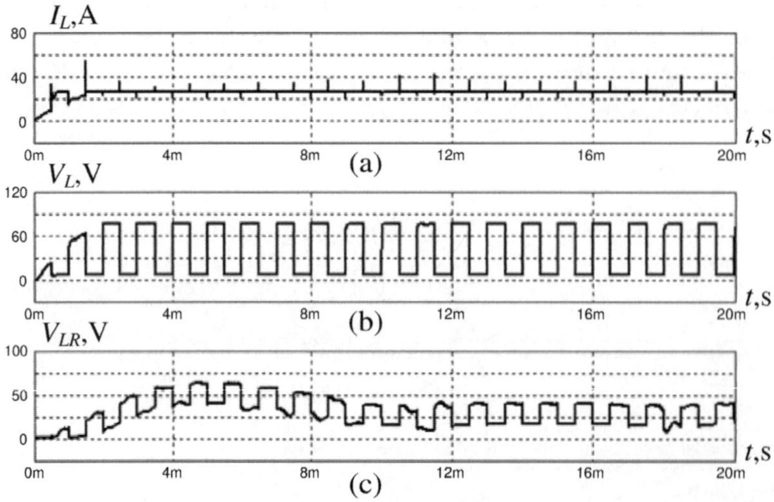

Figure 8. Load switching response with the NC (RL1 = 0.1 Ohm, RL2 = 1 Ohm, VLR = 10 V): (a) load current, (b) load voltage, (c) output voltage of linear regulator.

The figures show that the introduction of the NC is quite effective: in the same load resistance switching range a minimum voltage at the LR is greater than zero, so the LR operates in active mode, and the load voltage v_L has the required form.

4. Conclusion

The paper studied the impact of parameters of the nonlinear compensator (NC) on the stability of medium and high power solar array simulator (SAS) control system by the describing function method. The parameters of the NC, which preclude the self-oscillations, were defined. The following results were obtained:

- the introduction of the NC in the feedback loop of the switching regulator allows to extend the range of the periodically switched resistance of the SAS;
- using the describing function method a mathematical description of the proposed NC circuit was developed, this description allows to select the parameters of NC based on load switching frequency range of the SAS;
- the describing functions of the NC allow to explore the impact of the dynamic properties of the NC on the switching regulator stability.

The simulation shows that the introduction of the NC is quite effective: in the same load resistance switching range a minimum voltage at the LR is greater than zero, so the LR operates in active mode, and the load voltage has the required form.

Acknowledgements

This study was supported by the Ministry of Education and Science of the Russian Federation (Government Contract 14.577.21.0082, unique identifier RFMEFI57714X0082).

References

[1] Lloyd S H, Smith G A, Infield D G 2000 Design and construction of a modular electronic photovoltaic simulator *Proc. of the 8th Intern. Conf. on Power Electronics and Variable Speed Drives* (London, UK) pp 120–123

[2] Rahmani N, Mostefai M 2013 Experimental results of a photovoltaic simulator

IOP Conf. Series: Materials Science and Engineering **155** (2016) 012012 doi:10.1088/1757-899X/155/1/012012

Conference Internationale des Energies Renouvelables (CIER'13) (Sousse, Tunisie) volume **2**
[3] Agilent E4350B, E4351B Modular Solar Array Simulators Datasheet www.agilent.com
[4] Elgar Solar Array Simulators. www.elgar.com
[5] Utility model # 144248 Russian Federation. Elektricheskiy imitator solnechnoj batarei (Electrical solar array simulator) / V.N. Mishin, V.A. Pchelnikov, O.V. Bubnov, Ju.A. Kremzukov – 20.08.2014. Tomsk state university of control systems and radioelectronics
[6] Tkachev S B, Mizrah E A 2011 Upgrading of photovoltaic array simulator with two-circuit control element *Vestnik SibGAU* **1 (34)** pp 70–75
[7] Mizrah E A, Tkachev S B, Shtabel N V and Poymanov D N 2015 Design Principles Of Space Power System Primary Energy Sources Simulators *International Journal of Applied Engineering Research* (vol. **10**) (Dehli: Research India Publications) number 20 pp 41136–41142
[8] Lurie B J, Enright P J 2012 *Classical Feedback Control: With MATLAB and Simulink, Second Edition* (FL: CRC Press) p 556
[9] Mizrah E.A., Sidorov A.S. Modeling of solar array simulator with cascade connection of continuous and impulse power amplifier *Vestnik SibGAU* **1 (14)** pp 7–12
[10] Impram S T, Munro Neil 2010 Describing functions in non-linear systems with structured and unstructured uncertainties *International Journal of Control* (vol.**74**) (Taylor and Francis) issue 6 pp 600–608

Numerical simulation of low pressure gas metal arc plasma

I V Karpov[1,3], A V Ushakov[2,3], A A Lepeshev[1,3], E A Dorozhkina[2], O N Karpova[2], A A Shaikhadinov[1] and V G Demin[1]
[1] Siberian Federal University, Krasnoyarsk, 660041 Russia
[2] Reshetnev Siberian State Aerospace University, Krasnoyarsk, 660037 Russia
[3] Krasnoyarsk Scientific Center, Siberian Branch of the Russian Academy of Sciences, Krasnoyarsk, 660036 Russia

E-mail: sfu-unesco@mail.ru

Annotation. In this paper the model of the cathode spot of a vacuum arc is considered in the framework of the drift-diffusion representation. It is shown that the main source of electrons, providing the current transport of vacuum arc, are ionization processes in the cathode zone, while the emission processes are an auxiliary source of initial electrons. The effects associated with the distribution of electron density in the plasma column of cathode spot, in particular the formation of a "plasma reaction zone" and "ionization wave", are discussed.

For proper developing the techniques and methods for producing nanopowders in the plasma of low pressure arc, it is important to understand the processes occurring in the cathode spot and parameters that have the greatest influence on the physical and chemical properties of nanopowders [1-12].

In this paper the model of the cathode spot of a vacuum arc is considered within the drift-diffusion representation. All the studies showed that the arcs have much in common with the spark discharges, [13] however, the traditional models of Townsend avalanche-streamer mechanism of plasma production cannot be applied due to the short length of the cathode spot.

The system of differential equations has been solved in the *Comsol* software package by the finite element method.

Since all the processes in the cathode spot are defined with characteristic time intervals of 5-100 ns, then the solution of the problem of ionization is sufficient for solving the drift-diffusion Nernst-Planck-Einstein equations, supplemented by the Poisson equation, neglecting the processes of plasma formation, electrical neutrality, etc. That allows revealing the influence of the separate mechanisms of ionization on the structure and dynamics of the cathode spot.

$$\begin{cases} \dfrac{\partial n_e}{\partial t} + div\left[-D_e \nabla n_e - \mu_e n_e \vec{E}\right] = v_{ion}(E) n_e - v_{att}(E) n_e - c_{ee} n_e n_+ + S_{ph} + S_{un} \\ \dfrac{\partial n_+}{\partial t} + div\left[-D_+ \nabla n_+ - \mu_+ n_+ \vec{E}\right] = v_{ion}(E) n_e - c_{ee} n_+ n_e + S_{ph} + S_{un} \\ \Delta\varphi = -\left(\dfrac{e}{\varepsilon_0}\right)(n_+ - n_e); \vec{E} = -\nabla\varphi \end{cases},$$

where n_e, n_+ are concentrations of electrons and ions, respectively, φ is the electric potential. The source function for the electrons if defined taking into account the appearance caused by impact ionization with a frequency of $v_{ion}(E)$, losses after the adherence with a frequency of $v_{att}(E)$, and losses at the electrons and ions recombination with the coefficient of c_{ee}. The source function for positive ions is defined taking into account their appearance caused by the impact ionization and their losses at the electrons and ions recombination (factor c_{ee} is set for the latter value).

The influence of photoionization was neglected for revealing the influence of different sources of ionization on the structure and dynamics of the cathode spot (S_{ph} = 0, S_{un} = 0). The ions and electrons concentration flows, defined as $N_e = -D_e \nabla n_e - \mu_e n_e \vec{E}$, $N_+ = -D_+ \nabla n_+ - \mu_+ n_+ \vec{E}$, consider the diffusion and drift component respectively.

Since these intervals are not more than 10 ns, thus during this time, according to the equation shown above, the ions are displaced up to 1 μm [14], Therefore, the diffusion and ion mobility in the fields, which are not too high, can be considered as constant and equal to 3.7 10^{-3} [m^2/s].

Assuming that arc discharge contains no negative ions (as evidenced by numerous studies) the sticking coefficient can be neglected.

The coefficient of recombination of electron with simple ions (c_{ei}) has the main channel of recombination called dissociative. Constant value of c_{ei} equal to 10^{-13} m^3/s was assigned for simulation, since it varies insignificantly in the investigated range of the electric field intensity.

In this model nonstationary one-dimensional problem with a uniform field was solved. The electrons were emitted from the cathode and this process was constant in time. The emission of electrons was taken into account in the form of insignificant (10^{-16} m^{-3}) initial concentration on the boundary cathode-metal vapor. It was assumed that the emission electrons are the seed particles to start the main ionization processes. Ionization and recombination processes took place in the gap and were taken into account as a reaction rate. The initial concentration of the ions was defined according to the potential decrease across the gap and was located in the ionic cloud zone. The initial concentration of ions and electrons in the discharge gap was zero.

The boundary conditions for the cathode zone are presented in the form of:

$$-\vec{n}\left[-D_e \nabla n_e - \mu_e n_e \vec{E}\right] = 0,\ -\vec{n}\left[-D_+ \nabla n_+ - \mu_+ n_+ \vec{E}\right] = 0.$$

The boundary conditions for the ionic cloud zone were taken from the conditions of ionization and recombination, and also diffusion and drift processes in the discharge gap. They can be represented, for example, as follows:

$$-\vec{n}\left[-D_e \nabla n_e - \mu_e n_e \vec{E}\right] = N_{0e},\ -\vec{n}\left[-D_+ \nabla n_+ - \mu_+ n_+ \vec{E}\right] = N_{0+},$$

where N_{0e} and N_{0+} are time-constant values for the flow of ions and electrons.

Let's analyze the main stages of the development of the arc discharge. The figures show the development of the arc discharge at different points in time. Fig.1,*a* shows the change in the concentration of electrons emitted by the cathode in the process of moving to the ion cloud. The number of electrons in it gradually increases under the influence of impact ionization, and the radius of the cloud increases due to the diffusion. The bulk charge is too small for significant changing the electric field that is the initial phase.

Two electron density peaks are clearly seen during the development of the arc discharge. The first peak is the main and is observed in the cathode zone, while the second is observed near the ion cloud. The appearance of the second peak is caused by the first few electrons reached the ion cloud. Because of the diffusion some separate electrons reach it before all the other electrons. These electrons lead to a rapid increase in the concentration right at the ion cloud due to a high field intensity near ions and as a consequence, due to a high ionization rate. This peak rapidly exceeds the first one by magnitude and becomes noticeable on the background in the moment of time t = 2.8 ns. Nevertheless, since the rate of ionization processes is much higher than electron drift rate, the first peak begins to grow rapidly and they become equal at a moment of time of 4.2 ns.

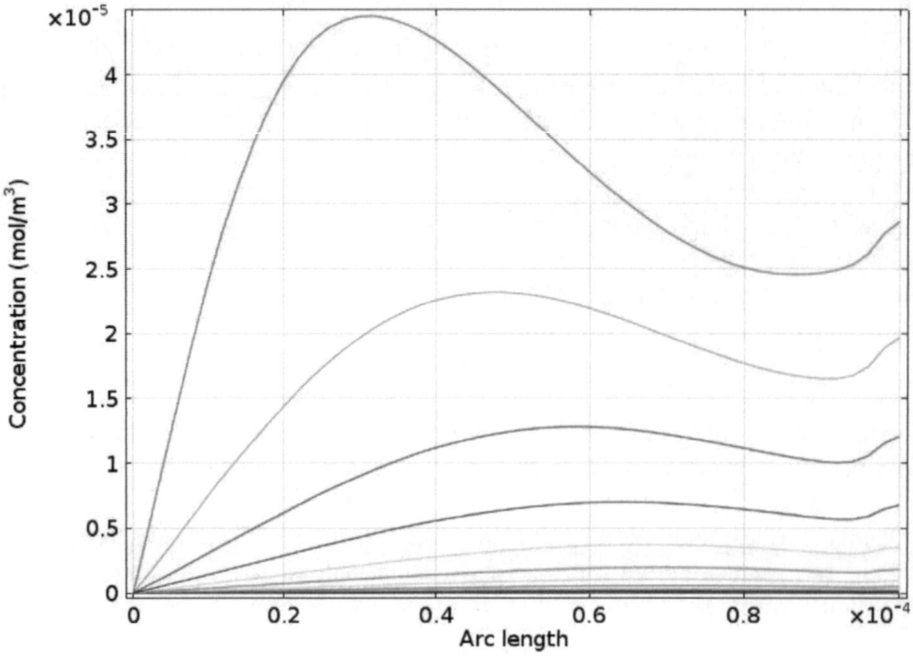

a

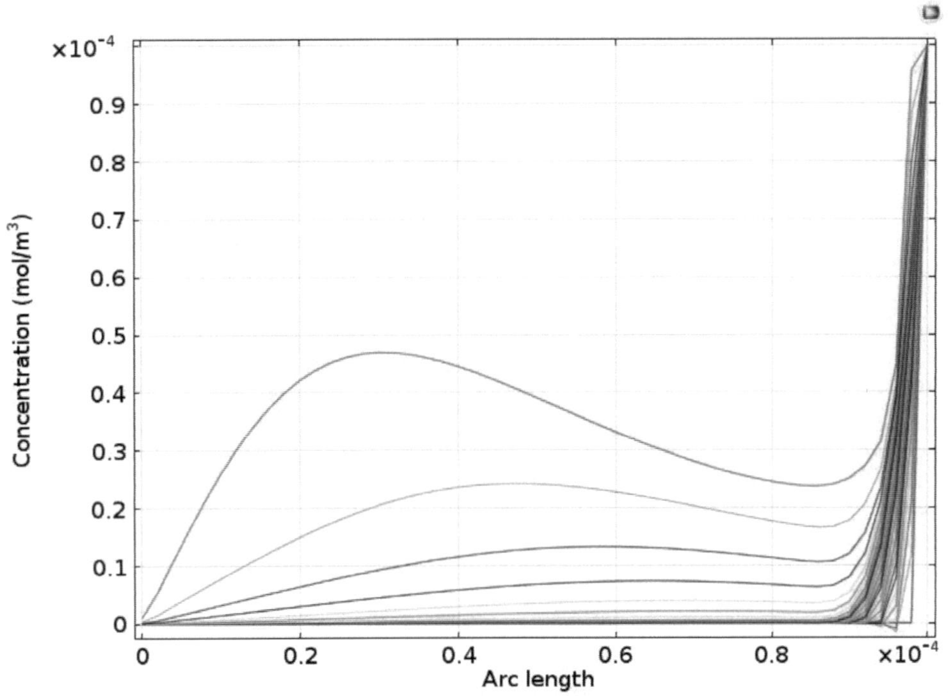

b

Figure 1. Electron (*a*) and ion (*b*) concentration in the cathode zone in the time intervals of 2 to 5 ns in increments of 0.1 ns

Due to intense ionization in the cathode layer, the uncompensated positive charge is gradually accumulated (Fig. 2), and begins to influence the external electric field, displacing it from the zone of increased concentration of electrons to the boundary of this zone. The zone with increased ionization is also moved there. This leads to increasing the concentrations of ionized particles before the boundaries (Fig. 1,*b*). Thus the zone of high concentration of charged particles changes its border again. Electron cloud starts to spread towards the cathode from the ion cloud. From this point it can be called "reflected" electron cloud. At this time the state of the zone with high concentration of electrons changed to the plasma state –the field is displaced from here, the bulk charge is practically zero.

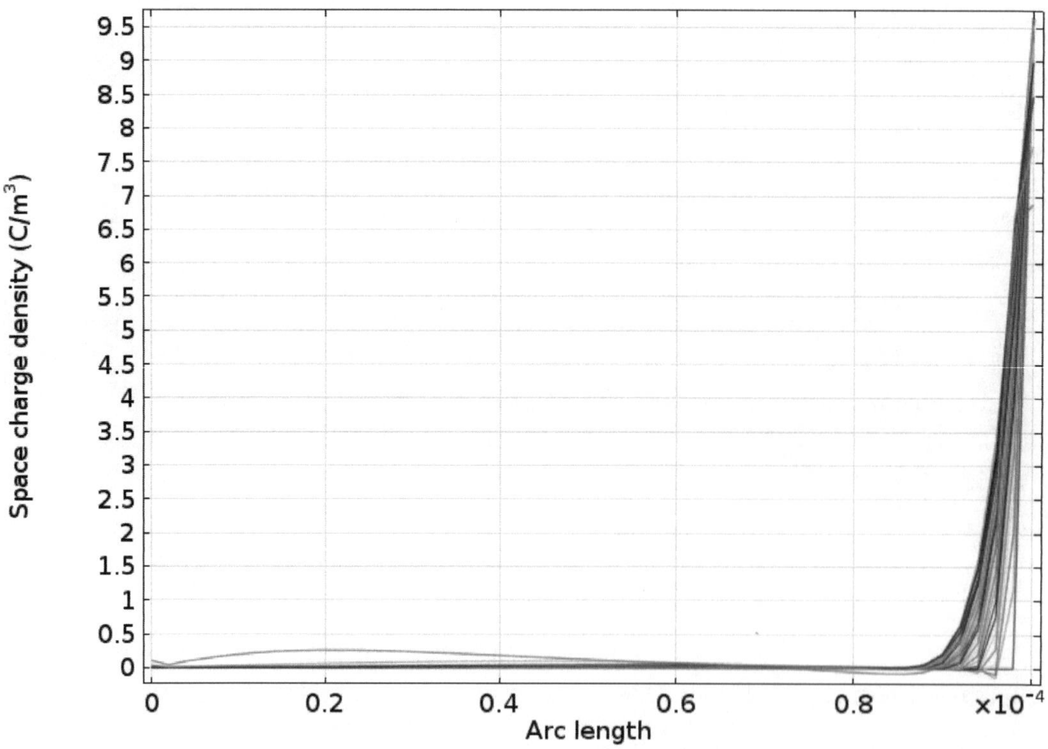

Figure 2. The diagram of distribution of the bulk charge [C/m^3] in the cathode zone at the moments of time of 2 to 5 ns in increments of 0.1 ns

The above-stated correlates the current understanding of the formed cathode spot structure, which is characterized by the plasma channel, and its field is displaced to the outer boundary of the channel [14,15]. The field peak in this case is located at the front boundary between the cathode zone and the ion cloud. If the arc is developed in the field formed by the electrodes, practically not having changed it, then since the formation of the plasma channel the conductive zone is developed at the expense of the newly formed plasma, deforming this field and forming a “plasma reaction zone” [16,17]. These peaks become new centers of the ionization processes where emission electrons are involved as seed electrons. A new plasma zone with high concentration of electrons ionization is generated as a result of ionization, which lengthens and narrows the plasma channel, displacing the electric field. The ionization wave moves as described above.

The rate of arc current increase is the rate of ionization wave, which can be much higher than a rate of ions move [18].

The field intensity in the plasma channel is less than in front of it. The field intensity it is too low and the impact of ionization in the channel does not occur. The key role is played by the loss of electrons due to recombination process.

In this paper the structure of the plasma column in the uniform field has been calculated within the computer model of the cathode spot. The principle advantage of this model is use of the least possible number of equations to describe all the basic parameters of the cathode spot of a vacuum arc discharge.

It has been shown that the field intensity of the plasma column is significantly reduced the column is surrounded by the bulk charge of the slow positive ions, providing high field intensity in front of it. At the stage of a plasma reaction zone ionization of the metal vapor is predominantly carried out in a thin layer in front of the ions cloud, while ion recombination predominates in the column. Ionization provides a constant elongation of conducting plasma column and moving the bulk charge layer on the outer boundary of the current state, i.e., towards the cathode. This, in turn, leads to a displacement of the local maximum of the electric field and to the displacement of the zone of intense ionization, and provides the movement of the so-called ionization wave.

Acknowledgements. The work was performed with a support of the State R&D Task of the Ministry of Education and Science of the Russian Federation (Task No. 11.370.2014/K).

References

[1] Ushakov A, Karpov I, and Lepeshev A 2016 *Vacuum* 133 25. doi:10.1016/j.vacuum.2016.08.007.
[2] Ushakov A, Karpov I, Lepeshev A, Petrov M and Fedorov L 2014 JETP Letters 99 99. doi: 10.1134/S002136401402009X.
[3] Ushakov A, Karpov I, Lepeshev A and Petrov M 2015 J. Appl. Phys. 118 023907. http://dx.doi.org/10.1063/1.4926549.
[4] Ushakov A, Karpov I, Lepeshev A, Fedorov L and Shaikhadinov A 2016 *Technical Phys.* **86**, 103. doi: 10.1134/S1063784216010230.
[5] Ushakov A, Karpov I and Lepeshev A 2016 *Technical Phys.* **86**, 260. doi: 10.1134/S1063784216020262.
[6] Lepeshev A, Karpov I, Ushakov A and Nagibin G 2016 *J. of Alloys and Compounds* **663** 631. doi:10.1016/j.jallcom.2015.12.168.
[7] Lepeshev A, Karpov I, Ushakov A, Fedorov L and Shaikhadinov A 2016 *Intern. J. of Nanoscience* **15** 1550027.
[8] Karpov I, Ushakov A, Lepeshev A and Zharkov S 2016 *Vacuum* **128** 123. doi: 10.1016/j.vacuum.2016.03.025.
[9] Fedorov L, Karpov I, Ushakov A and Lepeshev A 2015 *Inorg. Mater.* **51** 25. doi: 10.1134/S0020168515010057.
[10] Ushakov A, Karpov I, Lepeshev A, Petrov M and Fedorov L 2015 *Phys. of the Solid State* **57** 919. doi: 10.1134/S1063783415050303.
[11] Rudenko K, Miakonkih A, Rogojin A, Bogdanov S, Sidorov V and Zelenkov P 2016 *IOP Conf. Ser.: Mater. Sci. and Engineering* **122** 012029. doi: 10.1088/1757-899X/122/1/012029.
[12] Telegin S and Draganyuk O 2016 *IOP Conf. Ser.: Mater. Sci. and Engineering* **122** 012033. doi: 10.1088/1757-899X/122/1/012033.
[13] Korenyugin D, Martsinovsky A, Orlov K 2009 *Tech. Phys. Lett.* **35** 944. doi: 10.1134/S1063785009100204.
[14] Rakhovskii V 1976 *IEEE Transactions on Plasma Science.* **PS-4** 81. doi: 10.1109/TPS.1976.4316943.
[15] Mesyats G and Oks E 2013 *Tech. Phys. Lett.* **39** 687. doi 10.1134/S1063785013080105.
[16] Harris L 1982 *Cathodic processes. Vacuum arc.* (Moskow: Mir)
[17] Tsventoukh M, Barengolts S, Mesyats V and Shmelev D 2013 *Tech. Phys. Lett.* **39** 933. doi 10.1134/S1063785013110138.
[18] Daalder J. 1975 *Phys. D: Appl. Phys.* **8** 1647. doi: 10.1088/0022-3727/8/14/009.

IOP Conf. Series: Materials Science and Engineering **155** (2016) 012014 doi:10.1088/1757-899X/155/1/012014

Microstructures and Tribological Properties of Al-Cu-Fe quasicrystalline plasma sprayed coatings

A A Lepeshev[1,2], E A Rozhkova[1], I V Karpov[1,2], A V Ushakov[1,2], L Yu Fedorov[1,2], E A Dorozhkina[3], O N Karpova[3]

[1] Siberian Federal University, Krasnoyarsk, 660041 Russia

[2] Krasnoyarsk Scientific Center, Siberian Branch of the Russian Academy of Sciences, Krasnoyarsk, 660036 Russia

[3] Reshetnev Siberian State Aerospace University, Krasnoyarsk, 660014 Russia

E-mail: sfu-unesco@mail.ru

Annotation. The physical, mechanical, and tribological properties of quasicrystalline coatings based on the $Al_{65}Cu_{23}Fe_{12}$ alloy prepared by plasma spraying have been investigated. The specific features of the phase formation due to the competitive interactions of the icosahedral ψ and cubic β phases have been elucidated. A correlation between the microhardness and the content of the icosahedral phase in the coating has been determined. The decisive role of the quasicrystalline phase in the formation of high tribological characteristics of the coatings has been revealed and tested.

Quasicrystalline alloys, which are characterized by an unusual quasiperiodic structure and a crystallographically forbidden fivefold symmetry axis, possess a unique set of chemical, physical, and mechanical properties [1-7]. In this respect, noteworthy are quasicrystalline Al-Cu-*Me* alloys (where *Me* is Fe, Cr, Mn, or V) with a relatively small specific weight, which have high values of the chemical durability, elastic modulus, hardness, and wear resistance and a low value of the friction coefficient. The development of compositions and technologies for processing of these alloys has opened up wide possibilities for their use as antifriction materials in friction units. The high efficiency of friction pairs can be achieved both through the use of composites with a quasicrystalline alloy hardener and through the direct deposition of quasicrystalline coatings.

Among the most promising methods for fabricating units from quasicrystalline alloys is powder metallurgy, in particular, plasma spraying [8-21]. The use of technological capabilities of this method makes it possible, during the spraying of quasicrystalline coatings, not only to control their structure and phase composition but also to purposefully form composites with a specified ratio of the base material and quasicrystalline alloy components.

The purpose of this study is to investigate the processes of the structure and phase formation in quasicrystalline coatings under different technological conditions of their deposition and to perform tribological tests of the prepared coatings.

The investigations were performed on samples of coatings deposited on the outer surface of the copper rings using an oscillating plasma torch with coaxial powder injection into the cathode region [9, 10]. The temperature conditions of deposition were controlled by a thermocouple connecting the inner part of the ring and the outer surface. The ligature and initial powders of the $Al_{65}Cu_{23}Fe_{12}$ alloy for plasma spraying were prepared using the gas atomization method. The dispersion of the obtained powders was 25-106 μm.

The main parameters of the plasma spraying of quasicrystalline coatings based on the Al-Cu-Fe alloy are presented in below. Different thermal conditions of plasma spraying of the coatings were provided by varying the temperature of the copper substrate T_k.

- Arc current 200-250 A

- Power 13-17 kW
- Plasma-forming gas flow rate 0.008-0.0127 kg/s
- Carrier gas flow rate 0.08-0.12 kg/s
- Gas flow rate at the cross section 0.0065 kg/s
- Spraying distance 110-130 mm
- Powder flow rate 0.046-0.056 kg/s

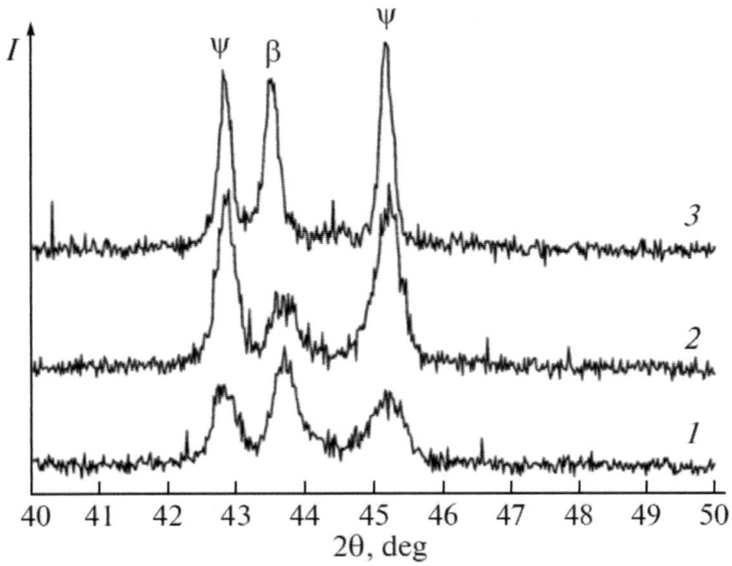

Figure 1. X-ray diffraction pattern of the $Al_{65}Cu_{23}Fe_{12}$ quasicrystalline alloy coatings deposited at different temperatures of the substrate T_k = (*1*) 500, (*2*) 650, and (*3*) 850°C.

The thickness of the deposited coatings varied in the range from 1.65 to 2.75 mm.

The X-ray structural and powder diffraction analyses of the initial powders and deposited quasicrystalline coatings were performed on a DRON-3 automated diffractometer (CuK_{α} radiation). The X-ray diffraction patterns were processed using the DRON-3 and ORIGIN-4 software programs.

The microstructure and morphological features of the quasicrystalline powders and coatings were examined on an MBI-15 optical microscope and a JEM-100C transmission electron microscope equipped with an EM-ASID-4 scanning adapter and an image processing system with the IMAGE-C software package.

The microhardness of the deposited quasicrystalline coatings was determined on a PMT-3 microhardness tester using the method of recovered indentation of a four-sided diamond pyramid.

The tribological tests were carried out on the upgraded SMT-1 friction machine according to the finger-disk scheme under conditions of dry friction.

The X-ray structural investigations of the initial powder and deposited coatings revealed that the coatings have a heterophase structure (ψ, β, λ, θ phases) with the competitive formation of the main phases, namely, the icosahedral ψ phase and the cubic β phase.

Figure 1 shows the X-ray diffraction patterns (filtered CuK_{α} radiation) of the quasicrystalline coatings deposited at different temperatures of the substrate. It can be seen from this figure that intense diffraction peaks lie in the range of angles 2θ = 42°-46° and that the structure of the coatings is heterogeneous and consists of a mixture of two main phases: the icosahedral ψ phase and the cubic β phase. The influence of the substrate temperature T_k on the phase formation and structure of the deposited quasicrystalline coatings manifests itself as a change in the intensity ratio of the diffraction peaks corresponding to the y and p phases, for example, (422222), (110), and (420024).

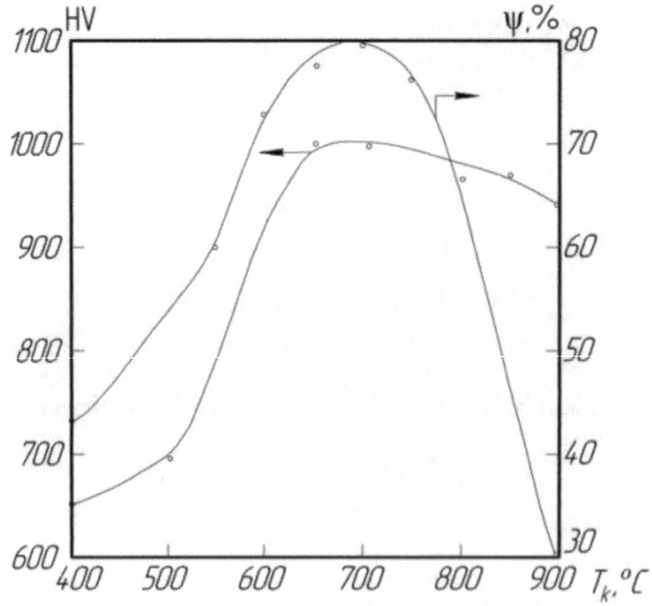

Figure 2. Dependences of the microhardness HV and the weight content of the icosahedral ψ phase in the coatings on the substrate temperature T_k during the plasma spraying.

The weight content of the icosahedral phase in the deposited coatings was determined from the X-ray powder diffraction analysis data and the results obtained in [10, 22]. The calculated dependence of the weight content of the icosahedral phase in the coatings on the deposition conditions is shown in Fig. 2. It can be seen that, with an increase in the substrate temperature, beginning from T_k ~ 400 °C, the fraction of the icosahedral phase increases, reaches a maximum near T_k ~ 700 °C, and then decreases to 33% at T_k ~ 880 °C.

Thus, the highest content of the icosahedral phase in the coatings (80%) is achieved upon their deposition on the substrates at a temperature of 600-750 °C.

The temperature conditions of the formation of quasicrystalline coatings are primarily responsible for their morphology. Figure 3 shows the cross sections of the coatings produced from the $Al_{65}Cu_{23}Fe_{12}$ alloy at different substrate temperatures T_k = 500, 650, and 850 °C. It can be seen from this figure that the morphological pattern (microcracks and pores, their concentration, size, and distribution) depends substantially on the conditions of the formation of the coatings. In particular, at the substrate temperature T_k = 500 °C (Fig. 3a), the coating is characterized by a large number of microcracks and cracks starting predominantly from the outer surface and, in some cases, extending through the entire thickness of the deposited layer. The pore formation nonuniformly occurs throughout the thickness of the coating, and the porosity reaches 20% near the substrate bottom and 10% on the free surface. In this case, an increase in the thickness of the coating leads to a decrease in the average pore size. In general, the pores have a rounded shape; however, near the inner surface of the deposited coating, there are pores that have an irregular shape and coalesce into microcracks [9, 11].

As the substrate temperature increases to T_k = 650°C, the morphological pattern of the cross section changes (Fig. 3b). The pores acquire a spherical shape, the average pore size decreases, and the pore volume distribution becomes more uniform. Moreover, as in the case of the substrate temperature T_k = 500°C, the coating also contains microcracks, even though to a lesser extent. Apparently, this is associated both with the specific features of the plasma spray technology and with the mechanical properties of the quasicrystalline alloy under investigation. Since the initial alloy has a high brittleness [4, 5], large internal stresses that are characteristic of rapid quenching lead not to plastic deformation but to the appearance of microcracks and to brittle fracture of deposited particles, which is clearly seen in Fig. 3 at the substrate temperatures T_k = 500 and 650 °C. The mechanism of the formation of long cracks arising on the side of the free surface of the deposited alloy coating is associated with the large shrinkage stresses generated during the rapid quenching under specified deposition conditions.

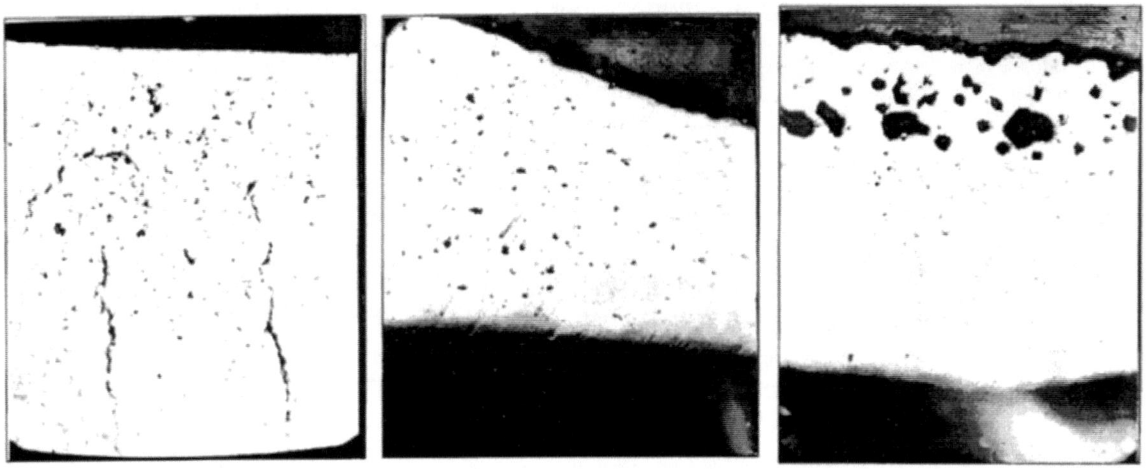

a b c

Figure 3. Morphology of the cross section of the coatings (magnification, ×90) deposited at different substrate temperatures T_k = (*a*) 500, (*b*) 650, and (*c*) 850°C.

At the substrate temperature T_k = 850 °C, the surface of the coating undergoes melting. This process leads to the fact that, in the bulk of the deposited coating, typical pores either are absent at all or have very small sizes, less than 1 μm. However, as can be seen from Fig. 3c, the alloy region near the free surface contains large pores of the bubble type with a size of 80 μm. The appearance of this layer can be associated with the intensive process of gas release through the molten region of the coating.

In order to better reveal the microstructure of the coating, these samples were subjected to selective etching. The cross sections of the samples of the $Al_{65}Cu_{23}Fe_{12}$ alloy coatings are shown in Fig. 4. It can be seen that the structure of these coatings is anisotropic and consists of uniform corrugated layers without foreign inclusions formed from molten particles, which is characteristic of the plasma spraying technology. The majority of the molten particles have a disk-like shape and are formed upon the deformation and solidification of the initial powder particles on the substrate, which were completely melted in the plasma. In addition, the coatings contain large particles, which were exposed to a high-temperature plasma jet but solidified before the collision with the substrate. The appearance of rounded small particles in the coating is caused by the spraying of larger droplets of the melt on the substrate. Furthermore, the deposited coatings contain particles of the mixed type, which combine features of several types of particles.

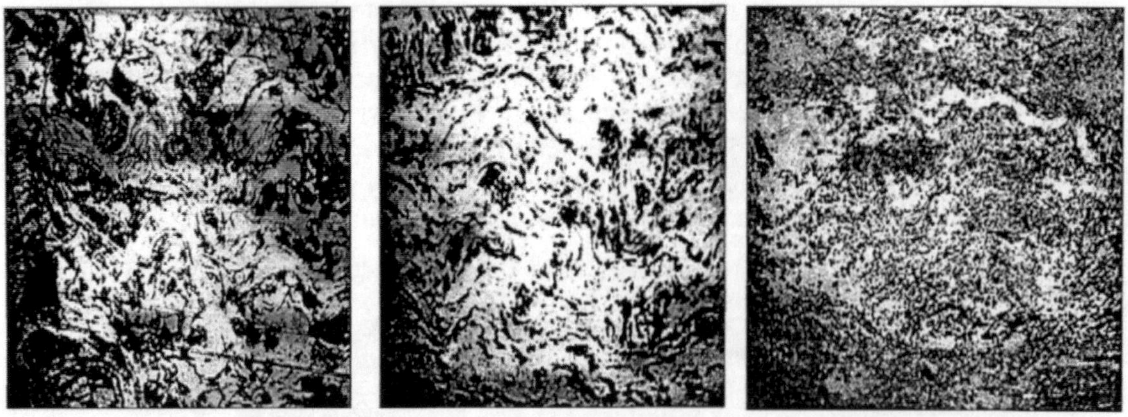

a b c

Figure 4. Microstructure of the $Al_{65}Cu_{23}Fe_{12}$ alloy coatings (magnification, ×630) deposited at different substrate temperatures T_k = (*a*) 500, (*b*) 650, and (*c*) 850°C.

The interparticle layers formed during the deposition contain pores with the percentage of 6-10% depending on the conditions used for the preparation of the coatings. In particular, the coatings deposited at the substrate temperature T_k = 500 °C are characterized by the lowest density. As the substrate temperature increases from T_k = 500 °C (Fig. 4a) to T_k = 850 °C (Fig. 4c), the corrugated structure of the coatings becomes denser, and the content of pores in the bulk of the material significantly decreases.

The microdiffraction pattern of the fracture surface of the deposited quasicrystalline coating consists of bright flat stepped regions. This pattern is characteristic of brittle fracture of the material. The cracking and fracture of the coating, as a rule, occur at the grain boundaries. A more detailed examination of the fracture surface revealed precipitates of individual powder particles and conglomerates forming the coating.

A comparative analysis of the microhardness was performed both on the initial quasicrystalline particles and on the deposited coatings based on them. The microhardness was determined on the polished surfaces of the samples with a roughness *Ra* = 0.34 by the diamond pyramid with an indentation load of 50 g for individual particles and an indentation load of 100 g for the deposited coating. According to the results of the measurements, the initial quasicrystalline particles have a microhardness of the order of 850-900 HV. Such high values of the microhardness are determined by the non-porous monolithic structure and the sufficiently high content of the icosahedral phase (~68%) in the bulk of the initial powders.

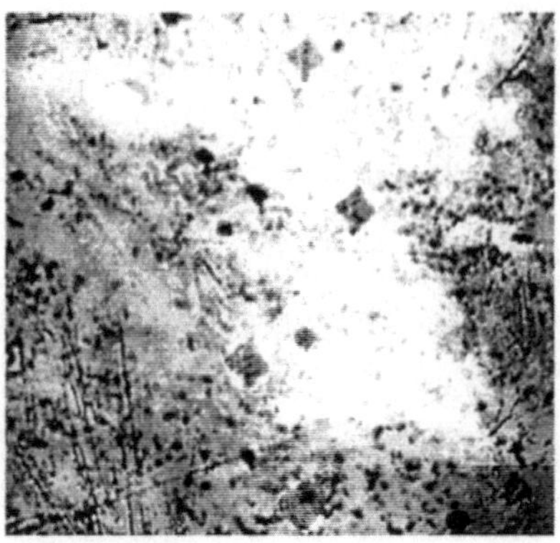

Figure 5. Measurements of the microhardness on the cross section of the coating deposited at the substrate temperature T_k = 650°C (magnification, ×630).

IOP Conf. Series: Materials Science and Engineering **155** (2016) 012014 doi:10.1088/1757-899X/155/1/012014

The microhardness of the coatings was examined on the samples prepared under different deposition conditions and with different contents of the icosahe- dral phase. Figure 2 shows the combined dependences of the microhardness and the content of the icosahe- dral phase in the coatings on the temperature conditions of their deposition. It can be seen that the microhardness of the material increases both with an increase in the substrate temperature and with an increase in the weight content of the icosahedral phase. In particular, the microhardness of the coatings with a 80-percent content of the icosahedral phase reaches values of the order of 1000 HV. With a further increase in the substrate temperature $T_k > 700$ °C, although the content of the icosahedral phase in the coatings decreases, their microhardness remains rather high. Apparently, this is associated with the formation of the molten layer and with the decrease in the porosity of the coatings.

Noteworthy is a significant increase in the brittleness of the Al-Cu-Fe alloy, which is evidenced by the tests on determination of the microhardness. After the indentation of a Vickers pyramid into the coating, radially directed cracks arise around the expression, which propagate mainly from the corners of the quadrangle.

The homogeneity of the properties of the deposited coatings was determined by analyzing the microhardness distribution in the cross sections of the samples. These measurements were performed on polished cross sections of the coatings with a step of 0.2 mm, beginning from the inner surface. The results of these investigations are presented in Fig. 5.

It can be seen that the structure of the indentations is identical almost over the entire thickness of the coating, which indicates a high uniformity of the deposited layers. The observed individual local overshoots of the values, most likely, can be explained by the fact that, during the measurements, the indenter falls into the interparticle space or into the region of pores of the deposited coating. The calculated values of the microhardness according to these indentations are presented in Fig. 6.

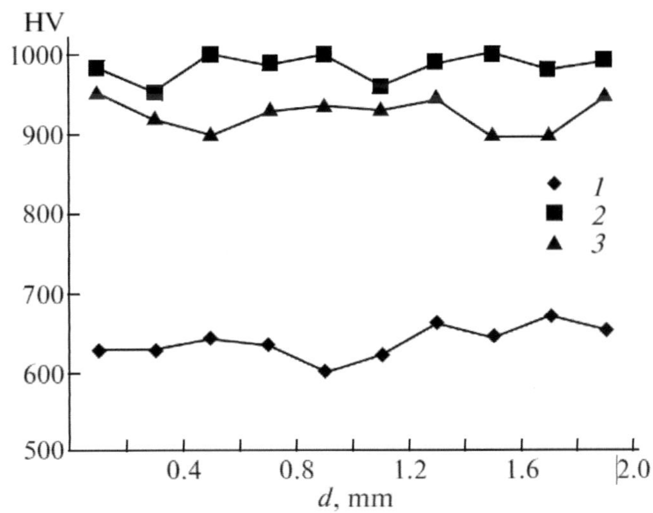

Figure 6. Change of the microhardness HV in the cross section of the sample, beginning from the inner surface of the coatings deposited at different substrate temperatures T_k = (1) 500, (2) 650, and (3) 850 °C.

The tribological (friction and wear) tests were carried out on the quasicrystalline coatings with the best combination of physical and mechanical properties, such as the content of the icosahedral phase, hardness, porosity, adhesion, etc.

The friction coefficient was determined using profilography with a Vickers pyramid under a load of 30 N for the Al-Cu-Fe alloy coatings deposited at different substrate temperatures

T_k. The performed investigations demonstrated that the friction coefficients of the deposited coatings differ only slightly from each other and their values lie in the range μ = 0.15-0.20 (Table 1).

Table 1. Friction coefficients of the quasicrystalline coatings prepared under different technological conditions of deposition

T_k, °C	400	500	650	750	850
Friction coefficient μ	0,17-0,20	0,18-0,20	0,15-0,17	0,15-0,18	0,16-0,19

The somewhat larger friction coefficients, which are characteristic of quasicrystalline coatings deposited at T_k = 400-500°C, can be explained by their higher porosity, lower content of the icosahedral phase, and microhardness.

The tests for wear resistance of the deposited coatings were carried out under dry friction conditions at different loads, wear rates, and wear times. The hardened steel ShKh15 was used as a counterbody.

It was found that the studied coatings possess a high wear resistance: the weight loss of the alloy was small and changed only slightly depending on the applied load, wear rate, and testing time (Fig. 7).

The morphological pattern of the coating surfaces after abrasion was smooth and did not exhibit grooves, adhesive wear traces, microcracks, and noticeable plastic deformation at the edges of the abrasion zone.

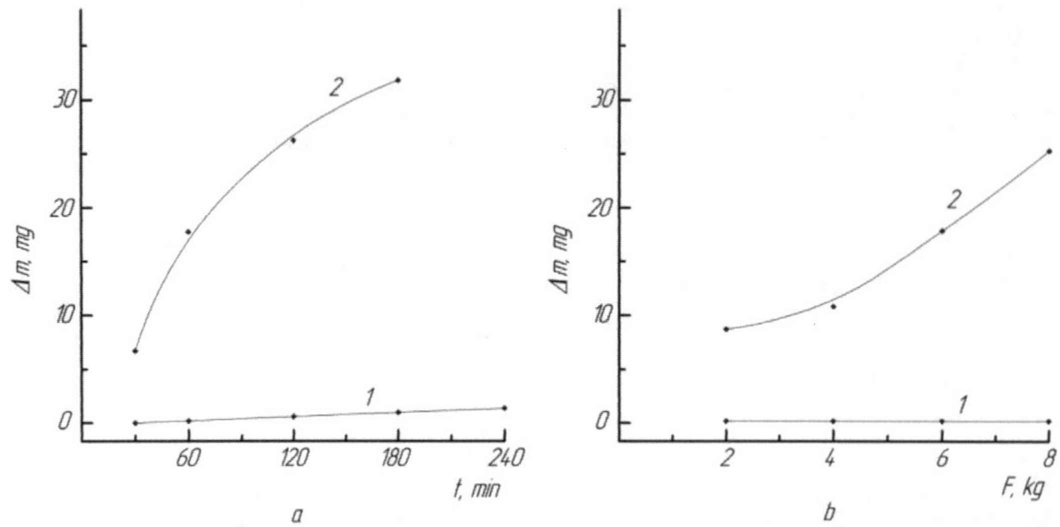

Figure 7. Dependences of the weight loss of (*1*) the $Al_{65}Cu_{23}Fe_{12}$ coating and (*2*) the titanium alloy VT 1-0 on the test time and (*b*) the applied load.

This behavior of the quasicrystalline coatings in testing for friction and wear differs significantly from the wear picture of the reference alloy VT 1-0 (Fig. 7, curve 2). The results of the tests performed for the VT 1-0 alloy under conditions similar to those used in the experiments with the Al-Cu-Fe coatings have demonstrated that, even in the early stages, an increase in the specific pressure leads to a sharp increase in the wear rate. The morphology of the worn surface of the titanium alloy is typical of severe plastic deformation with scratches, deep grooves, and traces of gripping and metal transfer. This eventually determined the low wear resistance of the titanium alloy VT 1-0 as compared to the quasicrystalline coatings.

The observed high level of wear resistance of the quasicrystalline coatings under dry friction conditions is associated with both the low value of the friction coefficient and the high hardness of the deposited alloy coating. Since the friction coefficients of the studied samples differ only slightly, the dominant role in the formation of high wear resistance, apparently, should belong to their high hardness.

Thus, from the practical point of view, the quasicrystalline coatings based on the Al-Cu-Fe alloy have important tribological characteristics under dry friction conditions. The technological principles underlying the preparation of such coatings are based on the justified choice of the chemical composition of the initial powders and the conditions of their deposition taking into account the regularities of the processes occurring in quasicrystalline alloys.

The morphological pattern of quasicrystalline coatings has a typical layered microstructure with a characteristic volume distribution of the pores and microcracks, which depends on the deposition conditions, in particular, on the substrate temperature T_k. The melting of the surface layer (T_k = 850°C) leads to a decrease in the porosity in the bulk of the coating and to an increase in the gas release through the molten layer.

The formation of the quasicrystalline coating is accompanied by the competitive interaction of two main phases, namely, the icosahedral ψ phase and the cubic β phase. As the temperature of the substrate increases, the fraction of the icosahedral phase first increases, reaches 80% at T_k = 700°C, and then decreases to 33% at T_k = 880°C.

It has been found that there is a correlation between the microhardness and the content of the icosahedral phase in the coating. The microhardness of the coatings increases with an increase in the content of the y phase and reaches HV = 900-1000 when the content of the ψ phase is 80% (T_k = 700°C).

Acknowledgments. The work was performed with a support of the State R&D Task of the Ministry of Education and Science of the Russian Federation (Task No. 11.370.2014/K).

References

[1] Sordelet D, Rozhkova E, Besser M, and Kramer M 2002 *Intermetallics* **10**, 1233.
[2] Sordelet D, Rozhkova E, Besser M, and Kramer M 2002 *Appl. Phys. Lett.* **80** 4735.
[3] Yang X, Kramer M, Rozhkova E, and Sordelet D 2003 *Scr. Mater.* **49** 885.
[4] Koster U, Liu W, Leibertz H, and Michel M 1993 *J. Non-Cryst. Solids* **153-154** 446.
[5] Noskova N and Ponomareva E 1994 *Fiz. Met. Metalloved.* **78** 34.
[6] Kronmuller H and Moser N 1988 *Z. Phys. Chem.* **157** 837.
[7] Bratkovskii A, Danilov Y and Kuznetsov G 1989 *Fiz. Met. Metalloved.* **68** 1045.
[8] Sordelet D, Kramer M. and Unal O 1995 *J. Therm. Spray Technol.* **4** 235.
[9] Lepeshev A, Sordelet D, Rozhkova E and Ushakov A 2011 *J. of Cluster Sci.* **22** 289. doi: 10.1007/s10876-011-0378-2.
[10] Ushakov A, Karpov I, and Lepeshev A 2016 Vacuum 133 25. doi:10.1016/j.vacuum.2016.08.007.
[11] Ushakov A, Karpov I, Lepeshev A, Petrov M and Fedorov L 2014 *JETP Letters* **99** 99. doi: 10.1134/S002136401402009X.
[12] Ushakov A, Karpov I, Lepeshev A and Petrov M 2015 *J. Appl. Phys.* **118** 023907. http://dx.doi.org/10.1063/1.4926549.
[13] Lepeshev A, Karpov I, Ushakov A, Fedorov L and Shaikhadinov A 2016 *Intern. J. of Nanoscience* **15** 1550027.
[14] Karpov I, Ushakov A, Lepeshev A and Zharkov S 2016 *Vacuum* **128** 123. doi: 10.1016/j.vacuum.2016.03.025.
[15] Lepeshev A, Bayukov O, Rozhkova E, Karpov I, Ushakov A and Fedorov L 2015 *Phys. of the Solid State* **57** 255. doi: 10.1134/S1063783415020249.
[16] Ushakov A, Karpov I, Lepeshev A, Petrov M and Fedorov L 2015 *Phys. of the Solid*

State **57** 919. doi: 10.1134/S1063783415050303.

[17] Ushakov A, Karpov I and Lepeshev A 2015 *Phys. of the Solid State* **57** 2320. doi: 10.1134/S1063783415110359.

[18] Fedorov L, Karpov I, Ushakov A and Lepeshev A 2015 *Inorg. Mater.* **51** 25. doi: 10.1134/S0020168515010057.

[19] Lepeshev A, Karpov I, Ushakov A and Nagibin G 2016 *J. of Alloys and Compounds* **663** 631. doi:10.1016/j.jallcom.2015.12.168.

[20] Rudenko K, Miakonkih A, Rogojin A, Bogdanov S, Sidorov V and Zelenkov P 2016 *IOP Conf. Ser.: Mater. Sci. and Engineering* **122** 012029. doi: 10.1088/1757-899X/122/1/012029.

[21] Telegin S and Draganyuk O 2016 *IOP Conf. Ser.: Mater. Sci. and Engineering* **122** 012033. doi: 10.1088/1757-899X/122/1/012033.

[22] Ebalard S and Spaepen F 1989 *J. Mater. Res.* **4**, 39.

Development of methods for equivalent transformation of GERT networks for application in multi-version software

M V Saramud[1], P V Zelenkov[2], I V Kovalev[3], D I Kovalev[4], I N Kartsan[5]

[1] Research engineer, Reshetnev Siberian State Aerospace University, Krasnoyarsk, Russia

[2] Associate professor, Reshetnev Siberian State Aerospace University, Krasnoyarsk, Russia

[3] Professor, Reshetnev Siberian State Aerospace University, Krasnoyarsk, Russia

[4] Researcher, Reshetnev Siberian State Aerospace University, Krasnoyarsk, Russia

[5] Associate professor, Reshetnev Siberian State Aerospace University, Krasnoyarsk, Russia

E-mail: kartsan2003@mail.ru

Abstract. The usage of information and control systems in various fields of production and research requires a certain level of reliability in the functioning of these systems. One of the most promising and already positively proven methodologies to ensure high reliability and fault tolerance of software is multi-version designing. The greatest value this factor have in the areas where control system failure can result in significant financial and material losses, as well as harm to human health and life.

Keywords: multi-version software, distributed information processing systems, Condor, check point.

1. Introduction

As has been repeatedly noted, multi-version software has a number of features, which allows to classify software on the basis of whether it is multi-version or not. One of the features is that scope of application of multi-version software is quite narrow, due to the fact that increasing the reliability of such an approach goes by software and resource redundancy[1]. As a result, large part of such programs used to control objects for which reliability is a key characteristic (such as the space industry). This fact, in turn, makes it difficult to obtain the actual technical characteristics of such systems [2]. On the other hand analysis of multi-version software could be based on distributed information processing systems. Currently, almost all of the big software systems are distributed. Distributed system – it is a system, in which processing of information focused not on one computer, but distributed between several computers. Often, in such systems the user's computer is controlled by a single operating system (OS), computer for collection of requests to process – the second, and computers-nodes can be controlled by a system different from the first two. In this case, these systems are called distributed heterogeneous data processing systems.

Such systems could be built with assistance of existing packs, for example project Condor. (http://www.cs.wisc.edu/condor).

Condor allows to use as part of a single cluster nodes, that not only differ in hardware, but which running under different operating systems, which makes it possible to use for computing the existing computer equipment and the existing communications thereby significantly reducing cost of creating the cluster. [3]

Condor has several modes of launch of calculating tasks: Standard, Vanilla, PVM, MPI, Globus, Java.

PVM, MPI, Globus and Java modes – it is compatibility support for programs, written using these libraries, so we will not consider them. The most interesting are the modes Standard (with the "backup") and Vanilla (without the "backup"). [4]

In «Standard» mode Condor makes checkpoints at a specified interval. Check Point – it is "snapshot" of the current state of the task. If migration of task is necessary (for example, a user has started to use a computer), then Condor creates an image of checkpoint, move it to another machine and resume calculation from the place where he stopped. If the calculating node finished its work with crash or communication breakdown with a node, when Condor puts on another node last stored image of the checkpoint task and resumes calculations. Thus, the task can continuously compute. over a long period of time. This mode corresponds to the mode complete a task with the periodic execution of backup of its status.

Vanilla mode allows to run random console applications, but does not allow to perform the migration and backup tasks. The figures bellow are functional diagrams of GERT-network modeling behavior of the system Condor when calculating tasks with a fixed duration in modes without backup and migration.

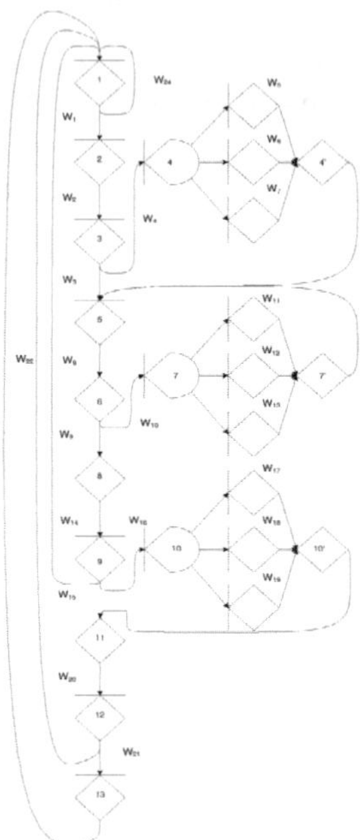

Figure 1. GERT-network, describing the work of CONDOR system in «Vanilla» mode

GERT-network shown in Fig. 1 consists of nodes corresponding to the start and completion of each operation and arcs representing the actual execution time of each operation.

Before you go to the search loops of the network lets produce some changes that significantly simplify network [5]. Suitable replacements are shown in Figs. 2-4.

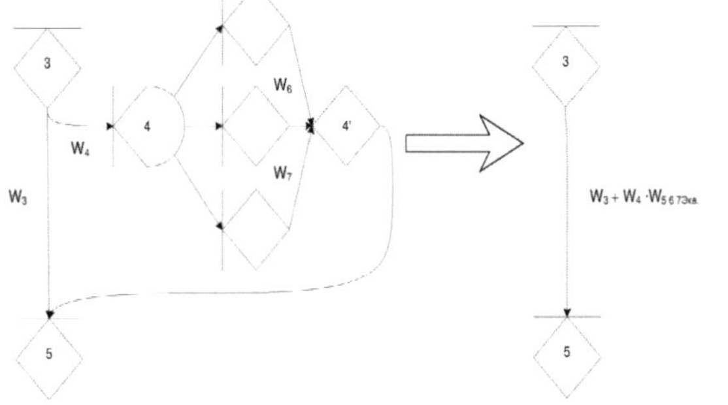

Figure 2. Application of equivalent transformation for nodes 3 and 5

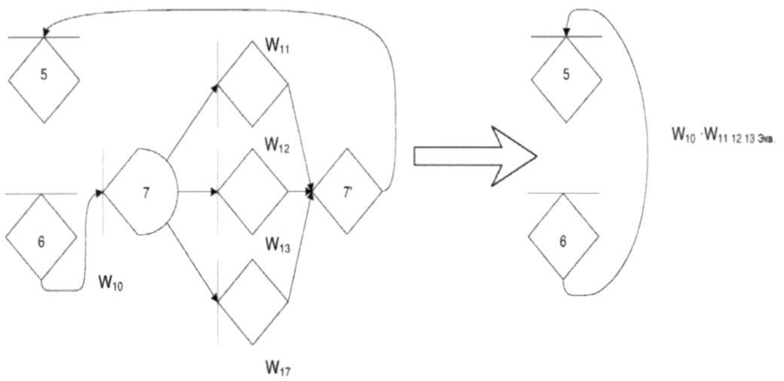

Figure 3. Application of equivalent transformations for nodes 5 and 6

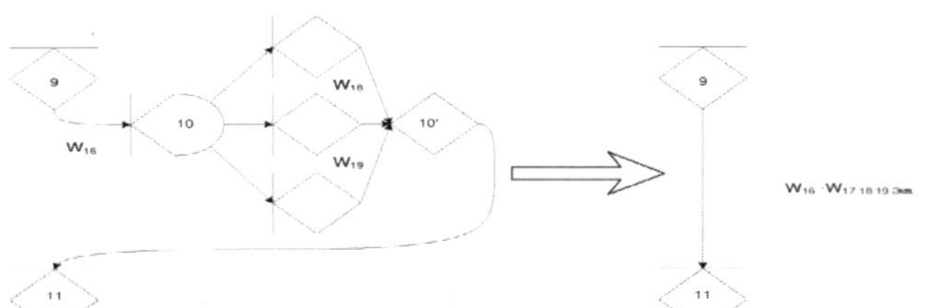

Figure 4. Application of equivalent transformation for nodes 9 and 11

The According to the network (see. Fig. 5) determine the appropriate equivalent coefficients of passing loops of the first and second order.

First order loops: W_{24}, $W_8 \cdot W_{10} \cdot W_{11,12,13eq.}$, $W_1 \cdot W_2 \cdot W_8 \cdot W_9 \cdot W_{14} \cdot W_{15}(W_3 + W_4 \cdot W_{5,6,7eq.})$, $W_1 \cdot W_2 \cdot W_8 \cdot W_9 \cdot W_{14} \cdot W_{16} \cdot W_{17,18,19eq.} \cdot W_{20} \cdot W_{22}$ и $W_1 \cdot W_2 \cdot W_8 \cdot W_9 \cdot W_{14} \cdot W_{16} \cdot W_{17,18,19eq.} \cdot W_{20} \cdot W_{21} \cdot \left(1/W_E \right)$

Second order loops: $W_8 \cdot W_{10} \cdot W_{11,12,13eq.} \cdot W_{24}$

Using topological equation, we get the following equivalent *W*-function for GERT-network describing the work without a backup system:

$$W_E(t) = \frac{W_1 W_2 W_8 W_9 W_{14} W_{16} W_{17,18,19eq.} W_{20} W_{21}}{1 - W_{24} - W_8 W_{10} W_{11,12,13eq.} - W_1 W_2 W_8 W_9 W_{14} W_{15} \left(W_3 + W_4 W_{5,6,7eq.}\right) - W_1 W_2 W_8 W_9 W_{14} W_{16} W_{17,18,19eq.} W_{20} W_{22} + W_8 W_{10} W_{11,12,13eq.} W_{24}}$$

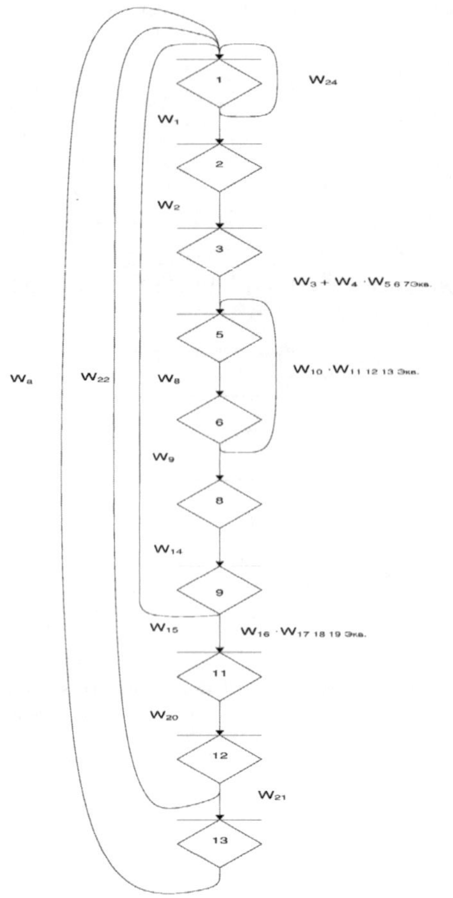

Figure 5. Equivalent GERT-network, which describe the work of CONDOR system «Vanilla» mode

3. Conclusion

Using these functions, obtaining the mathematical expectation and the dispersion becomes possible. As you can see, the methods of equivalent transformation of GERT-networks simplify the search for loops and network performance calculations.

References

[1] Kovalev P V, Laikov A N, Gritsenko S N 2009 Determination of reliability multi-version software using network analysis methods *Vestnik of Siberian State Aerospace University* **No. 1-2**, P 55-60

[2] Kovalev P V 2009 GERT-network analysis of multi-version software architecture *Uspekhi sovremennogo estestvoznaniya* **No. 9** P 161-164

[3] Kovalev I V 2005 Model evaluation time task on a cluster with a consistent and parallel architectures of data exchange *Sistemy upravleniya i informacionnye tekhnologii* **No. 3(20)** P 58-62

IOP Conf. Series: Materials Science and Engineering **155** (2016) 012015 doi:10.1088/1757-899X/155/1/012015

[4] Kovalev I, Zelenkov P, Ognerubov S 2015 The minimization of inter-module interface for the achievement of reliability of multi-version software *IOP Conference Series: Materials Science and Engineering* **No. 70 (1)**
[5] Kuznetsov P A, Kovalev I V, Zelenkov P V 2015 Dangerous failures in multifunctional systems *IOP Conference Series: Materials Science and Engineering* **No. 94**
[6] Kovalev I V, Zelenkov P V, Karaseva M V, Tsarev M Y, Tsarev R Y 2015 Model of the reliability analysis of the distributed computer systems with architecture 'client-server' *IOP Conference Series: Materials Science and Engineering* **No. 70 (1)**

Analyzing measurement errors for navigation parameters in on-ground short-range navigation systems based on pseudolites*

Yu L Fateev[1], V N Ratuschnyak[2], I N Kartsan[3], V N Tyapkin[4], D D Dmitriev[5], A E Goncharov[5]
[1] Associate professor, Siberian Federal University, Krasnoyarsk, Russia
[2] Associate professor, Siberian Federal University, Krasnoyarsk, Russia
[3] Associate professor, Reshetnev Siberian State Aerospace University, Krasnoyarsk, Russia
[4] Associate professor, Siberian Federal University, Krasnoyarsk, Russia
[5] Associate professor, Siberian Federal University, Krasnoyarsk, Russia
[6] Associate professor, Reshetnev Siberian State Aerospace University, Krasnoyarsk, Russia

E-mail: kartsan2003@mail.ru

Abstract. This article discusses coordinate measurement errors in short-range pseudolite navigation systems. An analysis of the components of the measurement error of radio navigating parameters has been performed; the geometric factor values have been calculated for various variants of constructing the system. The peculiarities of measuring the spatial orientation of objects using interferometric methods in on-ground short-range navigation systems have been ascertained.

Keywords: short-range radio navigation system, pseudo-satellite (pseudolite), geometric factor, coordinate measurement errors, spatial orientation.

1. Introduction

Modern traffic control systems – especially air and water traffic control systems – are challenged by the significant increase in the density of this traffic. In these circumstances modern high-precision positioning systems based on radio navigation are introduced at an ever-growing rate. Despite their increasing popularity, global navigation satellite systems (GNSS) are rarely used for autonomous navigating maintenance activities such as providing flight support services in the vicinity of aerodromes, supporting flight for unmanned aerial vehicles, assisting vessels in navigating in coastal waters and rivers, providing geodetic, cartographic and special services. The reason for this is the low noise immunity of GNSS which occasionally makes it impossible to comply with requirements of continuity and accuracy for navigation purposes.

One way to resolve this contradiction is by supporting the GNSS with a ground-based radio navigation system located in the vicinity of the navigation activities of a specified user. This enables the construction of a navigation field in a given area using ground-based pseudo-satellites (pseudolites) [1]. In such a radio navigation system, the high precision measurements for navigation parameters is proved by highly accurate geodetic coordinate

* The research had been conducted with financial support from the Russian Science Foundation grant (project №16-19-10089).

definition and mutual synchronization of the pseudolites, which are stationary navigation signal transmitters similar to the signals of GLONASS navigation satellites.

This arrangement for a navigation system significantly differs from local, regional or wideband differential GNSS subsystems which can be used only if there is a stable GLONASS signal, whereas the proposed navigation system based pseudolites can provide navigation services even in the absence of signals from a designated satellite.

2. Possible errors occurring during coordinate measurements of radio navigation systems based on ground-based pseudolites

The disposition of pseudolites within a specific area as well as their number in a short-range navigation system largely depends on the local terrain and the size of the coverage area. Thus, the area of an aerodrome would include the glide path of the aircraft; if we are speaking about river navigation then the main channel of the river would be the area covered by the navigation system. Despite this area being highly susceptible to various interference, it is necessary ensure high-precision continuous navigation services irrespective of the prevailing conditions. The limits of this are correspond to the boundaries in which the requirements for coordinate accuracy measurements and special orientation of angles are performed by the consumers of navigation services.

Various factors may influence the precision for measuring the coordinates of a moving object in a given area. These are some of the most widespread:

- The geometry of the pseudolite's location in relation to the navigation receiver (it is known as *the geometric factor*);
- Pseudo range measurement errors may be caused by the structure of the navigation signals and their selected processing method;
- Errors caused by the geodetic positioning quality of the pseudolite and the quality of mutual synchronization between transceivers and the Russian State Standard of the Coordinated Universal Time (UTC) scale of the state standard (similar to the ephemeris error in GNSS);
- Errors introduced by the signal propagation path;
- The power ratio of the desired received signal and interference.

The formation of a pseudolite short range navigation system arises a complex of tasks, including the selecting the optimal location for the pseudolites, the power output of the navigation signals, the structure of the navigation receivers surveillance schemes, which will provide a desired accuracy.

Measurements of coordinates of objects (*X*, *Y*, *Z*) in a short-range navigation system are performed using the standard range-difference (pseudorange) method:

$$R_i = \sqrt{(X_{mi} - X)^2 + (Y_{mi} - Y)^2 + (Z_{mi} - Z)^2} + C\Delta t, \qquad 1)$$

where X_{mi}, Y_{mi}, Z_{mi} are the coordinates of the i-th pseudo-satellite; X, Y, Z are the coordinates of the navigation receiver; C is the speed of light; Δt is the systematic error caused by discrepancy in time scales.

The value of Δt includes additional uncontrolled signal delays emerging on the signal propagation route.

The error in the measurement of the object's coordinates will depend on the hardware error of pseudorange measurement, the measurement error of the pseudolites' coordinates and errors in the path of signal propagation. For ground-based short-range navigation systems based on pseudolites using GLONASS signals the magnitude of the hardware pseudorange measurement error will be approximately equal to a similar error in the GNSS: it will be approximately 1.5 m [2].

The error of inaccurate coordinate determination for pseudolites and their mutual timing (analog ephemeris errors in GNSS) for a discussed navigation system will be significantly less than in the GNSS. This is the result of a high-precision definition of geodetic coordinates of the pseudolites and the possibility of their mutual synchronization using highly stable time and frequency standards. The error in the signal propagation path cannot be ignored due to rather small distances in the system.

Errors which depend on the properties and methods of navigation signal processing – so-called internal errors – are planned during the design stage for such a short-range navigation system. One advantage provided by the ground-based system over GNSS is the ability to plan external caused by geometric factors. It is possible to determine the location of a pseudolite in such a way, which would minimize its impact on an important navigation sector.

The practice of solving navigation problems demonstrates that the values of the geometric factor in the measurement of planimetric coordinates and altitude may significantly vary. Therefore it is necessary to individually assess its allowable values to ascertain any errors in the various components of coordinate positioning.

The essence of the geometric factor is as follows. To calculate the coordinates of an object, a system of nonlinear equations is used (1).The number of these equations is determined by the number of pseudolites. Nonlinear equations are characterized by a radius of curvature of the wave front, which is equal to the distance from the object to pseudolite. Since the radius of the wave front curvature is significantly larger than the range measurement error, the system of equations can be linearized at the point of signal reception. In this situation the conversion error will be close to linear. When performing linear transformation, the coordinate positioning error is described by a covariance matrix.

The covariance matrix may be obtained from a gradient matrix:

$$cov(r) = Gr^{T} Gr, \tag{2}$$

where Gr is the gradient matrix:

$$Gr = \begin{Vmatrix} \frac{\partial R_1}{\partial X} & \frac{\partial R_1}{\partial Y} & \frac{\partial R_1}{\partial Z} & \frac{\partial R_1}{\partial (C\Delta t)} \\ \frac{\partial R_2}{\partial X} & \frac{\partial R_2}{\partial Y} & \frac{\partial R_2}{\partial Z} & \frac{\partial R_2}{\partial (C\Delta t)} \\ \cdots & \cdots & \cdots & \cdots \\ \frac{\partial R_N}{\partial X} & \frac{\partial R_N}{\partial Y} & \frac{\partial R_N}{\partial Z} & \frac{\partial R_N}{\partial (C\Delta t)} \end{Vmatrix} = \begin{Vmatrix} \frac{X_{m1} - X}{R_1} & \frac{Y_{m1} - Y}{R_1} & \frac{Z_{m1} - Z}{R_1} & 1 \\ \frac{X_{m2} - X}{R_2} & \frac{Y_{m2} - Y}{R_2} & \frac{Z_{m2} - Z}{R_2} & 1 \\ \cdots & \cdots & \cdots & \cdots \\ \frac{X_{mN} - X}{R_N} & \frac{Y_{mN} - Y}{R_N} & \frac{Z_{mN} - Z}{R_N} & 1 \end{Vmatrix}, \tag{3}$$

where $\frac{X_{mN} - X}{R_{iN}} = \cos\alpha_i$; $\frac{Y_{mN} - Y}{R_{iN}} = \cos\beta_i$; $\frac{Z_{mN} - Z}{R_{iN}} = \cos\gamma_i$ is the direction cosines of the radius vector joining the *N*-th user *i*-th and the pseudolite; R_{iN} is the distance between the *i*-th pseudolites and *N*-this the user.

If it is assumed that the pseudorange measurements are of equal accuracy, it is possible to divide the measurement errors and transformations. In this case, the coordinate determination error will be equal to the product of the covariance matrix and the dispersion ranging; the resulting error can be estimated through the trace of the covariance matrix.

Generally, the covariance matrix contains non-zero off-diagonal elements which describe individual components of the coordinate correlation error. However, we can choose a system of coordinates in which the coordinates are uncorrelated errors, whereas the covariance matrix is diagonal. The elements of the main diagonal are the variance component coordinates; the resulting variance will be equal to their sum. At the same time, the sum of the diagonal elements of the covariance matrix for linear transformation does not change as it is a linear invariant tensor of rank 2.Consequently, the trace of a matrix – a ratio that shows what fold increases the variance of coordinates relative to the dispersion of the measured distance. The shape of the covariance matrix depends only on the relative position of pseudolites and navigation receiver, hence, giving it its name – the geometric factor.

The coefficients of the gradient matrix are the direction cosines in the pseudolite. Practically, the square root of the covariance matrix path, which shows the degree of increase in the standard deviation of the positioning error relative to the standard deviation range of the measurement error, is used:

$$\begin{aligned} \sigma_{x,y} &= \sqrt{cov(r_{11}) + cov(r_{22})}, \\ \sigma_z &= \sqrt{cov(r_{33})}, \end{aligned} \tag{4}$$

where $\sigma_{x,y}$, σ_z is, respectively, the standard deviation of the error in determining coordinates in the horizontal and altitude.

3. Calculating geometric factors for various pseudolite location configurations

Let's perform the calculation of the values of the geometric factor for a variety of pseudolite location configurations on the ground using the Math lab mathematical software environment [3].

3.1 Configuration No. 1

In the first configuration three pseudolites are placed on the corners of an equilateral triangle of side 50 km; one more pseudolite is positioned in the center of the triangle. Nominally, the pseudolites are situated on the earth and at the same altitude ($Y_m = 0$). Figs. 1*a*, 1*b* demonstrate the calculated values of the geometric factor for measurements of planimetric coordinates and altitude.

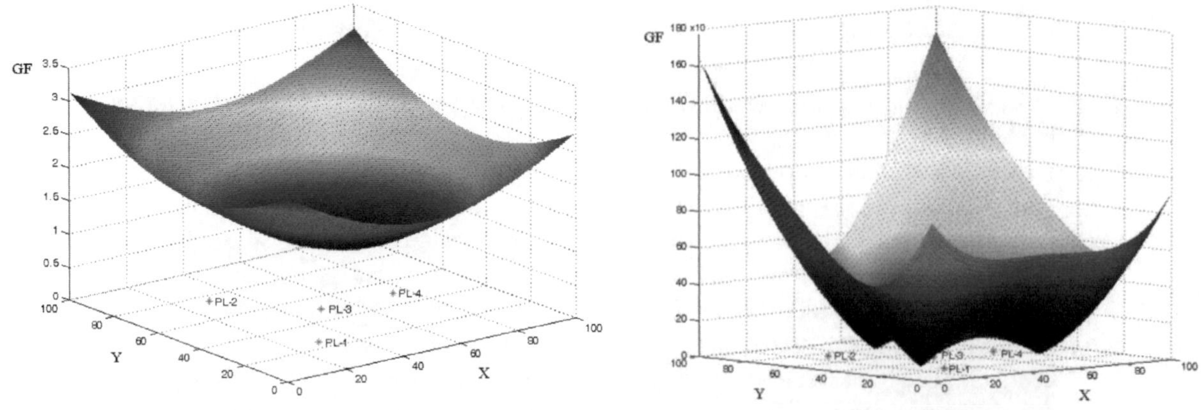

a) *b*)

Figure 1. Configuration No. 1: *a* is the geometric factor for measurement of planimetric coordinates; *b* is the geometric factor for altitude measurements

The geometric factor for measuring the coordinate's altitude exceeds the geometric factor for measuring planimetric coordinates. This can be explained by the fact that the pseudolites are located in one plane; as the object is brought closer to this plane, the altitude measurement error increases (Fig. 1*b*), whereas the planimetric coordinate error is almost unaffected by the object's altitude and is 1 ... 2.

For on-ground short-range navigation systems with the pseudolite positioned at the same altitude it is almost impossible to minimize the measurement error for the altitude with a large geometric factor. One option to solve this problem is by positioning the pseudolites on high masts, installing them on the tops of buildings, or by using the local terrain.

3.2 Configuration No. 2

Let's consider a configuration in which six pseudolites are situated at an altitude of 0 m on the sides of a square of side 60 km; two pseudolites are located in the center of the square at an altitude of 500 m (Fig. 2a).

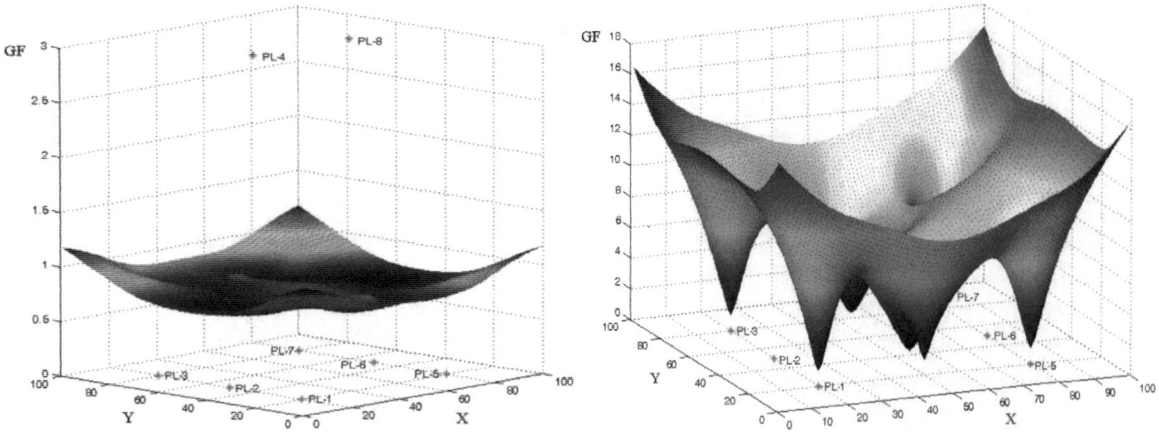

a) *b*)

Figure 2. Configuration No. 2: *a* is the geometric factor for measurement of planimetric coordinates; *b* is the geometric factor for altitude measurements

The geometric factor for measuring planimetric coordinates by shifting the location of the pseudolite configuration has not changed and does not exceed the value of 1 ... 1.3. Positioning the two pseudolites at an altitude of 500 meters significantly reduces the geometric factor for altitude measurements (10 in the short-range and 50 in the middle range (Fig. 2*b*).

3.3 Configuration No. 3

In the following configuration four pseudolites are situated at four corners of a square (the distance between them is 60 km); four more pseudolites are elevated to an altitude of 1,000 m (Fig. 3*a*).

The altitude of the suspended pseudolites can increase the area in which the value of the geometric adjustment factor is less than 10 (Fig 3*b*) while the geometric factor for coordinates in the short-range zone has not changed.

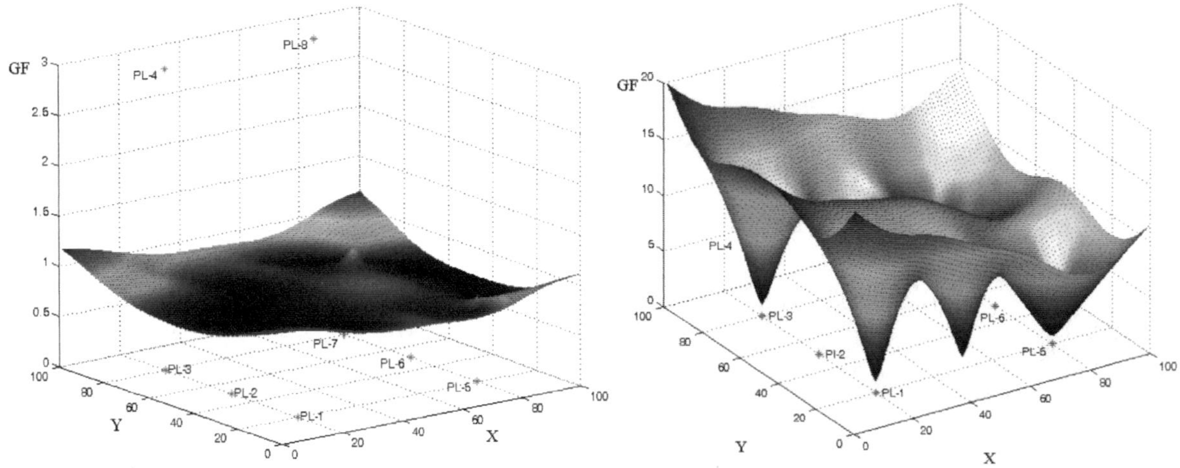

a) *b*)

Figure 3. Configuration 3: *a* is the geometric factor for measuring planimetric coordinates; *b* is the geometric factor for altitude measurements

Thus, the ground-based radio navigation system enables pseudolites to provide sufficient (for land and river transport) improvement for the precise measurement of planimetric coordinates. Air transport, naturally, requires higher accuracy for altitude measurements; these are defined by ICAO standards. In order to minimize the altitude measurement error, it is necessary to elevate one or more pseudolites to a certain height. In this case, the maximum value of the geometric factor and, respectively, the maximum altitude measurement error will occur at half the altitude of the elevated pseudolite.

Options for raising some pseudolites above the height of the rest of the configuration may include high masts or, if such are available, mountain peaks.

4. Measuring the spatial orientation of objects

Currently, it is possible to observe increasing interest to the expansion of the functionality in the consumer's navigation equipment. Among the various existing options for extra functions, determining an object's spatial orientation is a promising direction for research. In existing radio navigation systems the highly precise measurements for spatial orientation angles is provided only by interferometric methods, which involve the application of an antenna array [4–6]. In this case, we assess the difference in the path of the navigation signals

$\Delta R_{1,2}$ received by means of antenna diversity, through phase taper $\Delta R_{1,2} = \frac{\lambda \Delta \varphi_{1,2}}{2\pi}$. During this, the signal reception is performed by diverse antennas A_1, A_2 and A_3 which constitute the two vectors-bases of length B_1 and B_2 (Fig. 4). Not necessarily, these vectors-bases are perpendicular. More commonly, however, one vector-base is oriented parallel to the longitudinal axis of the object, the second is parallel to the transverse.

The phase shift of the pseudolite signal, received by two diverse antennas and the cosine of the angle between the vector-base and the vector-direction of the pseudolite can be expressed as

$$\cos \alpha_{1,2} = \frac{\Delta R_{1,2}}{B_{1,2}} = \frac{\lambda \Delta \varphi_{1,2}}{2\pi B_{1,2}}, \tag{5}$$

where λ is the wavelength of the pseudolite signal; α is the angle between the vector-base and the vector-direction of the pseudolite.

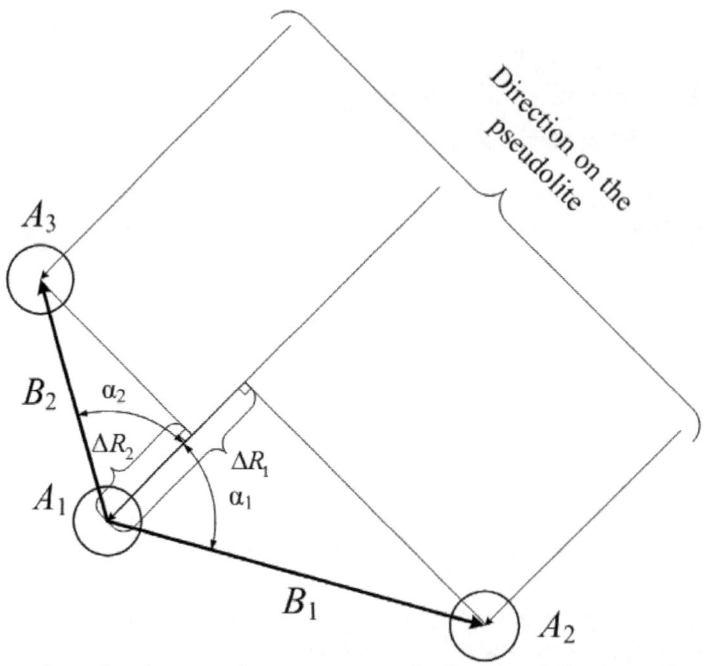

Figure 4. Position of navigation receiver antennas during spatial orientation of angular measurements

The direction cosines of the vector-base can be determined by an equation, based on the scalar product of vectors:

$$B \cdot \cos \alpha = k_x \cdot X + k_y \cdot Y + k_z \cdot Z, \tag{6}$$

There is a constraint equation between the components of the vector-base coordinates of the vector-base:

$$B_{1,2} = \sqrt{X_{1,2}^2 + Y_{1,2}^2 + Z_{1,2}^2}. \tag{7}$$

If there are two bases and considering (5) and (7), we shall write the equation for determining the direction cosine of two vectors-bases as such:

$$\begin{cases} k_{xi} X_{m1} + k_{yi} Y_{m1} + k_{zi} Z_{m1} = \Phi_{1i}, \\ k_{xi} X_{m2} + k_{yi} Y_{m2} + k_{zi} Z_{m2} = \Phi_{2i}, \\ B_{1,2} = \sqrt{X_{1,2}^2 + Y_{1,2}^2 + Z_{1,2}^2}. \end{cases} \tag{8}$$

where k_{xi}, k_{yi}, k_{zi} are the direction cosines in the pseudolite; $\Phi = \frac{\lambda \Delta\varphi}{2\pi}$ is the phase shift of the signals between the receiving antennas, expressed in units of length.

To find the position of a base-vector in space it is necessary to measure the phase shifts of signals from two pseudolites. Thus, the system of equations (8) comprises two linear equations based on the results phase shift Φ measurements and one equation of relation between the coordinates X, Y, Z.

The system of equations (8) is non-linear because of the nonlinear coupling equations. Consequently, due to the non-linearity, the accuracy of calculating orientation depends from the angular position of the vector-base.

Thus, as it can be observed from the features of this short-range navigation system based on pseudolites, positioned near the horizontal plane, in the system of equations (8) with the position of the vector-bases in the horizontal plane, the vertical directional cosine in the direction of the pseudolite k_y is equal (or is close) to zero.

From the linear part of the system of equations (8) we can only determine the planimetric coordinates of vector-base X and Z; the vertical component is obtained from a non-linear system of equations (7).

$$Y = \sqrt{B^2 - X^2 - Z^2} = \sqrt{B^2 - B_{plane}^2} \tag{9}$$

where B_{plane} is the projection of the vector-base on a horizontal plane.

Therefore, by using a vector-base along the longitudinal or transverse axes of the object it is impossible to determine the angles of pitch and roll.

The best solution to this contradiction arises by analyzing expressions (7), (8) and (9): it is the measurement of the angles of pitch and roll using a vertical base. The linear part of the system of equations (8) gives the planimetric coordinates for components of the vector-base; they increases with the pitch and roll angles. The vertical component is determined from the non-linear equation (7), while the sign for the vertical component is known in advance. Thus, we can determine the position of the vertical axis of an object on a vertically positioned base. At a certain course angle it is possible to determine the angles of pitch and roll. The measurement error, in this case, is comparable to the accuracy of determining the angle of course. An intermediate option is to position the arrangement of the bases at an inclined angle which will enable us to determine the angles of course, pitch and roll.

5. Conclusion

Thus we have demonstrated that ground-based navigation systems, containing pseudolites, are capable of significantly increasing the accuracy of measuring planemetric coordinates if compared to conventional GNSS: the error is reduced from 12 to 0.4 m. In order to measure an object's altitude using a pseudolite system, it is necessary to suspend part of the pseudolites in one configuration above the surface of the Earth; this variant, however, is not always possible.

In order to measure the spatial orientation of an object using a pseudolite-based system, it is necessary to arrange the receiver antenna system is such a manner that the vector-base is positioned vertically or inclined.

Thus, the possibility for selecting different locations for pseudolites enables us, considering the terrain, to adapt the navigation area for a designated area for specific navigation objects taking into account various features of the terrain, such as mountain peaks and hilltops, as well as to form the desired geometric factor in a given area. It is desirable to use 5 ... 10 pseudolites, arranged uniformly along the perimeter of the coverage area, while several pseudolites should be situated at various altitudes to improve the navigation signal above the ground. For a more accurate measurement of an aircraft's altitude it is recommended combine autonomous on-board barometers, altimeters and GNSS receivers with the receivers of short-distance navigation systems based on pseudolites.

References

[1] Pseudolites, Project URL: http://www.vedapro.ru/files/GPS2.pps

[2] GLONASS. Basic concepts and operation [*GLONASS. Printsipy postroeniia i funktsionirovaiia*] 2010 4th ed., Perov A I, Kharisov V N Eds. Moscow: Radiotekhnika

[3] Certificate No 2013660741 Russian Federation *The program for the automated calculation of radio-navigation parameter fields generated by grouping navigation satellites GLONASS devices: a certificate of official registration of the computer* Kremez N S, Dmitriev D D, Tyapkin V N, Fateev Yu L; applicant and copyright Siberian Federal University No 2013618834; appl. 03/10/2013; reg. 18/11/2013

[4] Fateev Yu L, Dmitriev D D, Tyapkin V N, Kartsan I N, Dmitriev D D, Goncharov A E 2015 Phase methods for measuring the spatial orientation of objects using satellite navigation equipment *IOP Conference Series: Materials Science and Engineering*. Vol. 94 doi:10.1088/1757-899X/94/1/012022

[5] Fateev Yu L, Dmitriev D D, Tyapkin V N, Kremez N S, Bondarev V N 2015 Phase ambiguity resolution in the GLONASS/GPS navigation equipment, equipped with antenna arrays *2015 International Siberian Conference on Control and Communications (SIBCON) Proceedings*

[6] Fateev Yu L, Dmitriev D D, Tyapkin V N, Garin E N, Shaidurov V V 2014 The phase ambiguity resolution in the angle-measuring navigation equipment *AIP Conference Proceedings* No. **12** (2014) 1611. P 12–14 doi: 10.1063/1.4893795.

Simulation modeling of functional adaptive interference nulling for multibeam hybrid reflector antenna systems*

I N Kartsan[1], V N Tyapkin[2], D D Dmitriev[3], A E Goncharov[4], P V Zelenkov[5], I V Kovalev[6]

[1] Associate professor, Reshetnev Siberian State Aerospace University, Krasnoyarsk, Russia

[2] Associate professor, Siberian Federal University, Krasnoyarsk, Russia

[3] Associate professor, Siberian Federal University, Krasnoyarsk, Russia

[4] Associate professor, Reshetnev Siberian State Aerospace University, Krasnoyarsk, Russia

[5] Associate professor, Reshetnev Siberian State Aerospace University, Krasnoyarsk, Russia

[6] Professor, Reshetnev Siberian State Aerospace University, Krasnoyarsk, Russia

E-mail: kartsan2003@mail.ru

Abstract. This paper considers the simulation of adaptive nulling mechanism patterns in hybrid reflector antenna systems with a 19-element feed element, in which the radiation pattern is formed as a cluster. Incidents of broadband and narrowband interference are studied in the article.

Keywords: hybrid reflector antenna; spatial interference filtering; adaptive algorithms; antenna radiation pattern;simulation modeling

1. Introduction

Multibeam antennas, built either on the basis of adaptive phased antenna arrays or hybrid reflector antennas have been among the most perspective directions for developing satellite navigation in Russia where the population distribution is uneven. Hybrid reflector antennas have a great advantage over multibeam adaptive phased antenna arrays when installed on space vehicles in terms of cost and mass reduction since the amplification coefficient for the antenna is very high (exceeding 35–40 dB) [1].

Among the main problems, which an engineer encounters when building adaptive multibeam hybrid reflector antennas is the development and research of methods and algorithms for object angular attitude to interference. In [2–5] an algorithm for hybrid reflector antennas amplitude-phase distribution synthesis had been developed. This algorithm considers both adaptive phased antenna arrays and reflector antennas.

In order to ascertain the effectiveness of the proposed methods it is necessary to conduct an experimental investigation. However, the production of hybrid reflector antenna models and their further testing using anechoic chambers and complex electronic test equipment is a complicated and expensive task. Mathematical modeling methods are an efficient substitute to the aforementioned methods. The significant reduction of cost, time and labor – essential to

* The research had been conducted with financial support from the Ministry of Education and Science of the Russian Federation (agreement № 14.577.21.0155, unique project identifier: RFMEFI57714X0155).

the research of complex electronic devices – are among the main advantages of mathematical modeling, which has made them very popular nowadays.

Modern computer-aided design software enables the simulation process to be significantly simplified. MATLAB is one of the most popular computing environment having multiple tools for complex mathematical operations, including complex numbers, vectors and matrices. MATLAB is also an excellent program for visualizing obtained experimental results [6, 7].

2. Description of computer models

We have developed several computer models, which imitate the functional adaptation process of the hybrid reflector antenna using MATLAB (R2011b). These models use Microsoft Windows XP / 7 operating systems. This adds to the versatility for testing the computer models that may be performed on most computers using standard repositories and software; this software is necessary for further editing of obtained results (images, tables, etc).

The computer models realize methods described in [1–5]. These methods include the synthesis of the amplitude phase distribution in hybrid reflector antennas, which forms a counter service area and nulls in the radiation pattern in the direction of at least three sources of interference. The other methods include a one for determining the direction of the interference signal source without using additional antennas and a method of data transmission by multiplexing the bandwidth applying perspective signal simulation methods.

The computer model of the system for adaptive forming of the amplitude phase distribution of the multibeam hybrid reflector antenna is designed for imitating the formation of a counter service area and nulls in the radiation pattern in direction of at least three sources of the interfering signal.

The model consist of a module for interference imitation that generates signals from different directions, a module for imitating desired signals that generates signals from different directions, a module for interference adaptation, and a module for building the service area counter.

The computer model imitates the following: the formation of a radiation pattern of the radiated antenna array in any direction within the limits of the rated angular sector, the formation of the service area counter, interference attacks and desired signals coming from different directions. The model also provides digital control of weighted coefficients; this enables us to do the following: adapt the radiation pattern, imitate the spatial selection of desired signals and interference, imitate adaptation to the interference situation in each beam, provide the desired adaptation requirements to interference during the adaptation process and immediately react to changes in the interference scenario.

The computer model for the system of determining the direction to the source of interference and spatial suppression of interference is designed to imitate interference from at least three sources. It activates the desired spatial methods for interference suppression.

The model consists out of a module for imitating desired interference with a given distributive law, which radiate from a given direction, a module for imitating desired signals, a module for imitating a system of signal suppression.

The computer model enables to imitate the formation of a radiation pattern of a radiating hybrid reflector antenna relay in any direction (within the limits of the rated angular sector). The following are also imitated with the help of the computer model: interferences of a given type with a given distribution law from different directions, desired signals, the algorithm for determining the direction of the source of interference, the work of the spatial selection system for desired signals, and digital control of weighted coefficients. This enables adaptive nulling in the radiation pattern.

There are a number of limitations accepted for the model: the main maximum of the hybrid reflector antenna radiation pattern is orientated towards the signal source in a direction determined by the user. The front of the wave from the desired signal source is considered to be flat. The direction to the interference source is characterized by azimuth θ and elevation φ, which are counted from the direction to the nadir. The amplitude of the interference signal is significantly greater than the amplitude of the desired signal; it is received by the side lobes of the radiation pattern.

For interference signals we have employed two types of interfering noises: narrowband (harmonic) and broadband.

3. Simulating the functional hybrid reflector antenna in various interference environments

In order to assess the quality of the functional adaptative nulling of a multibeam hybrid reflector antenna in environments with interference, we have simulated its performance under one broadband interfering noise that had the following characteristics:

- the magnitude of the interference exceeded that of the signal by 60 dB (3000 times in amplitude);
- the average frequency of the interfering noise was equal to f_0 of the central cluster; the width of the interfering noise band was 70 MHz;
- the interference attack direction was φ and θ, where φ was the elevation from the antenna plane; θ was the azimuth.

The output signal of hybrid reflector antenna without any adaptation is shown in Fig. 1.

There is a number of ways the interference can affect the system:

- only one cluster is suppressed;
- two neighboring clusters are suppressed with interference between them;
- three neighboring clusters suppressed with interference between them.

Let's consider all three variants.

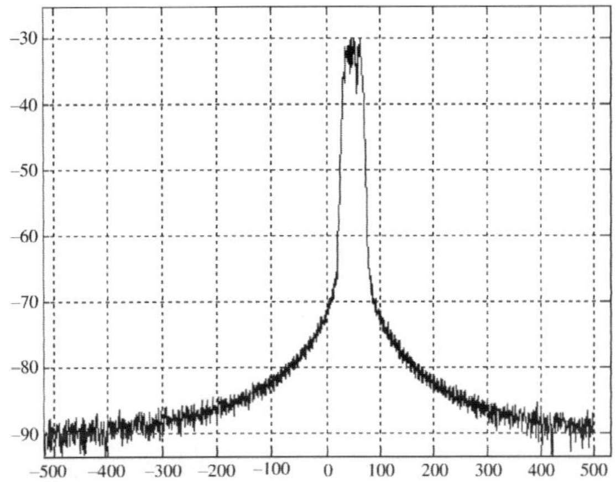

Figure 1. The output signal of a hybrid reflector antenna without any adaptive processing

A. Broadband interference attacking one cluster

The simulation of the hybrid reflector antenna system adaptive process to the attack of a broadband interfering noise on the fifth cluster had the same characteristics as shown above. The angular coordinates of interfering signal were: φ = 80°, θ = 210°.

Figs. 2 and 3 demonstrate null formation in the radiation pattern, which is formed by the adaptive system during interference attack to the main lobe of cluster 5. This results in the decrease of the antenna's power gain in the neighboring clusters: the broadband interference signal also suppresses them affecting their side lobes.

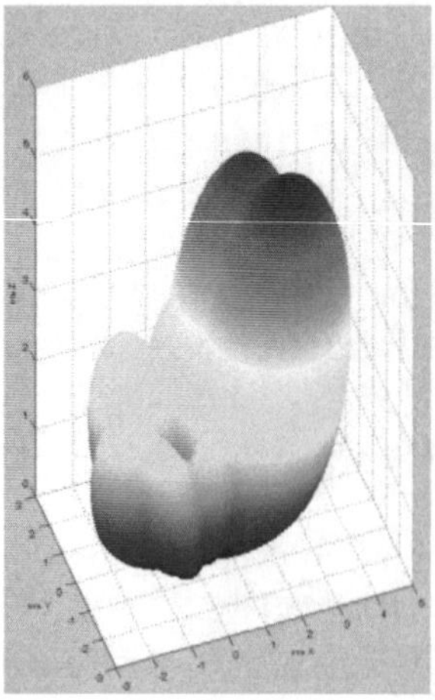

Figure 2. 7-beam radiation pattern of hybrid reflector antenna under broadband interference attack on the cluster 5

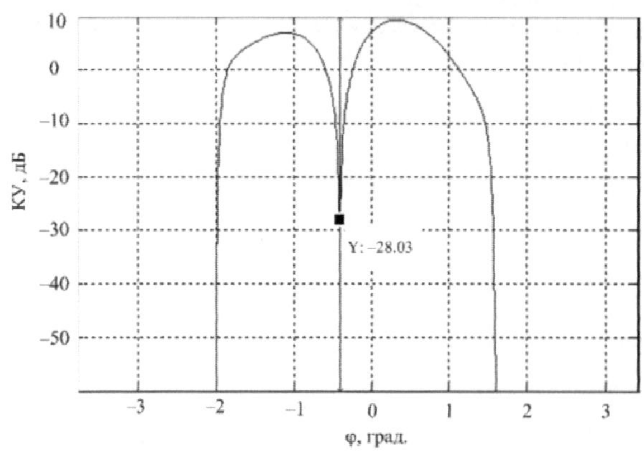

Figure 3. Cross-section of radiation pattern for cluster 5 with a vertical plane θ = 210° under broadband interference attack on cluster 5

As demonstrated in Figs. 2 and 3, during adaptive nulling in a hybrid reflector antenna to one broadband interfering signal there is an insignificant degradation of the whole radiation pattern. The antenna gain for cluster nulling is reduced by two times; the widening of the main lobe reaches 3°. The antenna gain is 28 dB. In result of the adaptive nulling the total service area was reduced (Fig. 4).

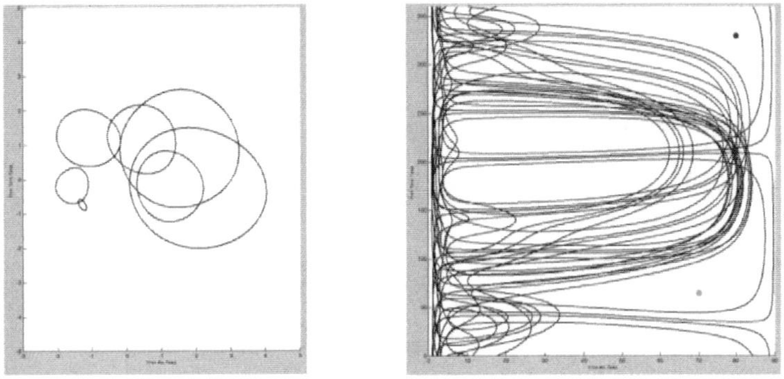

a) *b*)

Figure 4. Counter radiation pattern during cluster 5 suppression:
a) Level minus 3 dB; *b)* Level minus 35 Db

B. Broadband interference attack between two clusters

Let's simulate an adaptive nulling mechanism to one broadband interfering signal that impacts between clusters 5 and 6. The characteristics of the signal are identical to the ones in the previous situation with the exception that the angular coordinates of the signal are $\varphi = 80°$, $\theta = 270°$.

In Fig. 5 we can see a three dimensional view of a 7-beam radiation pattern for a hybrid reflector antenna, which is impacted by a broadband interference signal between clusters 5 and 6. A negligible decrease in the antenna gain can be visible in these two clusters. This decrease is explained by the fact that the main power of the interference signal is accepted by the side lobes. In Fig. 6 we can see a counter radiation pattern of a hybrid reflector antenna showing an insignificant reduction of the total service area. This can be compensated by a redistribution of the beams.

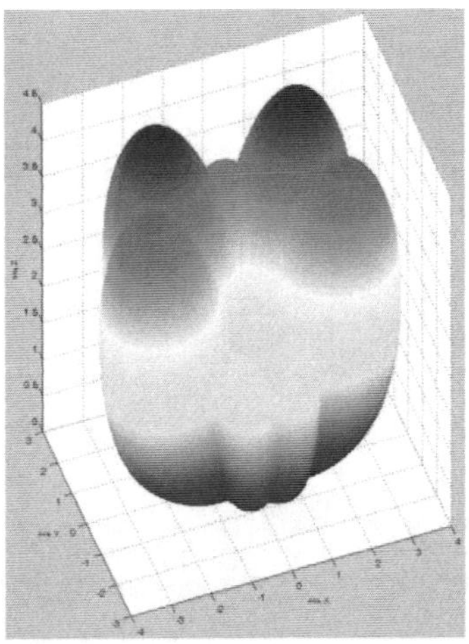

Figure 5. 7-beam radiation pattern of hybrid reflector antenna under broadband interference attack between clusters 5 and 6

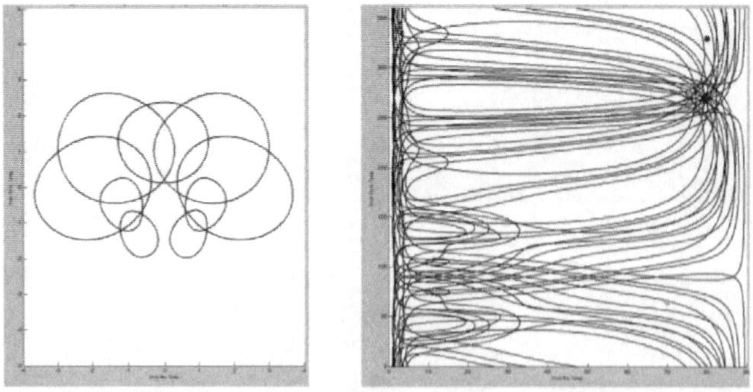

a) *b*)

Figure 6. Counter radiation pattern under broadband interference attack between clusters 5 and 6: *a* Level minus 3 dB; b) Level minus 35 dB

C. Broadband interference attack between three clusters

Let's now simulate an adaptive nulling mechanism to one broadband interference signal that impacts between clusters 1, 5 and 6. The characteristics of the signal are identical to the ones in the previous situation with the exception that the angular coordinates of the signal are $\varphi = 85°$, $\theta = 0°$.

Fig. 7 demonstrates a three dimensional view of a 7-beam radiation pattern of a hybrid reflector antenna, which is impacted by a broadband interference signal between clusters 1, 5 and 6. In Fig. 8 we can see a counter radiation pattern of a hybrid reflector antenna. The analysis of the figures shows that there had been a negligible degradation of the radiation pattern expressed in the decrease of the antenna gain for suppressed clusters and the widening of the main lobes of the radiation pattern. The total service area shows signal disruptions at signal level of minus 3 dB; if the levels are lower, it is possible to receive a signal.

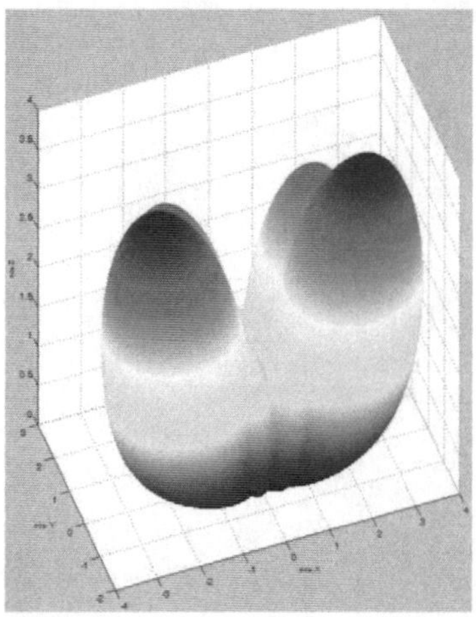

Figure 7. 7-beam radiation pattern of hybrid reflector antenna under broadband interference attack between clusters 1, 5 and 6

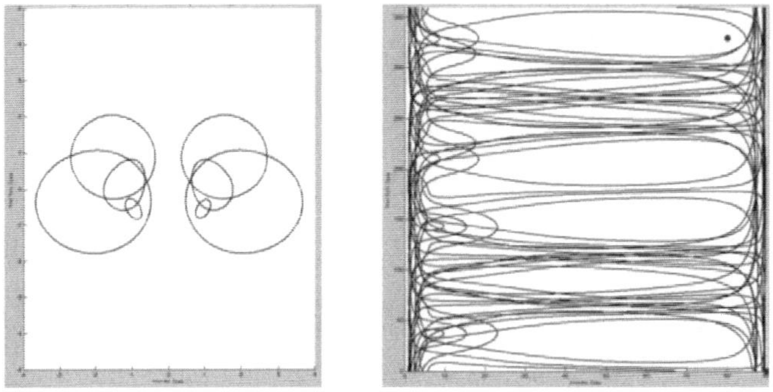

a) *b)*

Figure 8. Counter radiation pattern under broadband interference attack between clusters 5 and 6: *a)* Level minus 3 dB; b) Level minus 35 dB

As the calculation results for the radiation pattern of an adaptive hybrid reflector antenna show, the suppression of one broadband interfering signal in direction of the first side lobes of the radiation pattern and in direction of the slope of the main lobe had resulted in the interference suppression level being minus 35–40 db. The signal level in the rest of the service area remains unchanged. During interference suppression in direction of the main lobe of the radiation pattern a deterioration of the antenna gain level to 2 dB is observed; the level of interference suppression is minus 25 dB, which is enough for satellite communication.

D. Simulating hybrid reflector antenna performance under attack from one narrowband interfering signal

In order to assess the capabilities of adaptive hybrid reflector antennas for suppressing both broadband and narrowband interfering signals we have simulated adaptive nulling mechanisms to narrowband interfering signal with the following characteristics:

- the power of the interference exceeded the power of the signal by 72 dB (4000 times in amplitude);
- the narrowband interference frequency fell behind f_0 of the central cluster by 10 MHz in order to better illustrate and analyze the spectrogram;
- the interference attack was directed from a elevation level of φ and an azimuth θ.

Narrowband interference generally impacts only the cluster it strikes. This phenomenon, unlike that of broadband interference, is explained by the beams of the hybrid reflector antenna being subjected to frequency-division multiplexing. In this case, narrowband interference for neighboring clusters will have the characteristics of out-of-band interference that will be filtered in the receiving path.

In this aspect let's simulate the adaptive nulling mechanism for a hybrid reflector antenna to narrowband interference on the example of a single cluster. Let's look at two situations:

- interference strikes the main lode of the radiation pattern;
- interference strikes the side lobe of the radiation pattern.

In Fig. 9 we can see a narrowband interference spectrum at antenna array output.

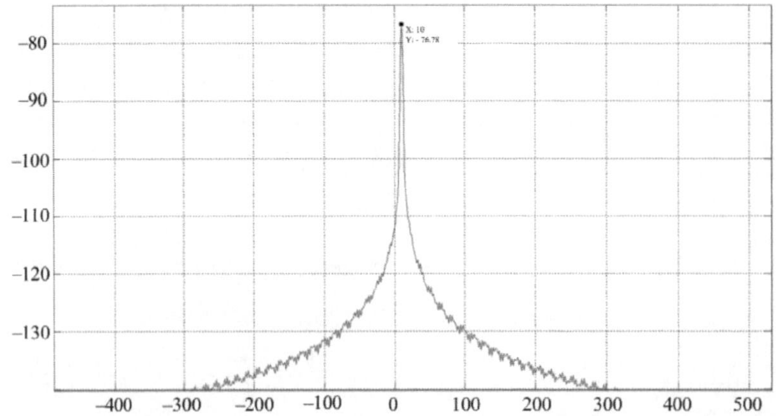

Figure 9. Narrowband interference spectrum at radiating antenna array output

During narrowband interference attack on the main lobe, its angular coordinates are: φ = 80°, θ = 210°. Adaptive nulling results in interference suppression. However a significant decrease in antenna gain results in direction of the source of interference.

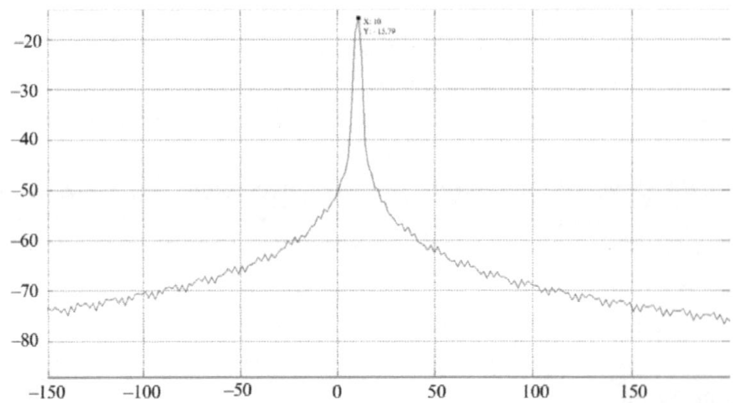

Figure 10. Narrowband interference spectrum attacking main lobe at radiating antenna array output after adaptive nulling

As seen in the figures, the interference cancellation ratio for narrowband interference on the main lobe of the radiation pattern has a satisfactory result of almost 61 dB (–15.79 – (–76.78) = 60.99).

The angular coordinates of narrowband interference attaking the main lobe are: φ = 18°, θ = 358°. In result of adaptive nulling even greater suppression of interference takes place (Fig. 11); a desired signal is emitted (zero frequency peak). In this situation the cancellation ratio for narrowband frequency is (–15.79 – (–89.2) = 73.41 dB. Such a significant difference in the cancellation ratio between narrowband and broadband interference is explained by the fact that the desired signal spectrum occupies a wide band and because further signal processing increases the signal-to-noise ratio in the signal path.

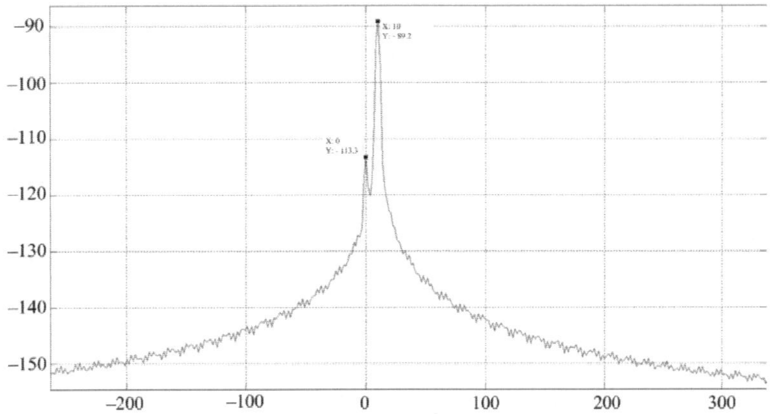

Figure 11. Narrowband interference spectrum attacking main lobe at radiating antenna array output

Thus, the simulation results demonstrated that an adaptive hybrid reflector antenna is capable of effectively suppressing narrowband interference with a significant cancellation ratio (up to 60 dB) under interference attack on the main lobe. The cancellation ratio during interference suppression to the side lobes of the radiation pattern can be increased to over 70 dB.

4. Conclusion

Thus, we have conducted a simulation of functional adaptive nulling for multibeam hybrid reflector antennas under interference using the developed computer models – systems for forming adaptive amplitude-phase distribution for multuibeam hybrid reflector antennas, which enable the realization of the developed methods of forming a counter service area and nulls in the radiation pattern in direction of at least three sources of interfering signals, together with a system for detecting the direction to the source of interference and the spatial suppression of the interfering signal.

We have discussed various scenarios of interference cancelling:

- simulating the suppression of broadband and narrowband interference with different directions of impact;
- suppressing a separate cluster:
- suppressing two and three clusters of the side lobes.

The conducted simulation for situations of interference suppression on the first side lobe, the slope of the main lobe and the main lobe has proved the capacity of the developed algorithms. During interference suppression in direction of the first side lobe of the radiation pattern and the slope of the main lobe, the level of interference suppression reached a maximum of minus 45–50 dB; the signal level in the remaining service area remained unaltered. The suppression of interference in direction of the main lobe of the radiation pattern showed deterioration in the antenna gain to 1 dB; the level of interference suppression reached minus 25 dB.

Since each beam of the adaptive hybrid reflector antenna, built according to a cluster scheme, forms a null independently from other beams, the long beams are provided with a cancelling for side lobes of the radiation pattern that can be attacked by interference.

References

[1] Tyapkin V N, Pershin A S, Dmitriev D D, Moshkina T G 2012 Adaptive multibeam antennas *Vestnik SibGAU* Issue **2 (42)** P 106–109
[2] Tyapkin V N, Dmitriev D D, Perschin A S 2013 The adaptive algorithm multibeam antenna, built on hybrid-reflector antenna *Journal of Siberian Federal University. Engineering & Technologies,* No. **7** Vo. **6** p 835–844
[3] Tyapkin V N, Dmitriev D D, Garin E N, Sokolovskii A V 2013 Controlling amplitude-phase distribution with an adaptive phase antenna array *Promyshlennye ASU i kontrollery* No. 4 P 66-71
[4] Tyapkin V N, Kartsan I N, Dmitriev D D, Goncharov A E 2015 Spatial filtering algorithms in adaptive multi-beam hybrid reflector *antennas International Siberian Conference on Control and Communications (SIBCON) – Proceedings*
[5] Serenkov V I, Kartsan I N, Dmitriev D D 2015 Synthesis methods for amplitude-phase distribution for hybrid reflector antennas *Vestnik SibGAU* Vol **16** No. 3 P 664-669
[6] Alkin O 2014 Signals and Systems: A MATLAB Integrated Approach. CRC Press
[7] D'iakonov V P 2010 MATLAB R2007/2008/2009 manual for radio-enginners (MATLAB R2007/2008/2009 dlia radioinzhhererov) Moscow DMK Press

Specialized software for the adaptive nulling unit of a hybrid reflector antenna*

I N Kartsan[1], V N Tyapkin[2], D D Dmitriev[3], A E Goncharov[4], P V Zelenkov[5], I V Kovalev[6]

[1] Associate professor, Reshetnev Siberian State Aerospace University, Krasnoyarsk, Russia

[2] Associate professor, Siberian Federal University, Krasnoyarsk, Russia

[3] Associate professor, Siberian Federal University, Krasnoyarsk, Russia

[4] Associate professor, Reshetnev Siberian State Aerospace University, Krasnoyarsk, Russia

[5] Associate professor, Reshetnev Siberian State Aerospace University, Krasnoyarsk, Russia

[6] Professor, Reshetnev Siberian State Aerospace University, Krasnoyarsk, Russia

E-mail: kartsan2003@mail.ru

Abstract. This paper considers the development of specialized software for a hybrid reflector antenna adaptive nulling unit; it conducts the calculation of a desired amplitude and phase distribution of the feed array in counter service area formation and interference cancelling.

Keywords: hybrid reflector antenna; spatial interference filtering, adaptive algorithms, specialized software

1. Introduction

Adaptive multibeam antenna systems are currently gaining significant popularity among designers of satellite communications equipment. Such systems enable the user to flexibly control traffic depending on the channel traffic load – this includes the formation of multibeam radiation patterns within the limits of the service area. Another advantage of these systems is their high level of noise immunity to jamming and industrial (undeliberate) interference by employing spatial interference suppression methods.

Adaptive multibeam antenna can be built either on the basis of phased antenna arrays with scanner beams, or multibeam antennas – built on the basis of hybrid reflector antennas. Phased antenna arrays with a numerical radiation pattern formation enables flexible control of the directions of the main maxima of each partial radiation pattern in a wide range of angles; it can also form deep nulls in the radiation pattern in direction of the source of interference.

However, phased antenna array have some major drawbacks: they are excessively large in size, their hardware – particularly the beamforming network – is quite complicated. All of these factors eventually sum up in the high cost of this type of system [1]. Adaptive hybrid reflector antennas have smaller scanning angles – a feature that is negligible for geostationary satellites and highly elliptical orbit satellites and can easily be compensated by a simpler

* The research had been conducted with financial support from the Ministry of Education and Science of the Russian Federation (agreement № 14.577.21.0155, unique project identifier: RFMEFI57714X0155).

design of the antenna. Antenna arrays with focusing quasi-optical reflectors and fewer elements are considerably more perspective than other types of arrays. Hybrid reflector antennas combine the advantages of reflector antennas and phased antenna arrays, enabling the formation of configurable multibeam radiation patterns and radiation patterns with special forms together with providing spatial signal and interference filtering.

The design of complex electronics requires the engineer to include modeling – a key stage of designing – into the development process. Modeling enables the engineer to ascertain the effectiveness of the proposed methods and algorithms, any technical solutions and schematics without being necessitated to construct a full-scale prototype of the tested device. In hybrid reflector antenna designing one of the most complicated devices is the adaptive unit – it is required for controlling amplitude and phase distribution of the radiating array of the antenna. Technically, the adaptive unit is an adaptive processer on the base of a signal processor; adaptive nulling is performed using specialized software.

In [2–3] a development of synthesis algorithms for amplitude and phase distribution in hybrid reflector antennas – designed for the adaptive processor – has been reviewed. The main idea of these algorithms is the following: if the signal/noise limit is exceeded, an adaptive nulling mechanism is launched in one of the beams. The algorithm then assesses the ability to suppress the interfering signal through possible amplitude and phase distribution options, which are cached in the system's memory. Since each beam in a hybrid reflector antenna is assigned to a specific service area, it is possible to detect the direction from which the interfering signal comes within a relatively small area. This reduces the time required for the system to select the best option for interference suppressing.

To assess the effectiveness of these algorithms we have developed a model of the receiving sector of an adaptive hybrid reflector antenna. As the basis for this model we have selected an adaptive unit that had been built on the basis of a new generation low-consumption multi-core signaling microprocessor 1892VM14Ia [6, 7]. Its integrated circuit is produced using CMOS technology with a minimal topological element size of 40nanometers. The microprocessor is a crystal system with high productivity; it includes two DSP kernels, two CPU ARM Cortex-A9, a system of correlators and integrated input/output buses. Phased antenna array control algorithms are introduced using specialized software. The microprocessor development board is connected to a personal computer via a JTAS emulator. Computer modeling allows the user to access all necessary tools using only one interface. The operational system for 1892VM7Ia integrated circuit is Linux.

MCStudio – an integrated software modeling environment – has been designed for the described integrated circuit; it provides a complete development and debugging cycle for the program. This environment is versatile since it can function on any IBM PC. The integrated design environment includes the following: an environment for designing programs for CPU and DSP kernels, a C++ environment for program adjustment in the source codes and a debugger for the debugging circuit module for the specified integrated circuit or the whole device.

2. Developing specialized software for the adaptive nulling unit of a hybrid reflector antenna

Specialized software enables the forming of a radiation pattern in any direction within the working degree of curvature. It also renders possible numerical control of weighted coordinates – this allows for adaptive nulling of the radiation pattern, providing it with a desired path, a given form and null formation in the direction of the interfering signal source, the spatial selection of the desired signal, i.e. the shifting of a partial single-degree beam within a small area (for interference cancelling), adaptive nulling in each partial beam to the interference environment, maintaining a given level of adaptive nulling to interference during the adaptive process and immediate reaction to changes in the interference situation.

The software for the adaptive unit (comprising a processor and memory) calculates and controls the phase-shifters and controlled amplifiers for the desired amplitude and phase distribution of the radiating array.

The programs key function is the minimizing the value of the objective function $F(x)$. Parameter values for achieving a minimal value of the objective function are chosen using successive iterations

$$F(\bar{x}) = \max\{f_1(\bar{x}), f_2(\bar{x}), f_3(\bar{x}), ..., f_k(\bar{x})\}.$$

For the first iteration we have taken an equal amplitude syn-phase distribution of feed element's array (each array comprises 7 to 19 feed elements). Next, the algorithm calculates the antenna gain for each beam on the basis of the equal amplitude distribution and determines the maximal value of the difference between the required and the available amplification ($\left|E(\theta,\varphi)_{\text{цель},i}\right|^2 - \left|E_i(\theta,\varphi)\right|^2$). The algorithm then changes the value of the amplitude and phase of each beam for the purpose of reducing the difference between the required and actual amplification. This process is repeated until the difference between the two for each station reaches its minimal measurement. The number of iterations can be programmed, or the algorithm will cease working as soon as the minimum in changes has been achieved.

The software for the adaptive nulling unit model of a multibeam hybrid reflector antenna has two functioning modes – a multi-beam mode and a counter beam mode.

For the first mode the adaptive nulling unit, solidary with the antenna system, covers the required service are with a set of narrow beams. Incorporating a multi-beam signal distribution allows the frequency resource to be reused multiple times: there is a spatial discrimination between the beams.

In the second mode the adaptive unit forms a radiation pattern identical to the service area of the system; power can be flexibly distributed within the limits of the area.

Both modes can visibly complement each other: for example, in conditions without any interference it is preferable to use a multi-beam mode – it will provide contact in a global service area. If interference is intense, it is best to switch to the counter mode with its capacity for counter radiation pattern formation.

The software for the adaptive nulling unit of the hybrid reflector antenna comprises a software module for beam formation in the given amplitude and phase distribution and adaptive nulling to interference, a software module for phase signal, delay and frequency detection and a software module for allocating desired data.

All modules are synchronized between each other are timed from one clock oscillator – the frequency of which depends on the frequency of data production by the analog-to-digital converter. The *P*-circuit of signal processing can be seen in Fig. 1. Here ADC is the analog-to-digital converter, FGEN is the beam forming circuit in a given amplitude and phase distribution and adaptive nulling to interference, NAVSRCH is the signal detection circuit, PHTRACK is the tracking circuit for signal phase, PRSTRACK is the signal delay tracking circuit, FTRACK is the tracking circuit for intermediate signal frequency, DEMOD is the code receiving and data demodulating circuit.

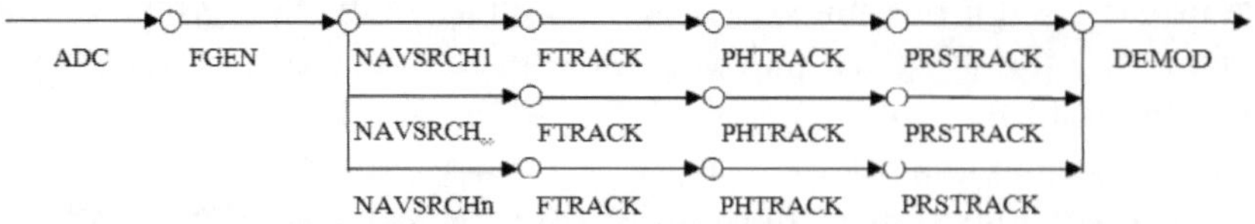

Figure 1. *P*-circuit for signal processing

A. A beam-forming module in a given amplitude and phase distribution and adaptive nulling to interference

A beam-forming circuit in a given amplitude and phase distribution and adaptive nulling to is realized using C++. Auxiliary c_complex and c_matrix classes realize mathematical operations on complex numerals, vectors and matrices.

The *P*-circuit for employing the adaptive function is shown in Fig. 2.

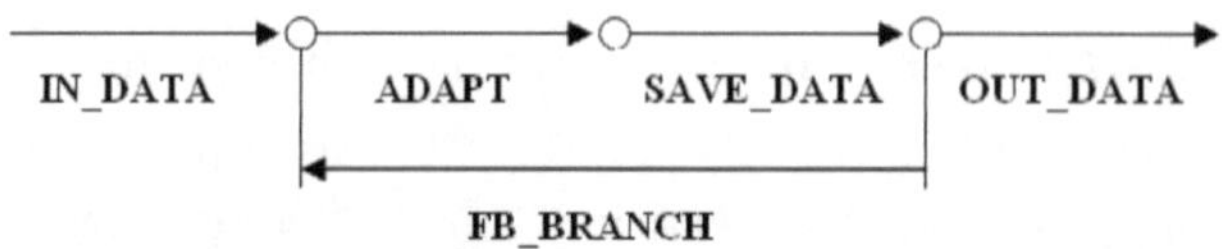

Figure 2. *P*-circuit for employing the adaptive nulling function

In Fig. 2 IN_DATA is the reading of input data in the complex view; input_vector_buffer is the production vector with antenna elements in a complex form; sum_signal is the total array signal in a complex form; inv_noise_matrix is the inverse correlative interference matrix; weights_buffer is the array weighted coefficients buffer; averaging is a parameter for determining the convergency speed for the adaptive algorithm; ans_container is a structure containing the calculated weighted coefficients and the inverse correlative interference matrix for further iterated adaptive nulling; ADAPT is the adaptive nulling; SAVE_DATA is the saving of intermediate weighted coefficients and the inverse interference matrix in the container; FB_BRANCH is the transfer of the container with intermediate data to the next iteration; OUT_DATA is the transfer of the calculations to the numerical processing unit for intermediate interference canceling.

B. Signal search module

Signal search is accomplished using syn-phase *I* and *Q* quadrate content of received signals. The process is divided into two types – full search and fine search.

Full search is performed in the absence of prior data on the frequency signals within the field of vision, data on delays, Doppler frequency shifts, receiver coordinates.

Search is performed within frequency ranges from minus 5 … plus 5 kHz with a step of 500 Hz. Thus, the search range in frequency ω_i is divided into 20 cells. Besides frequency search it is necessary to perform delay search. The detection criterion is selected depending on the desired probability of detection

$\sqrt{I_p^2 + Q_p^2} \geq h$, where h is the detection limit.

The duration of the accumulation of the analyzed signal depends on the signal-to-noise relation for the received signal and the desired probability of the correct detection; its range is between [1...10] ms.

Fine search is performed with any prior data on delays, receiver time and the last successful signal receiving session, the position of the receiver, etc. Any prior data narrows the search frequency range, consequently, saving time required for connection.

The search can be performed using correlators, matched filters, discrete-time convulsion, fast Fourier transform to frequency.

When using parallel correlators, the search range in frequency ω_i and delay τ_i is broken up into several areas, the limits of which correspond in time to the limits of signals shifting of the range-measuring code. Thus, the value of the code symbols in each area remains constant allowing us to perform a frequency search.

The *P*-circuit of the search is shown in Fig. 3. Here, INS is the input signal; ACUSELF is the circuit for accumulating the input signal in a selected frequency; CORPRS is the circuit for correlative processing with the code signal; SOLVEIQ is the circuit for decision-making considering the processed *I* and *Q*.

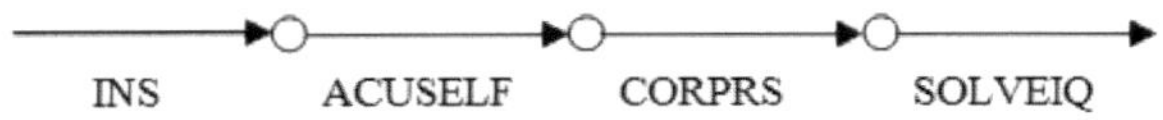

Figure 3. *P*-circuit of correlative signal search on in a selected frequency (CORFS)

When applying the circuit for phase tracking after the correlator, a phase discriminator is added. It automatically sets the frequency in which the signal has been detected though a feedback sequence.

The *P*-circuit for frequency tracking can be seen in Fig 4. Here, INS is the input signal; CORFS is the signal correlator circuit; PHTRACK is the signal phase tracking circuit; FILTER is the rectifier filter; FBBRANCH is the feedback branch in which the correction signal for the intermediate navigational signal frequency generator is formed; OUTS is the output signal.

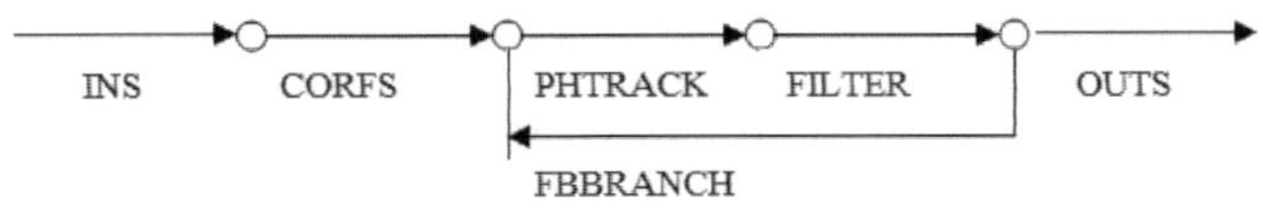

Figure 4. *P*-circuit for signal phase tracking

The delay tracking circuit also implies a discriminator, rectifier filter and feedback circuit for the formed correction signal. The discriminator uses the delayed and anticipated signals, which were formed in the correlator. The difference of this circuit from the phase tracking circuit is that the feedback circuit is connected to the clock generator with a 3-bit shift register.

The *P*-circuit for delay tracking is demonstrated in Fig. 5. Here, INS is the input signal; CORFSD is the signal correlator circuit for delayed and anticipated yields; DTRACK is the signal delay tracking circuit; FILTER is the rectifier filter; FBBRANCH is the feedback

IOP Conf. Series: Materials Science and Engineering **155** (2016) 012018 doi:10.1088/1757-899X/155/1/012018

branch in which the correction signal is formed for the clock generator; OUTS is the output signal.

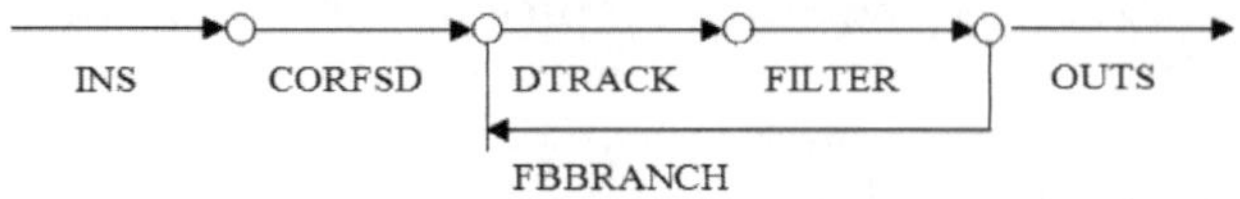

Figure 5. *P*-circuit for signal delay tracking

The frequency tracking circuit comprises a discriminator, rectifier filter and feedback circuit. The time required for signal accumulating in the discriminator circuit is 1 ms if a navigational signal correlator with search cells in a frequency of 500 Hz is used.

The *P*-circuit for frequency tracking is demonstrated in Fig. 6. Here, INS is the input signal; CORFSF is the signal correlator circuit; FTRACK is the signal phase tracking circuit; FILTER is the rectifier filter; FBBRANCH is the feedback branch in which the correction signal is formed for the navigational signal intermediate frequency generator; OUTS is the output signal.

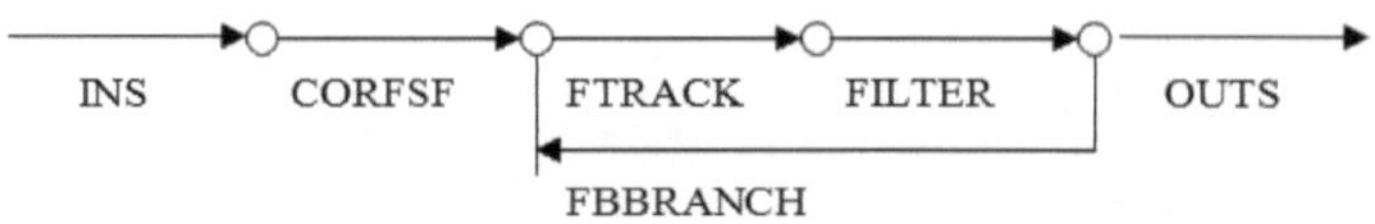

Figure 6. *P*-circuit for intermediate signal frequency tracking

C. Desired data selecting unit

Since the message symbols clocked with a frequency of 500 Hz are modulated by a code (clocked with a frequency of 1000 Hz), it is necessary to allocate the code symbols from the received packet before allocating the message symbols.

The *P*-circuit for data allocating is shown in Fig. 7. Here, INS is the input signal; PRSPULSE is the timed pulse of the range-measuring code allocation unit, PRSFORM is the forming unit; PRSPROBE is the pseudorandom number generator sequence symbol allocation unit; MOD2SUM is the module 2 adder unit; DATAPULSE is the data symbols frequency impulse forming unit; DATACCUM is the data symbols allocation unit, OUTS is the output signal.

The allocated data is transferred for further processing.

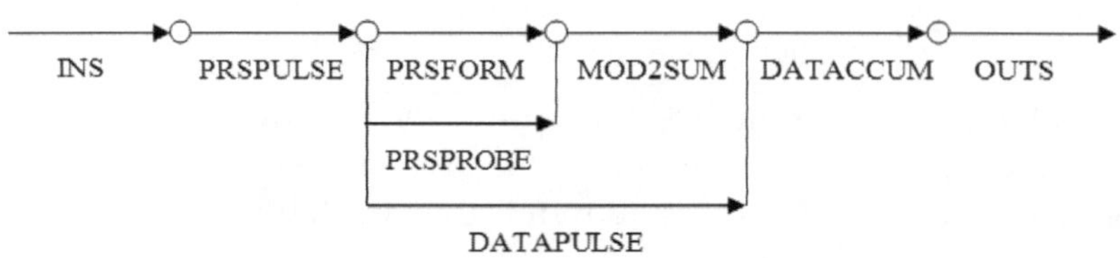

Figure 7. *P*-circuit for data allocation

3. Conclusion

Thus, we have described specialized software developed for the adaptive nulling unit of a hybrid reflector antenna. The software:

a) forms the radiation pattern of the radiating antenna array in a hybrid reflector antenna (within the working degree of curvature);

b) numerically controls the weighted coefficients enabling adaptive nulling of the radiation pattern, which sets the required direction and shape for the radiation pattern and forms nulls in direction of interference.

The software for the adaptive nulling unit may be used for conducting experiments and tests on adaptive hybrid reflector antenna models, testing adaptive algorithms in interference environments, controlling radiation patterns, forming counter service areas.

References

[1] Tyapkin V N, Pershin A S, Dmitriev D D, Moshkina T G 2012 Adaptive multibeam antennas *Vestnik SibGAU* Issue **2 (42)** P 106–109

[2] Tyapkin V N, Dmitriev D D, Perschin A S 2013 The adaptive algorithm multibeam antenna, built on hybrid-reflector antenna *Journal of Siberian Federal University. Engineering & Technologies* No. **7** Vo. **6** p 835–844

[3] Tyapkin V N, Dmitriev D D, Garin E N, Sokolovskii A V 2013 Controlling amplitude-phase distribution with an adaptive phase antenna array *Promyshlennye ASU i kontrollery* No. **4** P 66-71

[4] Tyapkin V N, Kartsan I N, Dmitriev D D, Goncharov A E 2015 Spatial filtering algorithms in adaptive multi-beam hybrid reflector *antennas International Siberian Conference on Control and Communications (SIBCON) – Proceedings*

[5] Serenkov V I, Kartsan I N, Dmitriev D D 2015 Method for the synthesis of amplitude-phase distribution of hybrid-mirror download *Vestnik SibGAU* Vol **16** No. **3** P 664-669

[6] Intergral circuit 1892VM14Ia. Basic information. Mikroskhema integral'naia 1892VM14Ia. Kratkaia informatsiia 2015 Moscow: OAO NPTs "ELVIS"

[7] Intergral circuit 1892VM14Ia. User's guide. Mikroskhema integral'naia 1892VM14Ia. Rukovodstvo pol'zovatelia 2015 Moscow: OAO NPTs "ELVIS"

Synthesis of an algorithm for interference immunity*

I N Kartsan[1], V N Tyapkin[2], D D Dmitriev[3], A E Goncharov[4], P V Zelenkov[5], I V Kovalev[6]

[1] Associate professor, Reshetnev Siberian State Aerospace University, Krasnoyarsk, Russia

[2] Associate professor, Siberian Federal University, Krasnoyarsk, Russia

[3] Associate professor, Siberian Federal University, Krasnoyarsk, Russia

[4] Associate professor, Reshetnev Siberian State Aerospace University, Krasnoyarsk, Russia

[5] Associate professor, Reshetnev Siberian State Aerospace University, Krasnoyarsk, Russia

[6] Professor, Reshetnev Siberian State Aerospace University, Krasnoyarsk, Russia

E-mail: kartsan2003@mail.ru

Abstract. This paper discusses the synthesis of an algorithm for adaptive interference nulling of an 8-element phased antenna array. An adaptive beam-forming system has been built on the basis of the algorithm. The paper discusses results of experimental functioning of navigation satellite systems user equipment fitted with an adaptive phased antenna array in interference environments.

Keywords: adaptive phased antenna array, adaptive nulling algorithms for phased antenna arrays, navigation satellite systems

1. Introduction

Navigation satellite systems accommodating the demands of various customer groups have been increasingly gaining popularity in the last years. Among the most important fields is aviation where precise navigational flight data, landing and piloting are provided by satellites.

Spatial selection methods are considered to be the most promising for interference cancelling. These methods are based on the usage of adaptive phased antenna arrays [1]. The theoretical part of their usage had been thoroughly studied by scientists both in Russia and abroad. However, the usage of phased antenna arrays in user satellite navigation equipment has a number of specific features that require further research (exploitation in three frequency ranges, low desired signal level and receiving desired signals from various directions).

In this paper we shall discuss the method for synthesizing an adaptive interference nulling algorithm for an 8-element phased antenna array. The algorithm became the basis for building a device for noise compensating. The analog part of the adaptive phased antenna array comprises 8 antenna modules with 8 radio links connected to them. Signals from the radio links arrive at the analog-to-digital converter and further at the digital section; it contains 12 independent signal processing channels (GLONASS L_1, L_2). Each channel comprises a beam-

* The research had been conducted with financial support from the Ministry of Education and Science of the Russian Federation (agreement from 28.11.2014 № 14.574.21.0131, unique project identifier: RFMEFI57414X0131)

forming circuit and a digital signal processing channel. The beam-forming arrangement is a weighted adder for signals from 8 analog highways. The weighted coefficients are calculated by a signal processor in accordance with the formation of the required radiation pattern.

Performed tests of the 8-element adaptive phased antenna array have demonstrated that the synthesized adaptive interference nulling algorithm is highly effective and enables suppression of both narrowband and broadband interference.

The interference suppression coefficient using spatial filtering methods was:

- 40–45 dB (for harmonic interference);
- 35–40 dB (for noise interference with a limited spectrum);
- 30–35 dB (for broadband interference).

The widespread usage of various electronic devices and the stable increase in their number has become a major issue in terms of the exploited frequency range limitability and the electromagnetic environment becoming constantly more complicated. This has become a problem for the functioning of electronic devices. In such situations, improving interference immunity in the user satellite navigation equipment has become an utmost concern.

In [2] it had been demonstrated that the choice of the effectiveness criterion with consideration for the weighted vector of the phased antenna array does not matter significantly. The choice of the control algorithm for building the radiation pattern of the array – the algorithm has direct impact on the speed of the transition process and on the complexity of implementing the whole system. In [3] an adaptive phased antenna array with primary equal amplitude distribution had been discussed. The effectiveness of its adaptive algorithm using a basis in the shape of a partial exponential Fourier exponential series had been determined in comparison with the optimal algorithm.

2. Algorithm synthesis

In this paper a synthesis of an algorithm for the 8-element circular phased antenna array adaptive nulling system has been performed.

The received signal in a general situation is described by a vector function that considers its description in time and space. Differences between the desired signal and the interfering signals are used to distinguish the former from the latter. From a mathematical position these differences are conveniently addressed through a dependence of signals and interference from parameters, which can be both spatial and time-frequent. Thus, the taken oscillation in the general case is

$$y(t) = x(t, \alpha, \beta) + n(t, \nu), \qquad (1)$$

where $x(t, \alpha, \beta)$ is the received desired signal vector with parameters α and β; α is the informative parameter vector (phase, delay time, Doppler frequency, etc.); β is non-informative parameter vector that are caused by signal fluctuation; $n(t, \nu)$ is the interference oscillation vector; ν is the interference parameter vector.

Let's assume that the interference is a vector accidental stationary process. The typical law of interfering signal oscillation is physically justified for most situations, since the interference normalizes in the comparatively narrowband interference paths of receivers.

The received signal phase in each antenna element of the array will be determined by the coordinates of this element, the direction from which interference and signals are transmitted and the angle of the beam.

In order to determine the vector of the amplitude and phase signal distribution it is necessary to know the direction from which the signal comes and the coordinates of the antenna element and the radiation pattern.

The direction from which the signal comes can be supposed as a single vector from a selected antenna element or phase center of the array to the signal source [4]

$$L(\psi,\gamma) = (\cos(\psi)\cos(\psi) \quad \sin(\psi)\cos(\gamma) \quad \sin(\psi))^{T}, \tag{2}$$

where ψ is the azimuth to the signal source (navigation satellite); γ is the angular altitude to the source of interference. The structure of the phased antenna array is determined by matrix A in which the Cartesian axles x, y, z are written in the k column of the i antenna element in relation to the phase center of the array, which matches the circumference center of the radius (R) on which the antenna elements are placed.

$$\mathbf{A} = R \cdot \begin{vmatrix} 0 & \frac{\sqrt{2}}{2} & 1 & \frac{\sqrt{2}}{2} & 0 & -\frac{\sqrt{2}}{2} & -1 & -\frac{\sqrt{2}}{2} \\ 1 & \frac{\sqrt{2}}{2} & 0 & -\frac{\sqrt{2}}{2} & -1 & -\frac{\sqrt{2}}{2} & 0 & \frac{\sqrt{2}}{2} \\ 0 & 0 & 0 & 0 & 0 & 0 & 0 & 0 \end{vmatrix}, \tag{3}$$

Let's suppose that the phased antenna array elements are not targeted; their position is shown in Fig. 1.

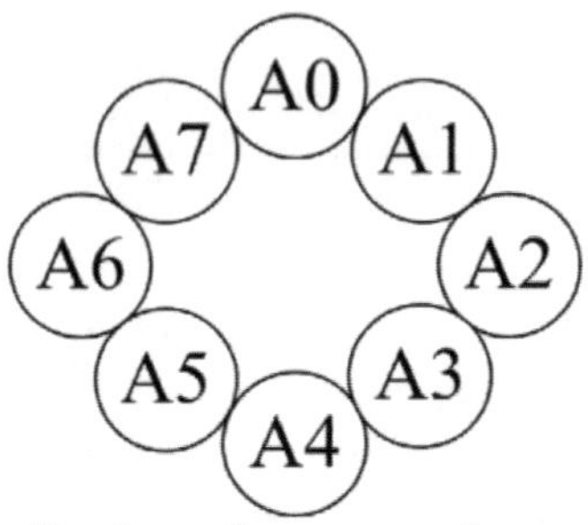

Figure 1. Distribution of antenna elements in the array

The vector from the phase center of the array (its coordinates for this case are $(0\ 0\ 0)^{T}$) to the k element a difference in the coordinates of the circumference center and array elements. In this case, it is just the k column of matrix **A**, i.e. $\mathbf{A}^{<k>}$. Thus, the vector of the amplitude and phase distribution through phase elements and setting beam direction to the i signal source shall be determined by the following

where $F_1(\psi,\gamma)$ is the amplitude and phase radiation pattern of

$$\mathbf{H}(\psi,\gamma) = \left(F_1(\psi,\gamma)\cdot\exp\left\{-2\pi j\cdot(R\mathbf{A}^{<1>}, L(\psi,\gamma))\right\} \ldots F_8(\psi,\gamma)\cdot\exp\left\{-2\pi j\cdot(R\mathbf{A}^{<8>}, L(\psi,\gamma))\right\}\right), \tag{4}$$

the k element of the array; $(R\mathbf{A}^{<1>}, L(\psi,\gamma))$ is the dot product.

Spatial processing algorithms (5) are typically modified in actual practice

$$\dot{\mathbf{Y}}_{\Sigma}(t)=\dot{\mathbf{Y}}^{T}\left[\dot{\mathbf{\Phi}}^{-1}\dot{\mathbf{X}}(\alpha)\right]=\dot{\mathbf{Y}}^{T}(t)\dot{\mathbf{R}}(\alpha), \tag{5}$$

where $\dot{\mathbf{Y}}_{\Sigma}(t)$ is the output signal of the adaptive antenna array; $\dot{\mathbf{Y}}^{T}$ is the vector of the input signal of the array; $\dot{\mathbf{\Phi}}^{-1}$ is the inverse correlative interference matrix; $\dot{\mathbf{R}}(\alpha)$ is the weighted coefficients vector. For this the requirements for a simpler construction of a spatial processing device are considered. The algorithm is modified in terms of multiplication of the vector-columns of the received $\dot{\mathbf{Y}}(t)$ and anticipated $\dot{\mathbf{X}}(\alpha)$ oscillations by a transforming matrix **A** [5]. The transformation of the vector-column can be determined by the formula

$$\dot{\mathbf{Y}}_{\text{пр}}(t)=\dot{\mathbf{A}}\dot{\mathbf{Y}}(t),\quad \dot{\mathbf{X}}_{\text{пр}}(\alpha)=\dot{\mathbf{A}}\dot{\mathbf{X}}(\alpha). \tag{6}$$

The correlative matrix $\dot{\mathbf{\Phi}}$ and the inverted $\dot{\mathbf{\Phi}}^{-1}$ are also transformed. The weighted vector-column $\dot{\mathbf{R}}(\alpha)$ undergoes transformation as well. The transformed values considering (6) shall equal:

$$\begin{gathered}\dot{\mathbf{\Phi}}_{\text{пр}}=M_{\Pi}\left[\frac{\dot{\mathbf{Y}}_{\text{пр}}(t)\dot{\mathbf{Y}}^{*T}_{\text{пр}}(t)}{2}\right]=\dot{\mathbf{A}}\dot{\mathbf{\Phi}}\dot{\mathbf{A}}^{*T},\\ \dot{\mathbf{\Phi}}^{-1}_{\text{пр}}=(\dot{\mathbf{A}}\dot{\mathbf{\Phi}}\dot{\mathbf{A}}^{*T})^{-1},\quad \dot{\mathbf{R}}_{\text{пр}}(\alpha)=\dot{\mathbf{\Phi}}^{-1}_{\text{пр}}\dot{\mathbf{X}}_{\text{пр}}(\alpha).\end{gathered} \tag{7}$$

The transformation structure of the spatial processing algorithm is identical to the transformed

$$\dot{\mathbf{Y}}_{\Sigma\text{пр}}(t)=\dot{\mathbf{Y}}^{T}_{\text{пр}}(t)\dot{\mathbf{R}}_{\text{пр}}(\alpha)=\dot{\mathbf{Y}}_{\text{пр}}(t). \tag{8}$$

When comparing algorithms (5) and (8) it is easy to see that they are totally equivalent providing that the transformation matrix **A** has an inverted $\dot{\mathbf{A}}^{-1}$. So, if we substitute in (8) values $\dot{\mathbf{Y}}_{\text{пр}}(t)$ and $\dot{\mathbf{R}}_{\text{пр}}(\alpha)$ from (6) and (7), the result shall be

$$\dot{\mathbf{Y}}_{\text{пр}}(t)=\dot{\mathbf{A}}^{T}(\dot{\mathbf{A}}^{T})^{-1}(\dot{\mathbf{\Phi}}^{-1})^{*}(\dot{\mathbf{A}}^{*})^{-1}\mathbf{A}^{*}\dot{\mathbf{X}}(\alpha)=\dot{\mathbf{Y}}(t)\dot{\mathbf{R}}^{*}(\alpha)=\dot{\mathbf{Y}}_{\Sigma}(t). \tag{9}$$

Matrix $\dot{\mathbf{A}}$ is used for building circuits with a main receiving channel and (M–1) compensating receiving channel with low-directivity features of the directivity. When selecting a transforming matrix it is necessary to consider the requirement for decoupling of the main and compensating channels for the desired signal. This is one of the possible examples for such a matrix:

$$\dot{\mathbf{A}}=\begin{Vmatrix}\dot{\mathbf{X}}^{*}_{1}(\alpha) & \dot{\mathbf{X}}^{*}_{2}(\alpha) & \dot{\mathbf{X}}^{*}_{3}(\alpha) & \cdots & \dot{\mathbf{X}}^{*}_{m-1}(\alpha) & \dot{\mathbf{X}}^{*}_{m}(\alpha)\\ -\dot{\mathbf{X}}^{*}_{1}(\alpha) & \dot{\mathbf{X}}^{*}_{2}(\alpha) & 0 & \cdots & 0 & 0\\ 0 & -\dot{\mathbf{X}}^{*}_{2}(\alpha) & \dot{\mathbf{X}}^{*}_{3}(\alpha) & \cdots & 0 & 0\\ \cdots & \cdots & \cdots & \cdots & \cdots & \cdots\\ 0 & 0 & 0 & \cdots & -\dot{\mathbf{X}}^{*}_{m-1}(\alpha) & \dot{\mathbf{X}}^{*}_{m}(\alpha)\end{Vmatrix}. \tag{10}$$

The transformed vector-column $\dot{\mathbf{Y}}_{\text{пр}}(t)=\left\|\dot{\mathbf{Y}}_{\text{пр}i}(t)\right\|$ of received oscillations $\dot{\mathbf{Y}}(t)$ switches on the voltage of the main channel

$$\dot{\mathbf{Y}}_{\text{пр}1}(t)=\sum_{i=1}^{M}\dot{\mathbf{Y}}_{i}(t)\dot{\mathbf{X}}^{*}_{i}(\alpha). \tag{11}$$

And the voltage of the compensating channels

$$\dot{\mathbf{Y}}_{\text{пр}ki}(t)=-\dot{\mathbf{Y}}_{i}(t)\dot{\mathbf{X}}^{*}_{i}(\alpha)+\dot{\mathbf{Y}}_{i+1}(t)\dot{\mathbf{X}}^{*}_{i+1}(\alpha) \tag{12}$$

IOP Conf. Series: Materials Science and Engineering **155** (2016) 012019 doi:10.1088/1757-899X/155/1/012019

The expected oscillation $\dot{\mathbf{X}}(\alpha)$ is transformed to

$$\dot{\mathbf{X}}_{\text{пр}}(\alpha) = \dot{\mathbf{A}}\dot{\mathbf{X}}(\alpha) = \left\| M \quad 0 \quad 0 \quad \cdots \quad 0 \right\|^T \tag{13}$$

where $M = \sum_{i=1}^{M} \dot{\mathbf{X}}_i^*(\alpha)\dot{\mathbf{X}}_i(\alpha)$.

The zero value of the elements of vector-column $\dot{\mathbf{X}}_{\text{пр}}(\alpha)$, beginning from the second indicate to the decoupling of the main and compensating channels for the desired signal. The latter is a mechanism for forming compensating channels with differential schemes. Due to this fact, a null is formed in the resulting directivity characteristics of the compensating channels. The angular position of this null matches the expected direction of the desired signal input α.

The expression for the transformed weighted vector-column of spatial processing $\dot{\mathbf{R}}_{\text{пр}}(\alpha)$ can be found using the following matrix equation

$$\dot{\mathbf{\Phi}}_{\text{пр}}\dot{\mathbf{R}}_{\text{пр}}(\alpha) = \dot{\mathbf{X}}_{\text{пр}}(\alpha). \tag{14}$$

It is convenient to solve, writing down $\dot{\mathbf{\Phi}}_{\text{пр}}$ and the vector-column $\dot{\mathbf{R}}_{\text{пр}}(\alpha)$ in the following manner

$$\dot{\mathbf{\Phi}}_{\text{пр}} = \begin{Vmatrix} \dot{\mathbf{\Phi}}_{\text{пр}11} & \dot{\mathbf{\Phi}}_{\text{пр}}^{*T} \\ \dot{\mathbf{\Phi}}_{\text{пр}1} & \dot{\mathbf{\Phi}}_{\text{пр}\kappa} \end{Vmatrix} \qquad \dot{\mathbf{R}}_{\text{пр}}(\alpha) = \left\| \dot{\mathbf{R}}_1 \quad \dot{\mathbf{R}}_2 \right\|^T, \tag{15}$$

where $\mathbf{\Phi}_{\text{пр}11}$ is the total interference and noise dispersion at output of the main receiving channel

$$\dot{\mathbf{\Phi}}_{\text{пр}11} = M_{\Pi}\left[\frac{\dot{\mathbf{Y}}_{\text{пр}1}(t)\dot{\mathbf{Y}}_{\text{пр}1}^*(t)}{2}\right]; \tag{16}$$

$\dot{\mathbf{\Phi}}_{\text{пр}1}$ is the vector-column of relative correlative moments of interference voltage in the compensating channels and main channel

$$\dot{\mathbf{\Phi}}_{\text{пр}1} = M_{\Pi}\left[\frac{\dot{\mathbf{Y}}_{\text{пр}k}(t)\dot{\mathbf{Y}}_{\text{пр}1}^*(t)}{2}\right], \quad k = \overline{2M}; \tag{17}$$

$\dot{\mathbf{\Phi}}_{\text{пр}\kappa}$ is the correlative interference matrix of the compensating channels

$$\dot{\mathbf{\Phi}}_{\text{пр}k} = M_{\Pi}\left[\frac{\dot{\mathbf{Y}}_{\text{пр}k}(t)\dot{\mathbf{Y}}_{\text{пр}k}^{*T}(t)}{2}\right]; \tag{18}$$

$\dot{\mathbf{R}}_1$ and $\dot{\mathbf{R}}_2$ represent the element of the weighted vector of the main channel and the vector-column of the weighted coefficients of the compensating channels respectively. Substituting (15–18) in matrix equation (14) we produce a system of two equations

$$\begin{cases} \dot{\mathbf{\Phi}}_{\text{пр}1\,1}\dot{\mathbf{R}}_1 + \dot{\mathbf{\Phi}}^{*T}_{\text{пр}1}\dot{\mathbf{R}}_2 = M \\ \dot{\mathbf{\Phi}}_{\text{пр}1}\dot{\mathbf{R}}_1 + \dot{\mathbf{\Phi}}_{\text{пр}k}\dot{\mathbf{R}}_2 = 0 \end{cases}, \tag{19}$$

Solving this system in relation to $\dot{\mathbf{R}}_1$ and $\dot{\mathbf{R}}_2$, we shall get

$$\begin{cases} \dot{\mathbf{R}}_1 = \dfrac{M}{(\dot{\mathbf{\Phi}}_{\text{пр}1\,1} - \dot{\mathbf{\Phi}}^{*T}_{\text{пр}1}\dot{\mathbf{\Phi}}^{-1}_{\text{пр}k}\dot{\mathbf{\Phi}}_{\text{пр}1})}, \\ \dot{\mathbf{R}}_2 = -\dot{\mathbf{\Phi}}^{-1}_{\text{пр}k}\dot{\mathbf{\Phi}}_{\text{пр}1}\dot{\mathbf{R}}_1 \end{cases} \tag{20}$$

From (20) we can see that the first element of the weighted vector is the real quality and functions as a normalization factor. Without quality loss for detection this factor can be taken as equal to zero, shifting to the normalizing weighted vector $\dfrac{\dot{\mathbf{R}}_{\text{пр}}(\alpha)}{\dot{\mathbf{R}}_1}$. According to (20), the latter shall be equal

$$\dot{\mathbf{R}}_\Delta = \frac{\dot{\mathbf{R}}_2}{\mathbf{R}_1} = -\dot{\mathbf{\Phi}}^{-1}_{\text{пр}k}\mathbf{\Phi}_{\text{пр}1}, \tag{21}$$

The output voltage of the processing device with a designated main channel is determined by the formula

$$\dot{\mathbf{Y}}_\Sigma = \dot{\mathbf{Y}}_{\text{пр}1}(t) + \dot{\mathbf{Y}}^{*T}_{\text{пр}k}(t)\dot{\mathbf{R}}^*_\Delta, \tag{22}$$

The device for compensating interference received by the side lobes of the antenna's radiation pattern is built with a synthesized algorithm; the device is shown in Fig. 2.

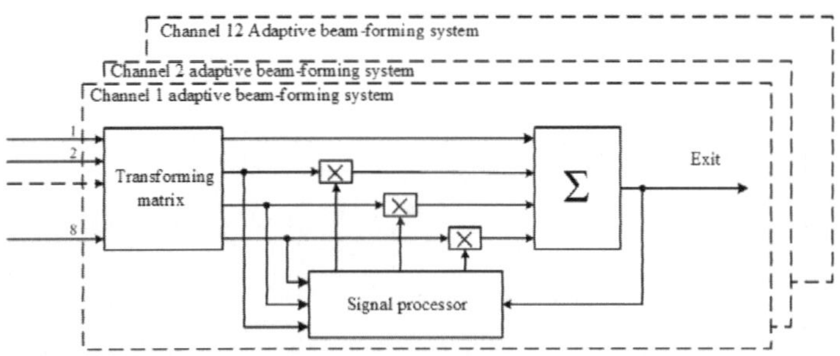

Figure 2. Adaptive phased antenna array beam-forming system

The analog sector of the adaptive phased antenna array comprises 8 antenna modules with 8 connected radio paths. Signals from the radio paths are transmitted to the analog-to-digital converter and then to the digital sector. The digital sector comprises 12 independent channels for processing L_1, L_2, GLONASS signals. Each channel consists of a beam-forming network and a digital signal processing channel. The beam-forming network is a weighted signal adder for signal from 8 analog paths. Weighted coefficients are calculated by the signal processor according to the desired radiation pattern. Particularly, during a high-precision measuring mode the weighted coefficients are selected to compensate the difference in the desired signal passage between antennas: thus, desired signals arrive at the adder with one phase. An example of such a mode for adaptive phased antenna arrays is discussed in [6].

Let's demonstrate the test results of a model of anti-interference navigation satellite systems user equipment with an 8-element phased antenna array. Comparison was performed on type MRK-33 user equipment.

The anti-interference test is performed in two modes of the navigation satellite system: signal grabbing and signal tracking of navigation satellite signals. It is necessary to calibrate the measuring circuits, using the method described in [7].

The evaluation criterion for interference immunity during natural tests is such a level of the interfering signal during which there is an absence of a solution to the navigation problem for at least four satellites that are in tracking and grabbing modes. Interference immunity is evaluated with and without using spatial filtering methods.

Increasing the interference level with a spacing of 1 dB from the original value, a value was recorded during which there was a loss in tracking (navigation data output stopped) – the level of suppressing the tracking scheme. Counting is begun from the point when one satellite is excluded from processing; it is continued until all prior visible satellites are excluded.

To evaluate the interference immunity of a phased antenna array in a grabbing mode, the interfering signal level is set at 10 dB higher than the level of tracking loss. An attenuator is used to gradually decrease the interference level with a spacing of 1 dB at output; grabbing of the navigational signal and navigational sightings were recorded.

In Figs. 3–5 the output spectra of signals at an intermediate frequency without any spatial processing (blue highlighting) is demonstrated; spatial filtering is shown in red. Approximately 100 000 samples were used to build the spectra.

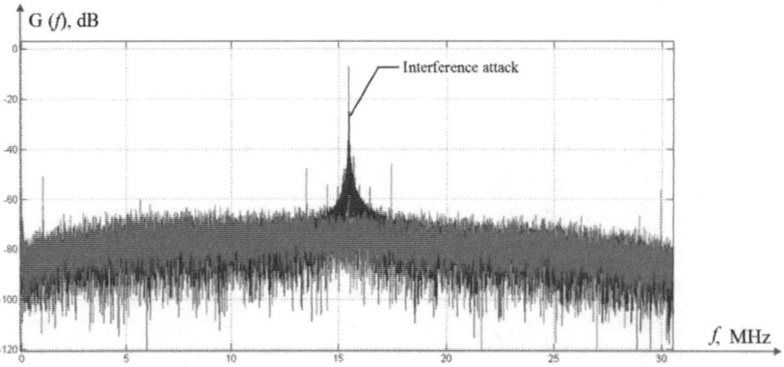

Figure 3. Output spectra of signals under narrowband interference attack: blue – without spatial selection; red – with spacial selection

For the experiment on the interference immunity of the user navigation equipment under harmonic interference attack, the initial level of interference power was set at 10 dB lower than the minimal level for narrowband interference for devices lacking any interference protection. This issue was discussed in [8].

Fig 4 demonstrates the output spectra of signals attacked by noise interference with a limited spectrum.

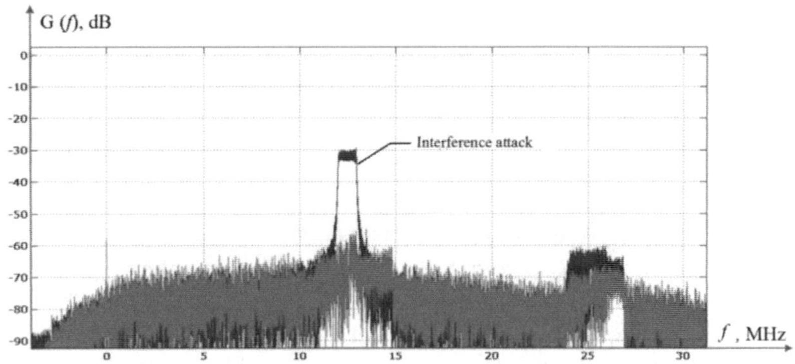

Figure 4. Output spectra of signals under noise interference attack with a limited spectrum: blue – without spatial selection; red – with spatial selection

Employing spatial selection methods during noise interference attack with limited spectrum (Fig. 4) provides an intereference suppression coefficient of 30 – 35 dB for user equipment.

In Fig 5 the input spectra of signals attacked by broadband noise interference is demonstrated.

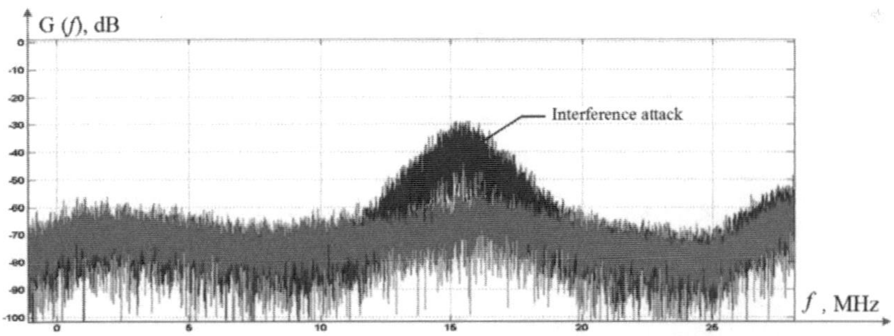

Figure 5. Imput spectra of signals unter broadband noise interference attack: blue – without spatial selection; red – with spatial selection

The analysis of the graphs in Fig. 5 shows that spatial filtering sustains the interference suppression coefficient for broadband interference at 25–30 dB higher for user navigation equipment with an adaptive phased antenna array than for equipment without any interference protection.

3. Conclusion

Thus, the performed natural experimentations of an 8-element adaptive phased antenna array demonstrate that the synthesized algorithm for adaptive nulling of phased antenna arrays is highly effective and is capable of suppressing both narrowband and broadband interfering signals.

The interference suppression coefficient using spatial filtering methods was:

- 40–45 dB for harmonic interference attack;
- 30–35 dB for noise interference attack with a limited spectrum;
- 25–30 dB for broadband interference attack.

References

[1] Electronic systems: basics of assembly and theory (Radioelektronnye sistemy: osnovy postroeniia i teoriia) 2007 Ed. 3 Ia D Shirman Ed. Moscow: Radiotekhnika

[2] Tyapkin V N, Dmitriev D D, Konnov V G, Fomin A N 2012 Method of determination of spectrum factors vector due to the likelihood ratio criterion *Vestnik SibGAU* Issue **3 (43)** P 76–79

[3] Tyapkin V N, Dmitriev D D, Garin E N, Sokolovskiy A V 2013 Operation of amplitude-phase distribution of adaptive phased array *Industrial Automatic Control Systems and Controllers* No. **4** P 66–71

[4] GLONASS. Basic concepts and operation [*GLONASS. Printsipy postroeniia i funktsionirovaiia*] 2010 4th ed., Perov A I, Kharisov V N Eds. Moscow: Radiotekhnika

[5] Almazov V B, Manzhos V N 1985 Gathering and processing radar data [*Poluchenie i obrabotka radiolokatsionnoi informatsii*]. Kharkov

[6] Sharfunova T G, Tyapkin V N, Dmitriev D D 2013 The measurement accuracy of navigation parameters in user equipment GLONASS satellite navigation system, equipped with an antenna array *Radiotekhnika* No. **6** P 22–25

[7] Tyapkin V N, Fateev Yu L, Dmitriev D D, Konnov V G 2012 Calibration of measuring path for ECM test of navigation equipment of customers of satellite navigation systems *Vestnik SibGAU* Issue **4 (44)** P 139–142

[8] Tyapkin V N, Dmitriev D D, Moshkina T G 2012 Potential interference immunity of navigation equipment of customers of satellite radio navigational systems *Vestnik SibGAU* Issue **3 (43)** P 113–119

Applying filtering for determining the angular orientation of spinning objects during interference*

I N Kartsan[1], Y L Fateev[2], V N Tyapkin[3], D D Dmitriev[4], A E Goncharov[5], P V Zelenkov[6], I V Kovalev[7]

[1] Associate professor, Reshetnev Siberian State Aerospace University, Krasnoyarsk, Russia
[2] Associate professor, Siberian Federal University, Krasnoyarsk, Russia
[3] Associate professor, Siberian Federal University, Krasnoyarsk, Russia
[4] Associate professor, Siberian Federal University, Krasnoyarsk, Russia
[5] Associate professor, Reshetnev Siberian State Aerospace University, Krasnoyarsk, Russia
[6] Associate professor, Reshetnev Siberian State Aerospace University, Krasnoyarsk, Russia
[7] Professor, Reshetnev Siberian State Aerospace University, Krasnoyarsk, Russia

E-mail: kartsan2003@mail.ru

Abstract. This article discusses the application of user navigation equipment of satellite radionavigation systems (GLONASS/GPS) for measuring the inclination of the aerial axis of rotation. The authors have demonstrated that the required accuracy of the angular inclination is achieved only in a relative phase mode, which is not always feasible. The application of filtering for measured parameters of navigation equipment has namely been used for planimetric coordinates. Filtering is performed using the second order Kalman filter. It has significant effect; thus, even at a speed of 1 rpm, there is no disruption in the autonomous coded mode; for the relative phase mode, the required accuracy is achieved at 3 rpm.

Keywords: GLONASS/GPS satellite radionavigation system, inclination from the aerial axis of rotation, second order Kalman filter.

1. Introduction

Currently, user navigation equipment of the GLONASS radionavigation system has found intense application for improving the mobility of aerial and space defense systems. As the demand for highly-accurate radiolocation data produced by mobile radiolocation devices grows, it becomes universal not only for topographic positioning, but for the precision of leveling antenna systems in triaxial mobile radiolocation devices, which greatly affects the instrumental accuracy for determining the altitude of aerial objects. The leveling error, using existing methods, usually exceeds 3 arc minutes; the altitude measurement error at a distance of 100 km is over 87 m.

It is preferable to position the antennas of user navigation equipment on the antenna system

* The research had been conducted with financial support from the Ministry of Education and Science of the Russian Federation (agreement from 28.11.2014 № 14.574.21.0131, unique project identifier: RFMEFI57414X0131)

of the mobile radiolocation unit. This is necessary to provide the radio coverage of the upper hemisphere and reduce the error of the multibeam signal detection. In this case, using single-antenna user navigation equipment, it is possible to determine angular velocity and deviations from vertical of the aerial axis of rotation.

A method for determining the angular position of a rotating object and errors for autonomous coded, deferential and phased modes for user navigation equipment functioning is described in [1, 2]. The accuracy analysis in [1] demonstrates that using single-antenna equipment in an autonomous coded mode does not allow us to determine, with a sufficient degree of accuracy, the inclination angle of the aerial axis of rotation, in comparison with conventional optical methods of leveling mobile radiolocation unit antennas. Differential and relative phase modes in user navigation equipment require at least two user antennas to be positioned on the mobile radiolocation unit antenna system – an operation that is not always possible. This adds to the relevance of filtering measured parameters in user navigation equipment for the purpose of increasing the accuracy for determining the inclination angle of the aerial axis of rotation.

2. Determining the angular orientation for spinning objects

Let's look at the algorithm for determining the inclination of an antenna's aerial axis of rotation. The position of the axis of rotation is determined by trajectory points; the current angular position is determined in the plane of rotation.

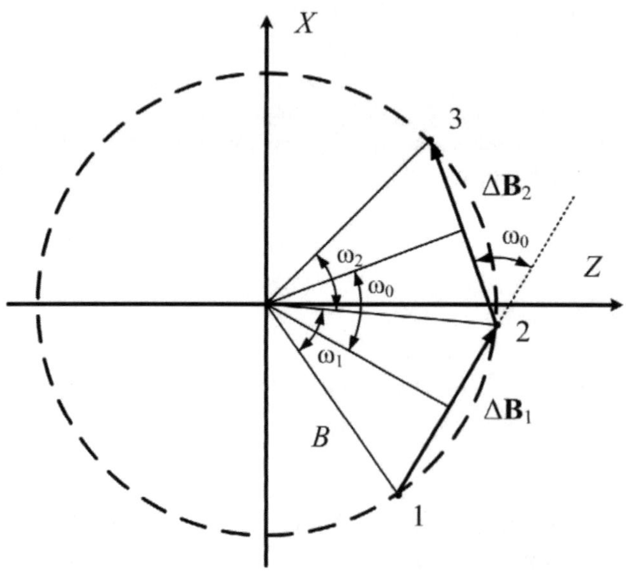

Figure 1. Determining the module of the angular velocity vector

The coordinate increment vectors are noncollinear and orthogonal to the axis of rotation, hence the coordinates of the axis of rotation are determined through a cross product of antenna coordinate increment vectors, as seen in Fig. 1.

$$\mathbf{n} = \Delta\mathbf{B}_1 \times \Delta\mathbf{B}_2. \quad (1)$$

Using expression (1), it is also possible to determine the direction of rotation. By definition, the angular velocity vector is directed so that from its end the rotation of an object is counterclockwise. In order to find the direction of the angular velocity vector, vector $\mathbf{\Delta B}_1$ must precede vector $\mathbf{\Delta B}_2$ in time [3, 4].

The **n** vector module equals:

$$|\mathbf{n}| = |\mathbf{\Delta B}_1| \cdot |\mathbf{\Delta B}_2| \cdot \sin \omega_0. \tag{2}$$

The angle of ω_0, as follows from Fig. 1, is equal to the following angle of pitch:

$$\omega_0 = \frac{\omega_1 + \omega_2}{2} = \arcsin \frac{|\mathbf{n}|}{|\mathbf{\Delta B}_1| \cdot |\mathbf{\Delta B}_2|}. \tag{3}$$

The module of the angular velocity vector:

$$|\omega| = \frac{\omega_0}{\Delta t} = \frac{\left(\arcsin \frac{|\mathbf{\Delta B}_1 \times \mathbf{\Delta B}_2|}{|\mathbf{\Delta B}|^2} \right)}{\Delta t}. \tag{4}$$

To determine the module of the angular velocity using formula (4) it is not necessary to know the length of the base. Moreover, the given method is functional, even if the base does not lie in the plane of rotation.

The proposed algorithm can be built using coded measurements in an autonomous mode, using a differential mode or angular phase observations. These modes have different errors for antenna coordinate increment measurements. (Fig. 2).

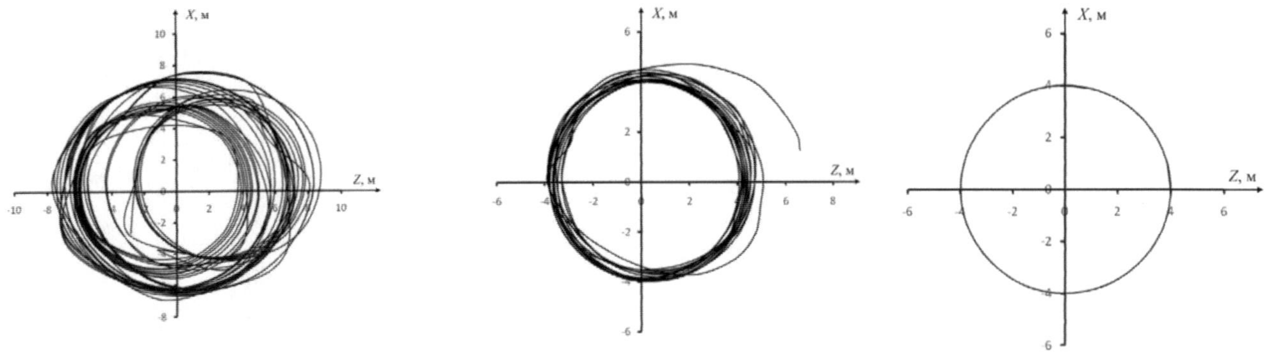

a) *b*) *c*)

Figure 2. Measuring coordinates *X*, *Z*: *a* – for an autonomous coded mode; *b* – for a differential mode; *c* – for a relative phase mode

During determining the vector of the axis of rotation, pictured in Fig. 3, the distance between the user's equipment antennas was 4 m; the speed of the antenna rotation was 6 revolutions per minute. An analysis of the demonstrated figures shows that the accuracy for measuring the inclination of the axis of rotation is much higher for differential and relative phase modes than for autonomous coded modes. Less irregular measurements have been observed in the third mode. For a more or less accurate determination of the inclination of the axis of rotation, it is necessary to conduct a large series of observations, i.e. to find the average of the measured parameters.

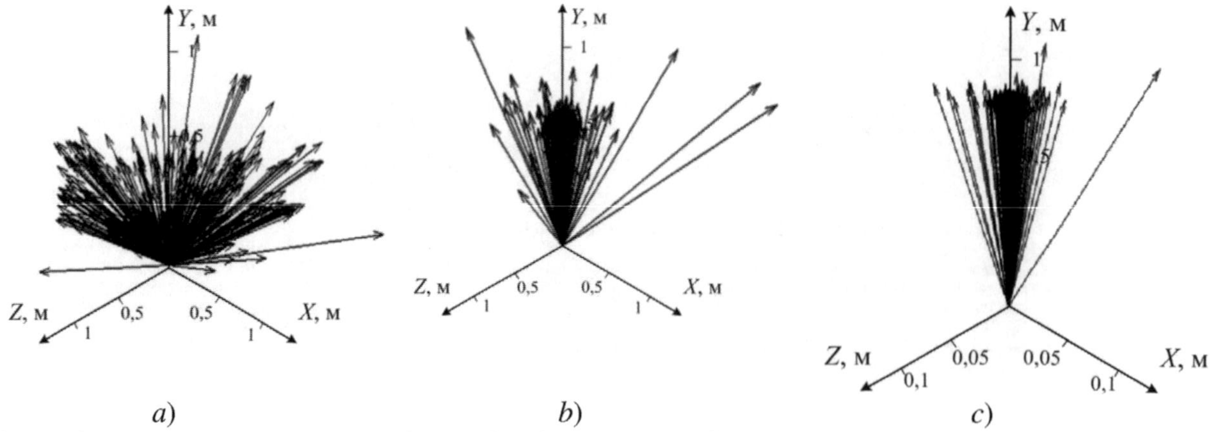

Figure 3. Determining the vector of the axis of rotation: *a* – for an autonomous coded mode; *b* – for a differential mode; *c* – for a relative phase mode

Fig. 4 demonstrates errors in measuring the inclination angle of the antenna from a vertical axis at different rotation velocities.

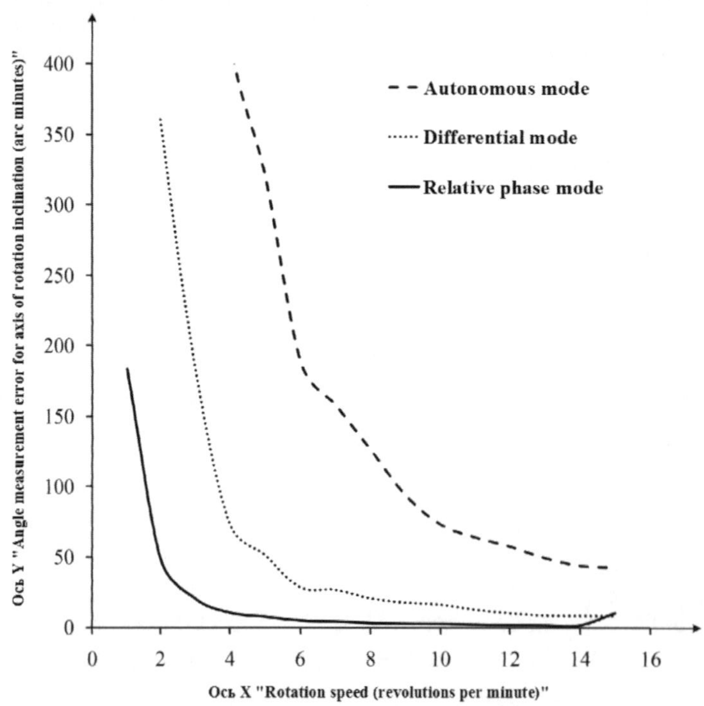

Figure 4. Results of modeling the measurement error for the angle of inclination for an aerial axis of rotation

The standard inclination is calculated, using the following formula:

$$\sigma_{Sy} = \frac{\sigma_{пл}}{|\Delta B|^2 \cdot \sin(\omega \cdot \Delta t)} = \frac{\sigma_{пл}}{4|B|^2 \cdot \sin^2(\omega \cdot \Delta t / 2) \cdot \sin(\omega \cdot \Delta t)}, \quad (5)$$

where $\sigma_{пл}$ is the standard deviation for measuring planimetric coordinates, which, in this case, are equal to: 0.7 m for an autonomous mode; 4 cm for a differential mode and 5 mm for a relative phase mode.

Analyzing the results of the measurements, we can see that in order for the algorithm to be accurate enough, it is necessary for the antenna to rotate at a high speed, especially in the automated mode. High precision is achieved only in a relative phased (or angular) mode; in an autonomous mode, even using a long arm, the accuracy of the measurements does not exceed one degree. On the other hand, to determine the deviation of the axis of rotation and its special orientation, it is possible to use only three trajectory points.

By using filtering for measurable parameters, it is possible to significantly increase the accuracy for determining the inclination of the aerial axis of rotation. For such problems it is practical to use the filtering of measured *X*, *Z* coordinates using the second order Kalman filter [5, 6]. During the development of the Kalman filter, the closest function has been selected: circumference $R^2 = X^2 + Z^2$.

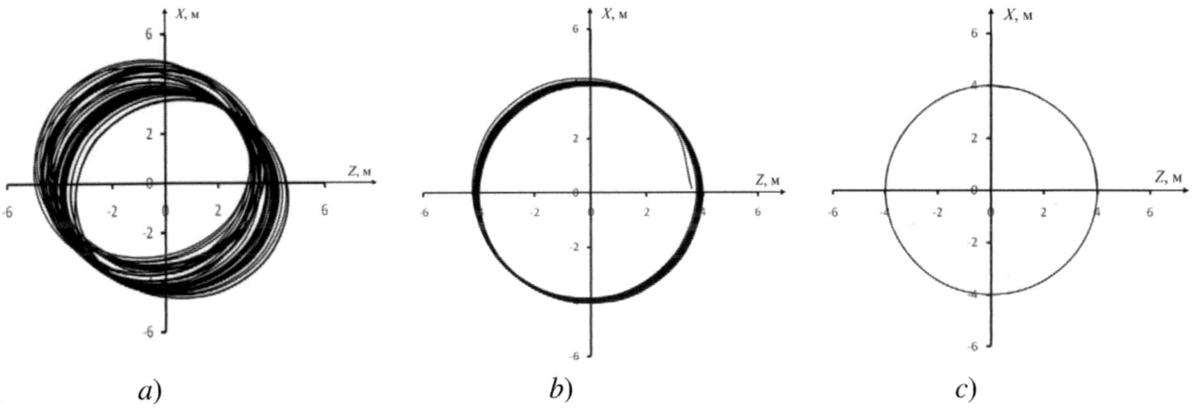

Figure 5. Measuring and filtering coordinates *X*, *Z*: *a* – for an autonomous coded mode; *b* – for a differential mode; *c* – for a relative phase mode

Fig. 5 demonstrates that in result of filtering the determined coordinates *X* and *Z*, the trajectory of the antenna path became more similar to a circle, without irregular measurements. Results in Fig 6 confirm a significant reduction in the error of determining the vector of the axis of rotation.

IOP Conf. Series: Materials Science and Engineering **155** (2016) 012020 doi:10.1088/1757-899X/155/1/012020

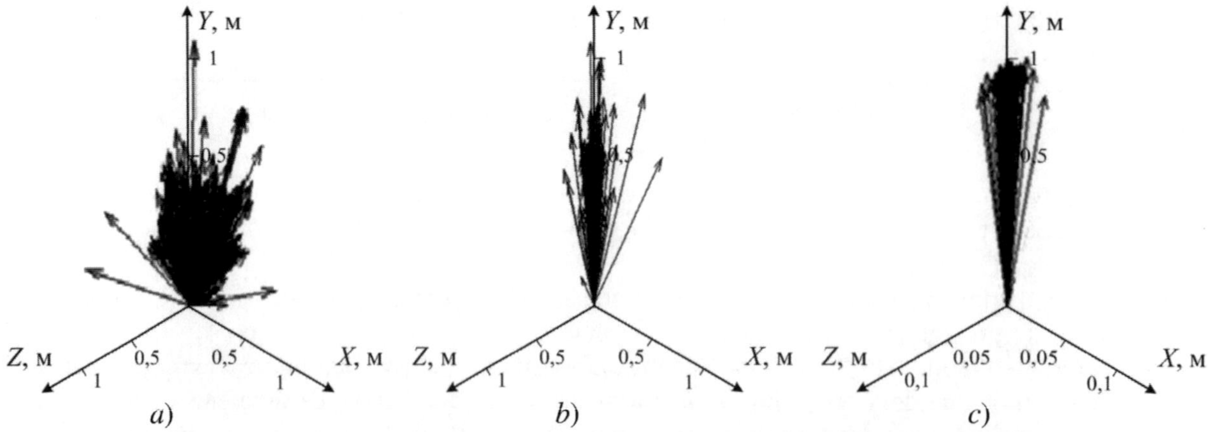

Figure 6. Determining the vector of the axis of rotation and its filtering: *a* – for an autonomous coded mode; *b* – for a differential mode; *c* – for a relative phase mode

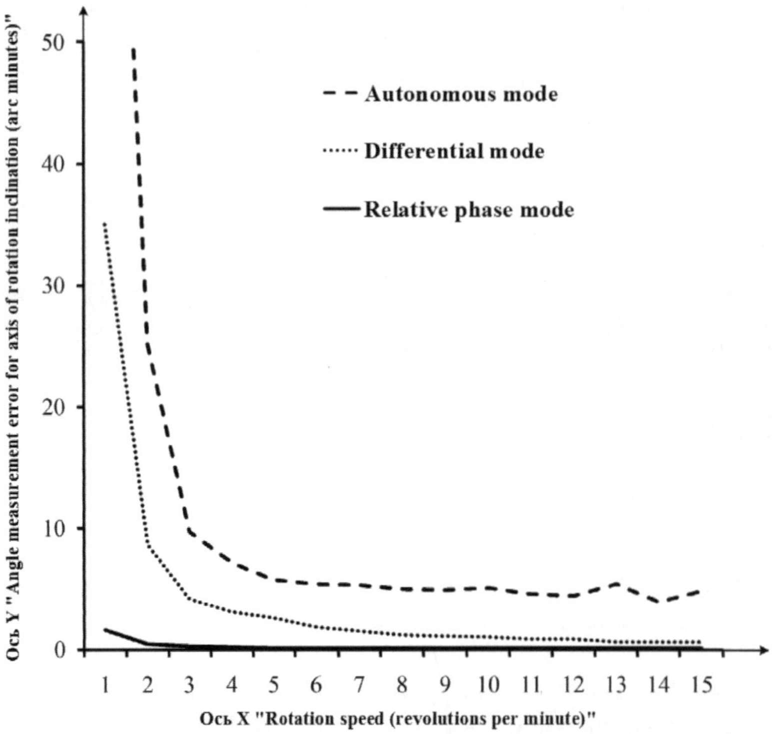

Figure 7. Results of simulating errors of determining the angle of the aerial axis of rotation inclination using a second order Kalman filter.

Fig. 7 demonstrates the results of simulating errors of determining the inclination of the aerial axis of rotation in various functioning modes of the user navigation equipment for radionavigation systems, filtering the measured coordinates with a second order Kalman filter.

Thus, based on the simulation results, we can conclude that filtering measurable planimetric coordinates yields a significant result. For instance, even at spinning speeds of 1 revolution per minute there are no failures in an autonomous mode; if the relative phase mode is employed, the desired accuracy may be obtained at a revolution of 3 rpm. In reality, the filtering level is selected, depending on the objects behavior.

4. Conclusion

The discussed method for determining the inclination of the aerial axis of rotation for mobile radiolocation devices, using user navigation equipment and filtering of measured parameters, enables us to perform immediate control over the antenna's position in space. The method significantly improves the maneuverability of mobile radiolocation devices and the accuracy of determining the altitude of aerial objects. It is possible to use not only differential and relative phase modes for operating the user navigation equipment, but the autonomous coded mode, enabling us to use serial and standard user navigation equipment for determining the inclination of the angle of the axis of rotation.

References

[1] Fateev Y L, Garin E N, Dmitriev D D 2012 Izmerenie uglovoi orientatsii vraschaiuscegosia o,"ekta [The measurement of angular orientation of a spinning object (in Russ.)] *Uspekhi sovremennoi radioelektroniki* No. **9** P. 69–75.

[2] Kartsan T I, Dmitriev D D, Kartsan I N, Tyaptin V N, Samoilov E A 2015 Integration of GNSS receivers and internal sensors *Reshetnev readings* Vol. **1** No. **19** P. 232–234.

[3] Kartsan I N, Tikhomovich A S, Dmitriev D D 2015 Determining navigation parameters of objects under the action of interference of various orgins *Vestnik SiGAU* **15(4)** P. 891–897.

[4] Iarovskii B M 1977 Spravochnik po fizike dlia inzhenerov i studentov vuzov [A handbook on physics for engineers and university students (in Russ.)] Moscow: Nauka.

[5] Balakrishnan A V 1988 Teoriia fil'tratsii Kalmana [Kalman's filtering theory (in Russ.)] Moscow.

[6] Fateev Yu L, Dmitriev D D, Tyapkin V N, Kartsan I N, Goncharov A E 2015 Phase methods for measuring the spatial orientation of objects using satellite navigation equipment *IOP Conference Series: Materials Science and Engineering. International Scientific and Research Conference on Topical Issues in Aeronautics and Astronautics (Dedicated to the 55th Anniversary from the Foundation of SibSAU)* P. 012022.

IOP Conf. Series: Materials Science and Engineering **155** (2016) 012021 doi:10.1088/1757-899X/155/1/012021

Efficiency of mobile applications for engineering language learners

N V Ivleva
Business Foreign Languages Department
Reshetnev Siberian State Aerospace University,
31, Krasnoyarsky Rabochy Av., Krasnoyarsk, 660037, Russian Federation

E-mail: natalie_ivleva@mail.ru

Abstract. Nowadays in the world of globalization and integration to survive in in it and be able to prove your competences and professionalism all educators are obliged to update their professional skills and knowledge. A big role of this update is to use cutting edge and innovative technologies. However, it is important to be rather careful and following certain recommendations to get the benefit from it.
Key words. Mobile applications, operating system, language learning, engineering language learners, mobile deliver.

1. Introduction

Nowadays not only is English language learning a crucial element of getting a career promotion and getting opportunities to work in international companies but also a great way to broaden your mind and be open to resources written in English which gives a specialist a competitive advantage among other workers in their business sector. Thanks to cutting edge technologies in almost all spheres of our life it became more flexible and adaptive to learn a foreign language through those technologies as well. There is a wide range of technologies for foreign language learning, particularly English. In this range we can include educational websites, social networks tailored to language practicing, massive open online courses (MOOCs) and, as a new element in language teaching, mobile applications.

Due to mass distribution of gadgets such as smartphones, tablets, netbooks, etc. the society reorientates to a new way of life that could be focused on modern technologies and many processes are implemented through gadgets they have to simplify their life. People pay online, participate in virtual events and workshops, arrange online meetings and conferences, play games in the virtual reality etc. As a result the IT market is forced to meet the society's needs and create all opportunities to computerize life processes not only through just a personal computer but also through mobile delivery. Consequently, more and more IT developers and designers are working in this area working on new projects and offering an extremely big amount of applications as this is demanded by the society and brings financial benefits for this business.

Mobile applications for foreign language learning are also developed by companies focused on the educational sector. It is an attractive area for the IT sphere as it brings good incomes due to popularity of foreign language learning in general. In our article we will stress on the English language as it is the global language and most popular among learners due to abovementioned reasons.

2. Categories of language learning mobile applications

Claire Bradin Siskin, an expert in English language teaching and linguistics, author of language teaching materials, categorized language learning mobile applications according to the following list which was specified by the author of this article:

- Built-in Applications – applications which are built in operating systems and can easily be used in the educational context for free without additional plugins and adaptation.

- Instructional Applications – applications which can tutor, discuss subject matters with students, and diagnose student errors with profoundly affect educational practice.
- Social Networking Applications - applications which give users a place to join an online community and establish a contact list, build photo galleries and share personal updates. It provides an opportunity for language learners to socialize with people all over the world and improve the language command.
- Repurposed Applications – applications which can be used in language learning, including voice search, voice email, postcard creation, audio recording, and children's games.
- Task Generators – applications aimed at creating tasks and tests which can be used by students within a study process.

Within all those categories it is possible to highlight programs which are widely spread on the territory of the Russian Federation among language learners. For example, Google Translate for Android which uses a voice mode and belongs to repurposed applications. Using this application learners can translate spoken words and phrases if a device has a microphone. In some languages it is possible to hear the translation spoken aloud. It allows learners to get the translation of a phrase or separate words and also gives them an opportunity to practise their pronunciation and language fluency. Another widely spread mobile application in the educational context is LinguaLeo. This resource is tailored to both mobile and desktop computer modes. Focusing on application feature it is crucial to highlight that this application allows to identify learners' language goals, boost the language competence through videos, doing a wide range of exercises. This platform also has inbuilt parameters to assess the study progress, provide awards and demonstrates effectiveness to learners which is essential from the pedagogical point of view. This mobile application can be referred to instructional applications according to its characteristics. As social networking is getting more and more popular in the Russian community it was the right way to introduce social networking applications into the educational process as they can improve learners' skills in a more engaging way. One of the most famous social networks in the Russian Federation and, as it was analyzed, in the Commonwealth of Independent States is VKontakte. Most population dealing with social networking creates accounts on this website and constantly communicates with their friends and followers through it. Users create communities depending upon their interests and invite others to join them. It became obvious for users to create communities for people who are into language learning and mastering other cultures. In such VKontakte communities people post inspiring quotations, lists of words grouped into categories upon a subject, various polls to identify learners' needs and many other useful materials which unite people into a global community of language learners. As an example of task generators which are attractive for language educators to put them into practice as well as for learners themselves is Voice Memos for iOS. This utility can be used for creating a dictionary with pictures and audio annotations. One more application from this category is Quizlet. It keeps learners engaged and motivated and gives them materials to help learn in a fun and more efficient way.

3. Engineering language learners.

It was covered in many researches that engineering students dramatically differ from other categories of students due to special characteristics typical only for them. Based on experiments implemented at the premises of the Reshetnev Siberian State Aerospace University, as a technical university, engineering students have analytical and cooperation professional skills. Moreover, intrinsically they are extremely curious that leads them to mastering everything new and undiscovered. This knowledge about engineering learners must be applied in all subjects taught to them as it must be considered to be a direct way to their efficiency as specialists in the future. Regarding the language learning process we must also apply it in the teaching practice. It means that all new methods and innovative technologies introduced in the area of teaching foreign languages should be immediately used for working with engineering students. It will motivate and involve them into a language study process, furthermore, it will help them meet their curiosity. At the same time it will all keep them engaged and inspired what is vital for successful education globally.

Mobile applications are one of the ways to support engineering students. It could be a good idea to use abovementioned categories of applications and at the same time always analyze which of them are more appropriate for engineering language learners. Depending on an engineering area it is possible to use Memrise as an additional resource on a regular base. Memrise is a language learning program that extends beyond vocabulary and language to things like history and science, but at its core it is a flashcard-style program that is augmented with memory tricks, images, and other useful tools to make learning a new language easier. Its focus is largely on memorization, but it is also designed to help you have fun learning the language you are trying to pick up. Memrise gamifies the process a bit, awarding you points and reputation as you learn, and the opportunity to compete against other users while you learn and complete activities. Memrise is free, web-based, and has iOS and Android applications so it gives a chance to take lessons on the go. It allows engineering students to choose the subject area they study in their program and be trained in those areas.

4. Developing for mobile delivery

According to Robert Godwin-Jones, professor of Virginia Commonwealth University, in order to take full advantages of the operating system capabilities of a mobile device, developers need to create an application using an approach consistent with that device's programming environment. For Apple iOS devices, such as iPhone, iPad, that means using Objective-C and Apple's XCode developers' tool. However, those applications do not run on Android devices, for which applications are written in Java running on a version of Linux. iOS applications are available exclusively from the Apple App Store, while Android applications are more widely available, from the official Android Market, Amazon's Appstore for Android, Handango, and other repositories. As is the case with iOS applications Android applications are free or affordable. Most applications are usually available in English only, few are localized for other languages. Other smartphone environments (Blackberry, Windows Phone 7, etc.) use different programming environments, all mutually incompatible. All of the different smartphone software companies make development tools available, which typically include a desktop phone simulator. Most are free, or available for a modest cost. Except for iOS development, which is Mac OS only, app development can be done on Windows or Macintosh machines, and, in some instances, on Linux.

Taking all abovementioned risks and problems which can occur in the process of usage, it is necessary to realize which device and software to use in the study process and recommend learners for self-study. Russian IT market is also developing at a quick pace. It tries not to lag behind other IT markets. Due to that for Russian learners there are no critical issues which could prevent from prompt usage of devices and, consequently, their applications. However, as educators, we should keep the following criteria for devices and applications used on the territory of the Russian Federation in mind:

1. Devices and operating systems learners use in their routine life.
2. Learners' possibilities to download applications on a paid basis.
3. If learners are a target audience of this developer or not.

5. Conclusion

As personal devices, smartphones are ideal for individualized formal and informal learning. The user determines which applications to acquire and how to use them. As language educators, we should encourage and assist the learner autonomy this enables and provide means for learners to combine formal and informal learning.

Learner autonomy is when students take control and responsibility for their own learning, both in terms of what they learn and how they learn it. It takes as its starting point the idea that students are capable of self-direction and are able to develop an independent, proactive approach to their studies.

In the field of higher education, learner autonomy is particularly important. Students may have limited classroom contact time for learning English but they may need to rapidly increase their knowledge and skills. It is therefore important for them to become self-reliant language learners who can continue learning efficiently outside the classroom.

References

[1] [EdVista]. (In Eng.). Available at: http://www.edvista.com/claire/pres/smartphones (accessed 29.09.2016)
[2] [Language Learning and Technology]. (In Eng.) Available at: http://llt.msu.edu/issues/june2011/emerging.pdf (accessed 29.09.2016)

IOP Conf. Series: Materials Science and Engineering **155** (2016) 012022 doi:10.1088/1757-899X/155/1/012022

Teaching critical thinking to engineering students through reading profession-oriented texts

N V Ivleva
Business Foreign Languages Department
Reshetnev Siberian State Aerospace University,
31, Krasnoyarsky Rabochy Av., Krasnoyarsk, 660037, Russian Federation

E-mail: natalie_ivleva@mail.ru

Abstract. Nowadays critical thinking is considered to be an essential learning outcome. To be able to use it in our life it is vital to obtain this skill through studying at school and university. In this paper there are some methods considered to drill this skill through reading within a translation course.
Key words. Engineering students, critical thinking, method, profession-oriented text, translation.

1. Introduction

Nowadays critical thinking plays a vital role in a teaching process. By critical thinking we mean the mode of thinking — about any subject, content, or problem — in which the thinker improves the quality of his or her thinking by skillfully analyzing, assessing, and reconstructing it. Critical thinking is self-directed, self-disciplined, self-monitored, and self-corrective thinking. It presupposes assent to rigorous standards of excellence and mindful command of their use. It entails effective communication and problem-solving abilities, as well as a commitment to overcome our native egocentrism and sociocentrism. Critical thinking development in a teaching process is considered to be an important element in the whole study process as it allows to teach students to be able to analyze, assess and judge carefully, to provide an extraordinary method in task completion and to be adaptable to various unexpected situations in their studies, personal life and also in the professional area in the future.

2. Critical thinking for engineering students

Critical thinking for engineering students, future engineers, is supposed to be more important as these specialists work in the areas of telecommunication, information and communication technologies, mechatronics and robotics, physics, machine building and other engineering areas which require accuracy and fidelity, clearness of actions, practical thinking, attentiveness and systematization. They all come from the ability to think critically to provide a reasonable argument and show a high level of professionalism. Consequently, it is crucial to introduce critical thinking activities in a study process and drill those skills within all courses given in the program starting with humanities and finishing with profession-oriented ones.

To analyze importance of critical thinking for engineering students and put it into teaching practice we should specify the definition of critical thinking from an engineer's prospective. According to Scriven (n.d.) critical thinking is the intellectually disciplined process of actively and skillfully conceptualizing, applying, analyzing, synthesizing, and/or evaluating information gathered from, or generated by, observation, experience, reflection, reasoning, or communication, as a guide to belief

IOP Conf. Series: Materials Science and Engineering **155** (2016) 012022 doi:10.1088/1757-899X/155/1/012022

and action. Taking this definition into account educators can design lesson plans, study programs and, globally, curricula with the focus on engineering students' critical thinking development.

3. Modern technologies in teaching critical thinking

Realizing how important critical thinking development is it is obvious that modern pedagogy must focus on it and dedicate an essential part of the whole study process to it. There is a wide range of methods and approaches which could be used for developing this kind of thinking. However, teachers should choose appropriate ones for a certain target audience depending on such factors as age, specialization, experience and practices.

This paper aimed at paving the ways to make engineering students achieve critical thinking skills through reading profession-oriented texts. Critical thinking and critical reading are closely connected with each other as critical reading is a form of language analysis that does not take the given text at face value, but involves deeper examination of the gist and details revealing that gist. It also provides the basis for thinking that leads to developing critical thinking skills.

Consequently, to develop critical thinking through reading we should choose a set of methods or approaches which could be tailored especially to engineering students paying attention to their characteristics mentioned above. Upon previous observations and researched done at the Siberian State Aerospace University (Krasnoyarsk, Russia) it was found out that engineering students work a lot with technical texts from course books which are rather challenging to be understood through traditional skimming. Moreover, the majority of engineering students study in the additional program named "Translation in the Professional Area". This program implies to provide translation skills and strategies to engineering students to be able to translate from Russian into English and from English into Russian in their professional area. To provide the engineering material through technical texts both in their main studies and additional ones it is crucial to choose methods or approaches which could help realize the idea of the context through critical thinking development. Below there are some methods and approaches selected which are suitable for engineering students and their specific of a study process.

"Know, want to know, learned (KWL)" method

KWL (Ogle, 1986) is an instructional reading strategy that is used to guide students through a text. Students begin by brainstorming everything they know about a topic. This information is recorded in the first column of a KWL chart. Students then generate a list of questions about what they want to know about the topic. These questions are listed in the second column of the chart. During or after reading, students answer the questions that are in the second column. This new information that they have learned is recorded in the third column of the KWL chart. In case the students got unanswered questions it is necessary to motivate them to find those answers themselves as their homework and present them next lesson.

The KWL strategy serves several purposes. It allows to elicit students' prior knowledge of the topic of the text, sets a purpose for reading and also helps students to monitor their comprehension in a proper way.

"Fishbone" method

Fishbone Diagrams (Ishikawa, 1943) (also known as Ishikawa Diagrams) can be used to answer the following questions that commonly arise in problem solving: What are the potential root causes of a problem? What category of process inputs represents the greatest source of variability in the process output?

IOP Conf. Series: Materials Science and Engineering **155** (2016) 012022 doi:10.1088/1757-899X/155/1/012022

Dr. Kaoru Ishikawa developed the "Fishbone Diagram" at the University of Tokyo in 1943. Hence the Fishbone Diagram is frequently referred to as an "Ishikawa Diagram". Another name for this diagram is the "Cause & Effect" or CE diagram. A completed Fishbone diagram includes a central "spine" and several branches reminiscent of a fish skeleton. This diagram is used in process improvement methods to identify all of the contributing root causes likely to be causing a problem. The Fishbone chart is an initial step in the screening process. After identifying potential root cause(s), further testing will be necessary to confirm the true root cause(s). This methodology can be used on any type of problem, and can be tailored by the user to fit the circumstances. In our case our user will be an engineering student who is going to develop critical thinking abilities through reading texts in the professional area.

Abovementioned methods were chosen to be the most effective methods for teaching engineering students through reading in terms of developing their critical thinking.

4. Experiment

The experiment on selecting the most suitable methods for engineering students was undertaken at the premises of the Reshetnev Siberian State Aerospace University (Krasnoyarsk, Russian Federation) in the frame of the course "Profession-Oriented Translation".

The experiment started with providing engineering students with profession-oriented texts which supposed to teach engineering students to translate such texts. However, it was important to teach them translate texts through prompt understanding of the gist and details to provide accurate translation. It was possible to implement it only through critical thinking. To check which methods better suit engineering students each lesson was dedicated to one method on critical thinking development through reading. They are "The basket of ideas", "False/true statements", "Cinquain poems", "Know, want to know, learnt (KWL)", "Fishbone". In total there were 5 lessons arranged. After each lesson the instructor made a survey on how much students understood after text analysis before starting to translate it. The second step was to check translation accuracy.

At the end of the set of lessons it became obvious that KWL and "Fishbone" methods and approaches work more efficiently with reference to provided translations as translations were closer to the original texts, gists and ideas were more accurately conveyed comparing to other methods and approaches.

5. Conclusion

Following the findings of the paper, it should be mentioned that curriculum developers, material designers, language and translation teachers are highly recommended to employ profession-oriented texts to help students achieve critical thinking skills and at the same time improve reading profession-oriented texts for better understanding and, consequently, prompt translation, though the results of this paper are not conclusive and more research studies are required in this regard.

References

[1] [The Critical Thinking Community]. (In Eng.). Available at: http://www.criticalthinking.org/pages/our-concept-and-definition-of-critical-thinking/411 (accessed 19.10.2016)

[2] [Critical Thinking for Engineers]. (In Eng.). Available at: https://sites.tufts.edu/eeseniordesignhandbook/2013/an-engineers-path-to-critical-thinking (accessed 19.10.2016)

[3] [The Engine Room for Continuous Improvement]. (in Eng.). Available at: https://www.moresteam.com/toolbox/fishbone-diagram.cfm

[4] Ogle, D.M. 1986. K-W-L: A teaching model that develops active reading of expository text. Reading Teacher 39: 564-570.

Tracking on the joint during the electron beam welding

V Braverman, V Bogdanov, V Belozertsev, N Uspenskiy
Reshetnev Siberian State Aerospace University,
31, Krasnoyarsky Rabochy av., Krasnoyarsk 660037, Russia
E-mail: braverman-vladimir@rambler.ru

Abstract. In the article the description of device, which provides automatic positioning of electron beam relative to joint of welded parts during welding, is given. Extremum seeking based on synchronous detection of sensor signal (X-ray or secondary emission) is realized in the device. Measurements are made when beam goes out of the channel following the welding direction.

The application of synchronous detection is possible due to the fact that during joint scanning with electron beam harmonics, carrying data about beam position relative to the joint appear in the joint sensor signal spectrum.

Introduction

Presently the problems of providing precise positioning of electron beam relative to the joint of welded parts are actual for long joints welding. Such by-effect phenomena as secondary electron emission and X-ray radiation are used as data sources of beam position. In case of welded edges burn-off absence, the dependence of secondary electron emission change and X-ray brems-strahlung intensity on the beam position relative to the joint is equal – extremal. This extremum is minimum which corresponds to the beam and joint coordinates coincidence. This allows applying of unified hardware to build joint tracking devices.

The analysis of secondary radiation under burn-off absence is important. The extraction of data about beam position relative to joint from welding channel is problematic because of high noise level. At the same time necessary data for welding process can be received in the immediate closeness to the channel. It can be realized by, for example, short-time taking the beam out of the channel, making measurements and taking the beam back to the welding zone [1]. It is obvious that maximum time of putting the beam out must provide minimal changing in the weld pool, and speed of beam shifting must provide minimal energy taken out of the weld pool to exclude edges burn-off.

In analyzed device extremum searching based on synchronous detection method of joint sensor signal is realized. The main idea of synchronous detection application is based on the following: under the joint scanning with electron beam in signal spectrum, harmonics with frequencies multiple to scanning frequency appear. First harmonic amplitude of scanning signal is proportional to beam deviation relative to the joint and its phase defines displacement direction.

The appearance of harmonics in sensor signal is connected with the introduction of periodic components into electron beam parameters. For example, mathematical expectation of beam position can be shown as sum of constant component (accidental beam deviation relative to the joint) and periodic component with given amplitude (joint scanning with beam).

Formalization of the Problem

Secondary electron emission and X-ray radiation are effect of primary beam interaction with material of welded parts. Therefore if we suppose that electron density of distribution obeys the normal distribution law, then under burn-off absence secondary electron current and X-ray brems-strahlung intensity can be defined with expression [2-5]

$$J(\varepsilon) = KI_b \left\{ 1 - \frac{1}{\sigma\sqrt{2\pi}} \int_{x_1}^{x_2} \exp\left[-\frac{(x-\varepsilon)^2}{2\sigma^2} \right] dx \right\}, \tag{1}$$

where J – secondary radiation parameter; I_b – beam current; σ – root-mean-square deviation of electrons from beam axis; ε – mathematical expectation of beam position; K – coefficient, characterizing secondary radiation nature; x_1 and x_2 – joint edges coordinates; $x_1 - x_2 = \delta$ – joint gap.

In Fig.1 the curves built with expression (1) are shown. From the presented expression (1) and curves we can see that extremum coordinate x coincides with joint coordinate $x = 0$ under $\varepsilon = 0$.

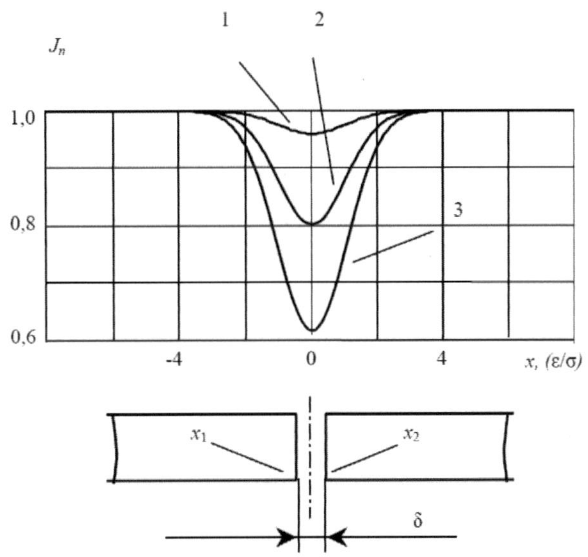

Figure 1. Analytical curves showing dependence of secondary radiations intensity on beam position relative to joint: $\sigma = \text{const} = 0.1\,\text{mm}$; $\delta = \text{var}$. Curves: 1- $\delta/\sigma = 0.1$; 2- $\delta/\sigma = 0.5$; 3- $\delta/\sigma = 1$, J_n - normalized parameter J

Extremal character of these dependences conditions the application possibility of known methods of extremum search for beam position relative to joint defining. One of these methods is synchronous detection method of joint sensor signal (secondary emission or X-ray). For this method realization searching beam shifting – joint scanning with electron beam is introduced. As a result mathematic expectation of beam position can be presented as:

$$\varepsilon = \varepsilon_0 + \varepsilon_m \sin\alpha, \tag{2}$$

where ε_0 – beam shift relative to joint; ε_m – searching beam shifting amplitude; $\alpha = \omega t$; ω –

frequency, t – time. Then, expression (1) becomes

$$J(\varepsilon) = KI_b \left\{ 1 - \frac{1}{\sigma\sqrt{2\pi}} \int_{x_1}^{x_2} \exp\left[-\frac{(x - \varepsilon_0 - \varepsilon_m \sin\alpha)^2}{2\sigma^2} \right] dx \right\}. \tag{3}$$

Having periodical component in parameter ε, sensor output signal (2) can be presented with Fourier series. Let us analyze the dependence of ε_0 on component b_1 with frequency ω. This component is defined as Fourier coefficient:

$$b_1(\varepsilon_0) = \frac{1}{\pi} \int_0^{2\pi} J(\varepsilon_0) \sin\alpha \, d\alpha. \tag{4}$$

In Fig. 2 plot showing function $b_1(\varepsilon_0)$ according to expression (3) is given. Thus, in joint neighborhood $b_1(\varepsilon_0)$ is proportional to mismatch of beam and joint positions, and the sign defines the mismatch direction.

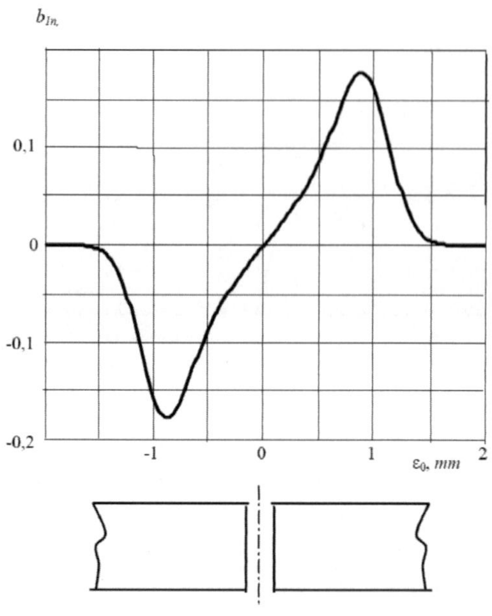

Figure 2. Dependence of first harmonic amplitude with frequency ω of sensor signal on beam position relative to joint: b_{1n} – normalized parameter b_1, $\sigma = 0.1$ mm; $\delta = 0.1$ mm; $\varepsilon_m = 1$ mm.

This fact makes data accessing about beam position relative to joint possible due to the application of synchronous sensor signal detection.

Structural device diagram

Functional device diagram is given in Fig. 3 [6, 7]. With the help of spike generator G and deflection system DS_Y the beam is occasionally taken out of the weld pool following the welding direction. At the same time generator CG and deflection system realize joint scanning with the beam (Fig. 4[430]). Simultaneously generator CG produces reference voltage, which comes to one of the inputs of synchronous detector SD, where sensor signals S and reference voltage multiply:

$$[b_1(\varepsilon_0) \sin\alpha] \sin\alpha = [b_1(\varepsilon_0)]/2 - [b_1(\varepsilon_0) \cos 2\alpha]/2 \tag{5}$$

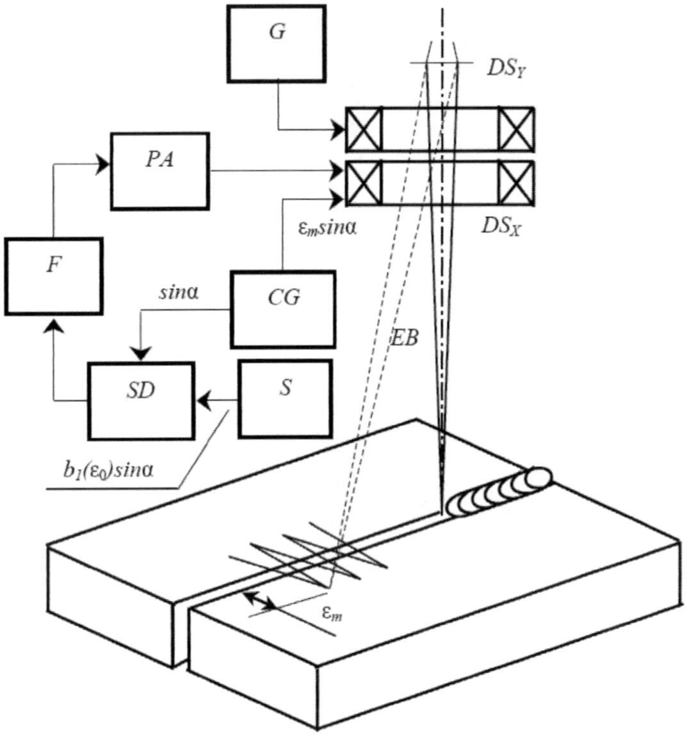

Figure 3. Structural diagram of the device for automatic beam control: S – sensor; CG – scanning beam generator; SD – synchronous detector; F – filter; PA – power amplifier; G – generator of spike; DS_X – deflection system for X-axis; DS_Y – deflection system for the Y-axis.

Expression (4) and oscillograms show that if there is beam shift $b_1(\varepsilon_0)$ from the joint, then synchronous detector output signal is a sum of direct component and variable component with frequency equal to 2ω (Fig. 4b). High frequency component is filtrated with pass F and direct current signal, proportional to beam shift through power amplifier, comes into deflection system. Beam position correction is realized. If beam and joint coordinates coincide $b_1(\varepsilon_0) = 0$, synchronous detector output signal is close to 0 (Fig. 4c).

IOP Conf. Series: Materials Science and Engineering **155** (2016) 012023 doi:10.1088/1757-899X/155/1/012023

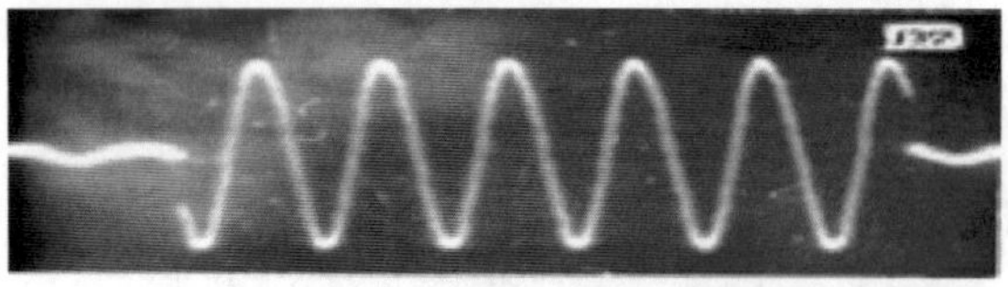

a)

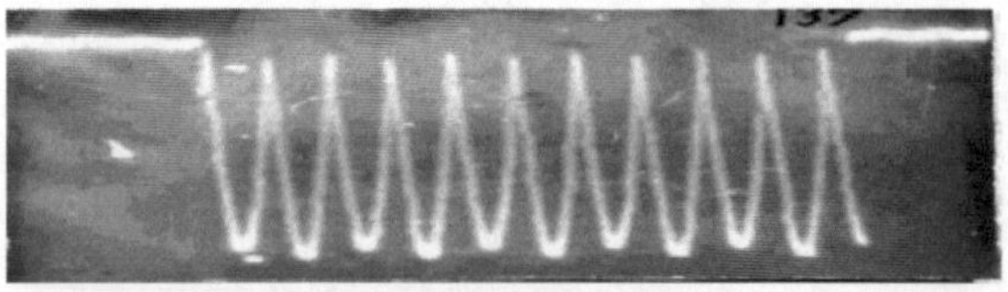

b)

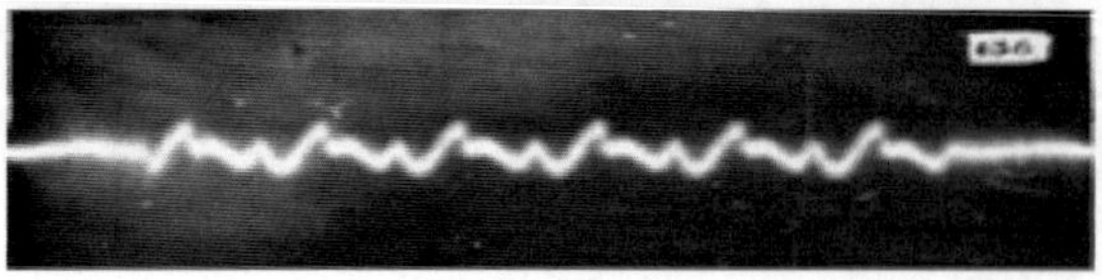

c)

Figure 4. Signals oscillograms: a) scanning current in deflection system DS_X; b) synchronous detector output signal under $\varepsilon_0 \neq 0$; c) synchronous detector output signal without mismatch of beam and joint.

Technical realization of the device is not problematic. Secondary electron collector is used as secondary emission sensor. Scintillation detector with photoelectron multiplier can be used as X-ray sensor (Fig. 5).

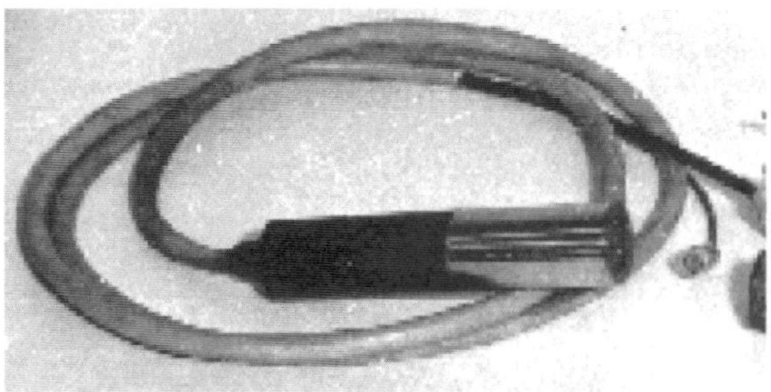

Figure 5. Scintillation detector.

IOP Conf. Series: Materials Science and Engineering **155** (2016) 012023 doi:10.1088/1757-899X/155/1/012023

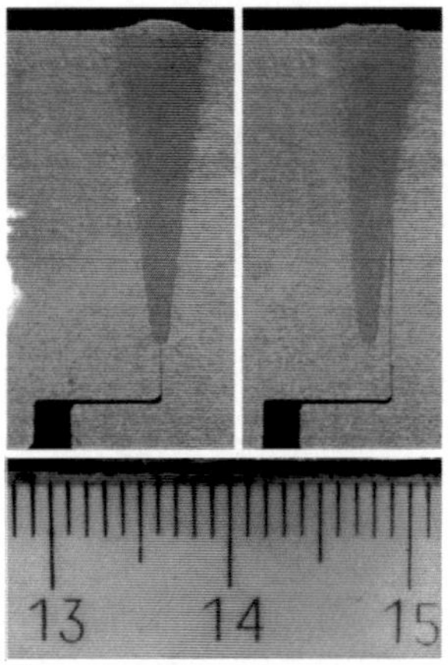

Figure 6. Welded joints microsections realized without tracking (on the right) and with automatic joint tracking (on the left).

At Fig. 6 joints microsections realized with electron-beam welding are shown.

Devices, designed in accordance with the given functional diagram, are utilized for electron-beam welding of large-dimension details in aerospace industry. Error of beam alignment with joint is not more than 0.1 mm.

Conclusion

1. Equal nature of secondary emission and X-ray radiations under electron-beam welding allows application of equal hardware to control welding process.
2. In conditions of electron-beam welding it is preferable to apply X-ray sensors for joint tracking, as X-ray radiation is the least sensitive to different destabilizing factors.
3. Synchronous detecting of joint sensor signal allows improving of device noise immunity due to the fact that interference risk at scanning frequency is low.

References

[1] Reichmann A, Leffler D, Bartel R 2004 State of the art of a FEP control unit of beam scanning for electron beam equipment 6th International Conference on Beam Technology, Halle (Saale)

[2] Braverman V 2008 Decelerating X-ray radiation during electron beam welding and its interrelation with the process parameters Vestnik SibGAU Vol. **3**

[3] Braverman V, Belozrtsev V, Litvinov V and Rozanov O 2008 Issues relating to controlling seam formation process in electron beam welding Vestnik SibGAU Vol. **2**

[4] Braverman V, Belozrtsev V, Goryashin N and Lelekov A 2009 Mathematical models for the process of changing the x-ray radiation from the beam position relatively the seam during the electron beam welding Vestnik SibGAU Vol. **2**

[5] Braverman V, Bashenko V and Bajakin S 1997 Device for control focus and depth of penetration on his own X-rays during electron beam welding in modulation of focus Welding Production Vol. **1**

[6] Braverman V, Shabanov V, Bashenko V and Bajakin S 1998 Method of electron-beam welding. Patent of Russia N2113954, International Classification Code B23K 15/00

IOP Conf. Series: Materials Science and Engineering **155** (2016) 012024 doi:10.1088/1757-899X/155/1/012024

Question-answering system

A A Stupina[1,2], E A Zhukov[2], S N Ezhemanskaya[2], M V Karaseva[1,2], L N Korpacheva[2]
[1] Siberian State Aerospace University, 31 "Krasnoyarskiy Rabochiy" av., Krasnoyarsk, 660037, Russia
[2] Siberian Federal University, 3 Vuzovskii Lane, Krasnoyarsk, 660025, Russia

E-mail: saa55@rambler.ru

Abstract
The paper gives a brief overview of the key problems of designing question-answering systems. It offers the approach to the determination of the question-answering system of giving an answer to the question based on a modified measure of one set to another inclusion. The results of the proposed approach in the problem of automatic selection of the answer to the question of several possible options are presented. These results showed that the system is able to give the correct answer (of 4 possible ones) in more than 50% of cases with a relatively small knowledge base.

Introduction

The information analysis given in the natural language, along with speech recognition and computer vision is one of the key directions of research in the field of artificial intelligence. Among the problems of natural language processing methods a special problem of development question-answering systems is of a great importance. According to some definitions a question-answering system is an information system that is able to get questions and answer them in natural language. In other words it is a system with a natural language interface. Conventionally, a question-answering system can be divided into highly specialized systems, i.e. those that are applied in any particular field and general question-answering systems with the capability of searching in the similar fields as well.

In the context of the conducted investigation a subclass of question-answering systems as a selection of the correct answer among the given finite set of answers, i.e., automatic test execution is considered.

Theoretical analysis

The development of the question-answering system is a very complex process that includes the application of a variety of NLP techniques (in English it means Natural Language Processing, i.e. natural language processing) as well as the data mining algorithms such as, for example, deep and recursive neural networks. In this investigation the algorithm based on searching the relevant answers in the knowledge base will be considered.

In general case one can point out 12 key objectives in the field of the QA-systems building developed in 2002 by a group of researchers [4].

1. Questions classification. It is considered as a selection of the question type (selection of one variant from several types; to insert an empty meaning into a phrase (a direct answer to the question, etc.) as well as a class in the field of knowledge; if there was a question, for example in physics or biology. This step is necessary in order to develop a more flexible

answer algorithm to the question because each class of the question may require its own method to find an answer.

2. Stage of processing and analysis. It includes semantic analysis of the question phrases, i.e. it identifies key words, splits a complex question into a few simple patterns, detects synonyms, idioms to recognize similar ones in meaning questions but presented in the different way, etc.

3. The ability of the system to see the context of questions allows it to make clarifying questions or alternatively, to find answers to the questions posed earlier.

4. Base of knowledge generation is one of the key stages in the QA-system development. The performance of the question-answering system in addition to the effectiveness of the text analysis methods depends on the quality of the text database; if it has no answers to the questions, the QA system can't find anything. It is essential the information is accurate and presented in different forms. This helps to ensure that the QA system is more likely to find the answer.

5. Stage of the answer selection and its evaluation. It means the direct information search in the base of knowledge that contains the answer to the question. The type of the required answer depends on the type of the given question.

6. Answer wording. It implies a simple selection of the text block from the base of knowledge and answers grouping from different sources including the reformulation of the text information based on the context of the question.

7. One should pay attention to such a property of the QA-system as the answer to the question in real time regardless of the complexity of the question and volume of the base of knowledge.

8. Support of multilinguistics. It is necessary to take into account the property of the multilingual support while the system development since questions could be asked in different languages taking into account the peculiarities of the language,

9. Interactivity is necessary in the case when either the system correctly understands the question or the answer does not satisfy the user. In this case it is necessary to support dialogue with a user to clarify the information that is necessary for the question and answer.

10. Ability for consideration. The requirements for the derivation of the new knowledge on the basis of the existed belongs to the QA-systems of a higher level and considered to be one of the most promising in the in the field of this investigation.

11. User's profile. Having the information about the user (his vocabulary, speech manner) the system can increase the performance in search of answers based on the individual approach to each user.

12. And, finally, one can distinguish the stage of the combined question-answering system development to use the advantages of certain systems, eliminating the weaknesses through the strengths of other systems.

The investigation considers a simplified QA-system based on the search of answer as close as possible in accordance with the selected metric to the base of knowledge.

Technology

First of all it is necessary to generate a base of knowledge to develop a QA-system. The base of knowledge can be local (books, scientific papers placed in local repository) and based on web technologies. The access to the base of knowledge through the Internet taking into account modern technologies does not cause any problems and it is preferable from the practical point of view. However, in cases with the large amounts of knowledge and limited Internet traffic one should pay attention to the local method of information location.

In the context of this investigation the basis for the base of knowledge development could be books, scientific papers and the school curriculum tests from different fields of knowledge.

First of all it is necessary to allocate so-called knowledge from a source of potentially useful information in the assembled list. By knowledge here we mean a Text phrase that includes brief and specific information about any fact. Preferably if the sentence has the information about only one fact and this information is unambiguous. This kind of phrase is proposed to generate the following way:

- on the basis of tests; knowledge is a question of the test in combination with a correct answer;
- on the basis of books, papers and other text information; that means the extraction from the text separate, complete declarative sentences having at least one subject and predicate in its structure. Also a paragraph of a few sentences corresponding to the property described above can be grouped into knowledge. It is possible to apply more complex algorithms to select informative and meaningful sentences; that is one of the objects of the further investigation solutions to this problem.

Further from each resulting phrase so-called "stop words" are excluded [2] and lemmatization [2] of words is performed to bring knowledge to the universal form. Then it is necessary to form a "bag of words" on the basis of the available text information [2]. And finally, it is necessary to present each type of knowledge in the form of a binary vector [2] of length equal to the volume of a "bag of words". In this case j-th value of the vector is 1 if the j-th word from a "bag of words" is found in the knowledge at least once, and 0 otherwise. In other words, each type of knowledge is a variety of unique elements from a "bag of words".

The search for the answer to the test question is proposed to realize using the following algorithm.

1. Let we have some question Q and a limited set of n_A answers $A = \left\{a_j,\ j = \overline{1, n_A}\right\}$ to choose the right one.
2. For each question Q, n_A phrases are formed: it is a combination of the question and the j-th answer from a set A .
3. As in the case with the base of knowledge, each combination of question-answering phrase undergoes the pre-processing; it means the "stop-words" excluding, lemmatization, and representation in the form of binary vector. However, note that each phrase in combination is processed separately. Thus it is possible a binary vector will appear; it consists of only zeros for a question or answer phrase for the case these phrases includes words not found in the previously generated "bag of words". Further the combination A will denote a pair of binary vectors.
4. Now it is necessary to evaluate the truth degree of each pair based on the available base of knowledge among all pairs $\left\langle Q, a_j \right\rangle$ for $j = \overline{1, n_Q}$. For this purpose it is necessary to introduce some measure of proximity of the pair to the base of knowledge. In the framework of the given investigation it is proposed to apply the measure modification of inclusion of one set into another; this modification was described by Seminum B.I. to evaluate the degree of the pair truthfulness [1].

Let we have n_K knowledge and n_A variants of answers for each question Q. The matching level (a measure of inclusion) for a pair $\langle Q, a_j \rangle$ and some knowledge K is proposed to be present with the formula (1):

$$d(\langle Q, a_j \rangle, K) = [d'(Q, K) + d'(a_j, K)] \cdot (d'(Q, K) > 0) \cdot (d'(a_j, K) > 0), \quad (1)$$

where $d'(Q, K)$ and $d'(a_j, K)$ K is the measure of the inclusion of question Q and answer a_j into the knowledge K.

5. The answer which is considered as correct corresponds to the maximum value of a inclusion measure among the whole knowledge from the base of knowledge (2):

$$Correct\_Answer = \arg_j \max_{j,i} d(\langle Q, a_j \rangle, K_i), \; i = \overline{1, n_A}, \; j = \overline{1, n_K} \quad (2)$$

Experiment

A problem of the QA system development for answering questions using school curricula tests for the 8th grade of the United States school was selected as an object for testing the proposed algorithm [8]. We have a training sample (2 500) and a test sample (21 298). The samples are a set of "question + 4 choices of answers" where the information about the correct answer for the given question is in the training sample. It is worth noting a very specific feature of the sample test questions that the question phrase is quite noisy, i.e. it contains words that may not be applied to the question. This approach to the generation of the test samples significantly reduces the opportunity to respond to the questions independently without any help of machines as the process of understanding the meaning of the question takes quite a long time. Thereby the higher requirements to the question-answering system are imposed i.e. to present the ability of the system to weed out uninformative components of the question.

The investigation was performed using three approaches to the base of knowledge generation:

1. only a training sample: a set of phrases that means a combination of a question + a correct answer;
2. a set of textbooks from different fields of knowledge (natural science, physics, chemistry, astronomy, etc.) presented on the portal ck12 [5];
3. a combination of the first and second approaches.

For the experiment let's answer all questions at random and we will receive the answers quality that is equal to about 25%. This result is expected.

First let us investigate the algorithm using the base of knowledge based on the training samples, i.e., having only 2500 informative phrases so called a base of knowledge (Table 1):

Table 1. The result of answers to the questions by the system using training samples as a knowledge base

Number of phrases	Run time	Quality of the algorithm
2 500	4 min	34.1%

As you can see in Table 1 the system was able to answer nearly a third of all new tasks having only the information about the correct answers to 2500 questions.

The following investigation stage is the base of knowledge use formed on the basis of textbooks (Table 2). For the beginning we use a textbook that contains the general information but covering a wide range of topics.

Table 2. The result of answers to questions by a system using a textbook as a base of knowledge

Number of phrases	Used books	Run time	Quality of the algorithm
124 798	Life Science concepts for middle school [7]	17 min	42.3%

It is clear that only one textbook has greatly improved the quality of the system of answers to the questions. The computational time has significantly increased.

Then a training sample was included into the existing base of knowledge (Table 3):

Table 3. The result of answers to questions by a system using a textbook and a training sample as a base of knowledge

Number of phrases	Used books	Run time	Quality of the algorithm
127 298	Life Science concepts for middle school [7]	22 min	42.7%

The inclusion of the training sample into the structure of the base of knowledge has not brought a significant improvement of the algorithm quality. That means that a textbook "Life Science concepts for middle school" contains almost the whole amount of information available in the training sample as well.

Now let's add a bit more information to the base of knowledge; that is let's include two subject-oriented textbooks: Biology [3] and Earth Science Concepts For Middle School [6] (Table 4):

Table 4. The result of answers to the questions by a system using three textbooks and a training sample as a base of knowledge

Number of phrases	Used books	Run time	Quality of the algorithm
270 535	Life Science concepts for middle school [7], Biology[3], Earth Science Concepts For Middle School [6]	47 min	53.6%

Again it can be noted that the use of additional training materials can significantly improve the quality of the algorithm. In this case the system could give answers probably to the subject-oriented questions the information was not included into the general textbook.

It should be noted that the use of the combined approach with only one book gives the results comparable with the use of the search engine Lucene [9] to apply it to Wikipedia [10].

Results

Despite the fact that the proportion of the correct answers of the QA system exceeded 50% a bit, there is a reason to consider the proposed approach correct and requiring further investigation.

Among the prospects of investigation one can point out:

- including a larger number of the more subject-oriented scientific literature into the base of knowledge;
- application of vectorization methods of computation to optimize the run time speed of the algorithm;
- application of the different methods of literature parsing to identify the more specific and informative knowledge;
- development of the new metrics and the application of other well-known metrics for evaluation the conformity degree of the question-answering phrase to the knowledge.
- application of the cluster analysis algorithms and classification for pre-identification of the field of knowledge to which the system answers the question.

References

[1] Semkin B I 2009 About the relationship between the average values of two inclusion measures and similarity measures *Bull. Botanicheskogo sada-instituta DVO RAN* Vol. **3** p. 91-101.

[2] Bird S, Klein E, Loper E 2009 Natural Language Processing with Python. O'Reilly Media 497 p.

[3] Brainard J Biology 2016 CK-12 Foundation

[4] Burger J, Issues, Tasks and Program Structures to Roadmap Researchin Question Answering (QA) / C. Cardie, V. Chaudhri, R. Gaizauskas, S. Harabagiu, D. Israel, C. Jacquemin, C-Y LinS. Maiorano, G. Miller, D. Moldovan, B. Ogden, J. Prager, E. Riloff, A. Singhal, R. Shrihari, T. Strzalkowski, E. Voorhees, R. Weishedel. / Q&A Roadmap Paper, 2003.

[5] CK-12. – CK-12 Foundation, 2016 Access: http://www.ck12.org/, free. – A title screen.

[6] Desonie D. Earth Science Concepts For Middle School CK-12 Foundation, 2016.

[7] Harwood J., Wilkin D. Life Science concepts for middle school / J. Harwood, D. Wilkin, CK-12 Foundation, 2015

[8] Kaggle: The Home of Data Science (The Allen AI Science Challenge's page). – Kaggle inc, 2016 -. -Access: https://www.kaggle.com/c/the-allen-ai-science-challenge/, free. – A title screen.

[9] Lucene: Ultra-fast Search Library and Server. - The Apache Software Foundation, 2016-. – Access: https://lucene.apache.org/core/, free. – A title screen.

[10] Wikipedia: The Free Encyclopedia. – Wikimedia Foundation, 2016 -. – Access: https://en.wikipedia.org/, access. – A title screen.

IOP Conf. Series: Materials Science and Engineering **155** (2016) 012025 doi:10.1088/1757-899X/155/1/012025

Control by technological mode parameters with an intellectual automated system

A A Stupina[1,2], A A Shigina[1,2], A O Shigin[3], M V Karaseva[1,2], L N Korpacheva[2]

[1] Siberian State Aerospace University 31 "Krasnoyarskiy Rabochiy" av., Krasnoyarsk, 660037, Russia

[2] Department of Economics and Information Technology of Management, Siberian Federal University, 3 Vuzovskii Lane, Krasnoyarsk, 660025, Russia

[3] Department of Mining machines and complexes, Siberian Federal University, 95 "Krasnoyarskiy Rabochiy" av., Krasnoyarsk, Russia

E-mail: saa55@rambler.ru

Abstract

The scheme of functional correlation of input and output parameters of the "drilling rig – rolling cutter bit – rock" system is proposed. The necessity to apply drilling rigs of the automated intellectual system with the adaptive element for a quick time response of the investigated system on the change of physical and mechanical rock properties and the maintenance of the adjusted operating parameters of the technical system in an optimal ratio.

Introduction

The control systems play an important role in the automation and control by technological processes in the mining industry. Nowadays the problem of such systems development is quite relevant since the increased efficiency of control is impossible without the automated control systems (ACS) based on the information technologies and advanced mathematical models [1, 2].

But modern automation control systems of the drilling process do not react to the changes of the interrelation properties of the object (rocks), do not adjust the operation modes and do not compensate the disturbances of the complex technical systems "drilling rig – rolling cutter bit – rock" functioning (hereinafter refers as the object of control or R-B-R) that reduces its efficiency [3]. The insuring the efficiency requirements (timeliness) and accuracy of the information transferring is a fundamental requirement for the control quality improving by the object and process.

Intelligent automated control systems

The urgency of the investigation in the field of intelligent ACS is that in the conditions of increase of requirements to the characteristics of the information exchange in ACS significantly increasing the demands on the quality of the transmission systems functioning [4]. A structure that interacts with the external environment and in the process of obtaining necessary information from it generates the purpose of the action and analyses the impact on the system (physical and information) is applied for the generalized intellectual systems. The main elements of the control system in this case are an intellectual converter and basic control system.

In the case of using the control system of artificial intelligence as intellectual converter of the implemented expert systems, situational management, management of structural dynamics of complex technological [5] and other intellectual systems and their elements. A mathematical model of intellectual control system includes three parts: intellectual converter of the control object and control systems (computing, transforming and performing units). The intellectual converter alters the information about the external environment and the object of control and transforms it into the signals of impact on the controllable devices of the system. The converter uses a unit of decision-making for the generation impact on the controllable system by the object.

In general the ability for adaptive changes of the state is the most important quality of the elements and environment. To achieve the adaptive state one should keep in mind that in the cases of the system motion under the direct influence of the external driving forces (the signal impact), the direction of adaptive motion of the system is predicted, and when the system moves under the indirect influence of external forces it is required to interrupt the periodic adaptive movement to determine the direction of the correcting action. An adaptive element as a unit of the electromagnetic type (an adaptive rotationally-feeding mechanism) designed for the simultaneous smoothing of random shock loads of the drilling process and getting rapid feedback on the time and magnitude of impact must be included in the hardware system for its implementation on the basis of the drilling rig intelligent automated control [6].

The control by the parameters of the drilling mode is characterized by a large number of input and output variables. A system R–B–R as an object of control can be represented in the form of relations (Figure 1). The specifies parameters for the adaptive element of the system R–B–R is information about the changes of physical and mechanical characteristics of the rock expressed by the values of drilling speed V_d, axial forces P_{af}, rotation frequency n_r at the current moment and the change of drilling speed ΔV_d for a discrete period of time. The output parameters of the adaptive element are the current I and its changes ΔI at the given discrete period of time that indirectly reflects the index of drill ability and its changes. The input parameters of the automated intellectual control systems (AIS) are the output parameters of the adaptive element, the drilling speed and its changes, axial force and rotation frequency at the given moment of time. At the output of the AIS applying the calculation algorithms the correction of the axial force and rotation speed as well as the information about the strength and structural properties of the rock are worked out. The output parameters of the regulator are correction values of the axial forces ΔP_{af}, rotation frequency Δn_r, calculated values of the drill ability index of I_d and its changes ΔI_d in this discrete period of time.

The adjusted operating parameters axial force $P_{af}+\Delta P_{af}$, rotation frequency $n_r + \Delta n_r$ are the input control impact for the system R–B–R. To generate a control impact for the control object corresponding to the algorithm of its operation it is necessary to have a control unit (a controller) that performs functions without the direct human activity. A controller in the system R–B–R is an intellectual system with an adaptive element.

The input parameters that characterize technical capabilities of the drilling rig and drilling conditions are divided into:

- controllable ones that are quickly adjustable (P_{af}, n_r);
- controlled ones that don not depend on the drilling process (design parameters K_r of the drilling rig: weight, dimensions, types of drivers of the basic mechanisms) and dependent on the drilling modes (diameter and depth of the hole P_d, vibration parameters of the rig: amplitude, frequency B_r);

- uncontrolled random: parameters of the destructible environment P_{en}.

A complex of indicators characterizing physical and mechanical properties of the rock as an object of destruction while drilling in the form of random functions (strength threshold of the rocks under uniaxial compression, fracture, abrasion, cut, etc.) belongs to the parameters of the destructible environment P_{en}.

The output parameters characterize physical results of the destruction process of the rock and they are divided into:

- observable (V_d, n_r, P_{af}, resource of a bit T);
- unobservable: quickly computed (the performance of the rig for a change P_{ch}; the energy intensity of the drilling process E_d, power of the rotationally-feeding mechanism N and not calculated quickly (cost for drilling one meter of the hole C_d).

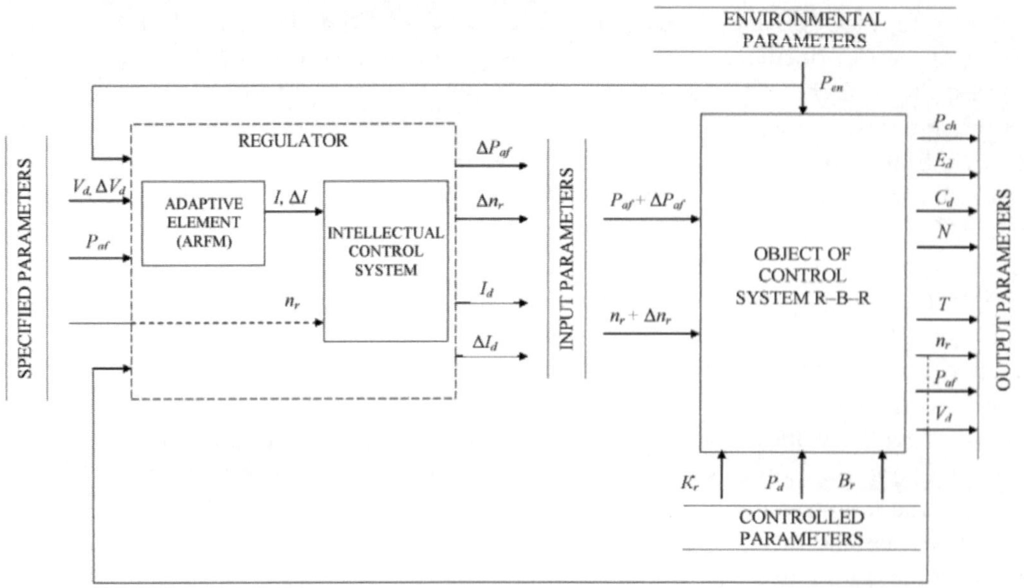

Figure 1. The proposed scheme of functional relations of input and output parameters of the system R–B–R

It should be noted that the energy intensity of drilling E_d can be regarded as a function of input parameters (P_{af}, n_r) and uncontrolled random disturbances (P_{en}) and they should have their actual values for the feed and rotation mechanisms of the drilling rig.

For optimal system output parameters R–B–R it is necessary to apply adaptive rotationally-feeding mechanism [7] when there is no significant time delays to adjust short-term random disturbances. The system R–B–R continuously receives the input control impacts in the form of adjusted values as the axial force and rotational speed resulting in the correspondingly changing output parameters. As the output parameters the controlled parameters (V_d, P_{af}, n_r, T, E_d, N, P_{ch}, C_d) can be taken. The desire to optimize the drilling process led to attempts to design models of this process based on theoretical calculations and empirical regularities characterizing the performance of bits and the efficiency of destruction of the rock [8]. For the mathematical formulation of the problem of the optimizing the studied technological process it is proposed to apply mixed criteria as a function of primary parameters. To improve the optimal control of the drilling process in non-steady modes of operation the problem of multi-criteria parametric optimization with the mixed criteria is stated [9] since the studied object of control in operation mode is continuously influenced by stochastic external influence, in particular, unpredictable changes of the rock properties causing critical impact shock loads and affects negatively the operation of the whole technical system.

The most valuable functionals of the possible relations among the parameters of the selected groups are:

$$Z_i = \varphi_i(U, P_{en}, P_d, N, K_r, V_d, B_r), \quad (1)$$

$$U_i = \varphi_i(P_{en}, P_d, N, K_r, V_d, B_r). \quad (2)$$

Functionals of the form (1) give a general solution. Functionals of the form (2) express the conformity control by the process of rolling cutter drilling. If the input parameters U_i correspond to the extreme values of the optimization criterion Z_i for each set of parameters the equation (2) describes an optimal control by the drilling process.

The intellectual automated control system by the technological process of drilling rock operates according to the following algorithm. The external environment influences the functioning process of the object. This system involves an auxiliary adaptive component [4], sensor, computer, and controller. The object is also affected by disturbances that are independent from the control system: random load disturbance (vibration, dust levels, temperature, error devices, failure in the control system). The adaptive element helps to smooth out these unpredictable disturbances.

For the analysis of input information about changing of the object properties the impact sensors (specified unit) in the computer send the information signals about the speed changes of the technological process and the current in the stator of the adaptive mechanism (specified impact). In the computer these information signals are converted to the control ones (the information about the actual characteristics of the object) with a controller (controllers, control units) intended to smooth short-term variations and to implement the process of control and software unit containing the developed computational technique (implementation of the control algorithm). Then control signals are sent to the operating unit realizing the decision made and of helping the appropriate operating parameters changing (automated adjustment). These techniques determine the predicted resource of tool, unit costs for the technological process corresponding to the actual values of the operating parameters and properties of the object. This information determines the optimum speed of the process and operating parameters (output parameters). To improve the quality characteristics of the system the actual values are compared with the optimal ones and automatically changed with the corrective units. The feedback realizes fast transmission of the information (one-hundredth of a second) about the current operating conditions of the control object from the control object to the control part. After the corrective action the adaptive element works in the newly defined modes. The calculated values are displayed on the dashboard using the visualization module intended to demonstrate the simulation results and the subsequent control of the operator.

Conclusion

To make a conclusion we can note the necessity to use AIS adaptive element for quick timely response of the system in drilling machines with to modify the properties of the object of influence and the subsequent adjustments and maintaining the operating parameters of the control object in the optimal ratio. The proposed algorithm of the optimal control system "drilling rig – rolling cutter bit – rock" is based on the use of correctitude values of axial force and rotation speed derived in the controller using computational techniques that assess the resource, productivity and economic efficiency of the process that allows to control current values of the operating parameters, properties of the rock mass, as well as changes in its strength, structural characteristics and to optimize the process of the system operation using the optimal values of the input characterized by the permissible value of thrust and the optimum value of rotation frequency of the operating unit. Further the application of such a AIS will reduce the operating costs for the drilling process in conditions of uncertainty and to improve the functioning of the technical system.

References

[1] Zaitsev K S 2003 Using group technology to plan data processing in computer-aided control system *Automatic and Remote Control*. Vol. **9** № **64** pp. 1507-1512.

[2] Shigina A A, Stupina A A and Shigin A O 2014 Functioning analysis of automated intellectual system with adaptive element *Modern problems of science and education*. Vol. 2, № 52. URL: http://www.science-education.ru/116-12783.

[3] Stupina A A, Shigina A A and Shigin A O 2013 The analysis of efficiency of the multi-parameter system. *Vestnik SibGAU* Vol. **2** № **48**. pp. 94-100.

[4] Stupina A A, Shigina A A, Shigin A O, Karaseva M V and Ezhemanskaja S N 2014 Automated intellectual system with the short-duration nature of feedback. *Life Sci J*. Vol. **11**, № **8s**. pp. 302-306.

[5] Ohtilev M Ju, Sokolov B V and Jusupov R M 2006 Monitoring and control intellectual technologies of difficult technical objects structural dynamics. Moscow: Nauka. 410 p.

[6] Shigin A O, Shigina A A, Gilev A V, Vokhmin S A and Kurchin G S 2015 Development of automated closed-loop drive upon designing of roller bit drilling rig for open pits. *Journal of Theoretical and Applied Information Technology*. Vol. **80** No.**1** pp. 95-104.

[7] Shigin A O 2013 Adaptable rotary-feed mechanism of rock drilling machine for reduction of unpredictable loads in complicated-structure rock drilling. *Gornyi Zhurnal*. № **7** pp. 79-83.

[8] Stupina A A, Shigin A O, Shigina A A, Karaseva M V and Korpacheva L N 2014 Control and Management by Resource of Rolling Cutter Bits in Drilling Rock Massif. *Middle-East J. Sci. Res*. Vol. **21** № **1** pp. 84-90.

[9] Stupina A A, Shigina A A and Shigin A O 2015 Mathematical formulation of technological processes optimization problem. *IOP Conf. Series: Materials Science and Engineering*. TIAA. IOP Publishing Ltd., Bristol, UK. Vol. **94** conf 1. doi:10.1088/1757-899X/94/1/012001.

IOP Conf. Series: Materials Science and Engineering **155** (2016) 012026 doi:10.1088/1757-899X/155/1/012026

Fuzzy clustering of EEE components for space industry

V I Orlov[1], D V Stashkov[2], L A Kazakovtsev[1,3], A A Stupina[1,3]
[1] Reshetnev Siberian State Aerospace University 31 "Krasnoyarskiy Rabochiy" av., Krasnoyarsk, 660037, Russia
[2]JSC "Sinetic", 127 "3 Internacionala" street, Novosibirsk, 630009, Russia
[3]Department of Economics and Information Technology of Management Siberian Federal University, 3 Vuzovskii Lane, Krasnoyarsk, 660025, Russia

E-mail: levk@bk.ru

Abstract. One of the most important problems of the space industry is obtaining reliable methods of automatic grouping (clustering) of specialized EEE components for using in space systems. The main purpose of automatic grouping of EEE components on a set different parameters is the most legible splitting group of EEE components into several homogeneous production batches produced from a single bath of raw materials. The Expectation Maximization algorithm first time applied for the classification of EEE components.

Introduction

Particular belonging of EEE components to a single production batch and a single batch of raw materials is the major necessary condition to increase significance of further actions realization carried out within multistage tests of EEE components [1, 2, 3].

Determining the belonging of a set of EEE components to a set of production batches is a clustering problem clustering (unsupervised learning) [4], there is no reference selection for application of approaches to the problems of classification (supervised learning), a priori the number of clusters (batches of the raw materials) is unknown, however, this number is bounded above.

The problem is even more complicated because of the possible existence of outliers (i.e. copies of the EEE components that do not belong to the main production and batch of raw materials or made with essential aberrations).

At the solution of such clustering problems, it is possible to use two approaches:

1. Hard clustering. Each object belongs to only one cluster. In this case if there is no clear boundary between clusters in the experimental data, the result will not be satisfactory (numerical characteristics of belonging to a cluster will be indistinct).

2. Fuzzy clustering. A set of probabilistic characteristics of belonging of the object to each cluster is the result of the algorithms work of this type.

At application of both approaches, classical methods [5] demand that the number of clusters was set before an algorithm starts.

Thus it is necessary to touch several options with various estimated parameter (number of clusters) for the further choice of the best decision.

For decision making on the result quality of a clustering algorithm (i.e. analysis of compliance of sets of objects and clusters), various quality evaluation criteria [6] are used.

Probability assessment of a regularity of the object reference to each cluster is result of work of such criteria.

The result of use of fuzzy clustering methods is defining probability assessment of belonging of each object to each cluster.

Thus need to use additional criteria for checking and confirmation (deviation) results of the algorithms disappears.

At present, there are a lot of methods for data clustering and classification [5, 7, 8].

The review of the methods mostly often used in practice is provided in [5].

EM algorithm

An EM algorithm (Expectation Maximization) is among the most popular methods. It is successfully applied to statistical problems of analysis of inexact data for cases when some statistical data are absent or for cases when function of credibility has a form which is not allowing convenient research techniques but allowing serious simplifications at introduction of the additional “unobservable” (“hidden”) variables [9].

Such problem definition (clustering of multi-dimensional data which have normal distribution with the hidden data) is used by us for solving the problem of splitting a set of EEE components into several production batches produced from a single batch of raw materials.

Let the density function on a set X have a form of a mixture k of distributions (we assume that these distributions are Gaussian):

$$\rho(x)=\sum_{j=1}^{k}\omega_j\rho_j(x),\ \sum_{j=1}^{k}\omega_j=1,\ \omega_j\geq 0,$$

where $\rho_j(x)$ is a likelihood function of *j*th component of these mixture of distributions, ω_j is its prior probability.

Let likelihood functions belong to parametrical family of distributions of $\varphi(x;\theta)$ and differ in values of parameter $\rho_j(x)=\varphi(x;\theta_j)$ only.

The problem of unmixing (fuzzy or indistinct clustering) is to estimate a vector of parameters $\Theta=(\omega_1,...,\omega_k,\theta_1,...,\theta_k)$ having selection of X^m of random and independent observations of the mix $\rho(x)$, knowing number k and function φ.

The idea of an algorithm is as follows. We introduce an auxiliary vector of the latent (hidden) variables G having two remarkable properties:

- on the one hand it can be found if vector values Θ of parameters are known;
- on the other hand searching maximum of likelihood strongly becomes simpler if values of the latent variables are known.

The EM algorithm consists of iterative repetition of two steps:

On an E-step, we calculate the expected value (expectation) of a vector G of the latent variables based on the current approximation of a vector Θ of the parameters. Let us designate $\rho(x;\theta_j)$ the probability density of the fact that an object x is received from *j*th component of our mixture of distributions. The conditional probability is

$$\rho(x,\theta_j)=\rho(x)P(\theta_j\,|\,x)=\omega_j\rho_j(x).$$

Let us introduce the designation $g_{ij} \equiv P(\theta_j \mid x_i)$. It is an unknown posterior probability of the fact that the object x_i is received from the *j*th component of the mixture of distributions. These values are used as the latent variables.

Note that $\sum_{j=1}^{k} g_{ij} = 1$ for any $i = 1,..,m$ since it means the total probability of belong of an object x_i of one of the k formulation constituent. From Bayesian formula

$$g_{ij} = \frac{\omega_j \rho_j(x_i)}{\sum_{s=1}^{k} \omega_s \rho_s(x_i)} \text{ for all } i, j.$$

On the M-step the problem of maximizing likelihood (maximization) is solved and there is the following approximation of Θ vector on the current values vectors of G and Θ.

Let us maximize a logarithm of the complete likelihood:

$$Q(\Theta) = \ln \prod_{i=1}^{m} \rho(x_i) = \sum_{i=1}^{m} \ln \sum_{j=1}^{k} \omega_j \rho_j(x_i) \to \max_{\Theta}$$

Solving an Lagrange optimization problem with restriction ω_j, we have:

$$\omega_j = \frac{1}{m} \sum_{i=1}^{m} g_{ij}, \; j = 1,..,k$$

$$\theta_j = \arg\max_{\theta} \sum_{i=1}^{m} g_{ij} \ln \varphi(x_i, \theta), \; j = 1,..,k \,.$$

Thus the M-step comes down to calculation of components weights ω_j as arithmetic averages and to a parameter estimation θ_j a solution of k independent optimization problems. Note that the separation of variables was possible thanks to successful introduction of the latent variables.

The repetitive process stops according to a stop criterion which is set in advance (we choose a metric $\rho(\theta_1, \theta_2)$ and number ε in advance). Process stops on m step if $\rho(\theta^{(m)}, \theta^{(m-1)}) < \varepsilon$.

It is experimentally established that the main EM algorithm has the strong instability according to input data. For example in case for four-component mix of normal distributions at a sample of 200-300 observations the replacement of only one observation by another can cardinally change the total estimates received by means of the EM algorithm [9].

The main EM algorithm as well as the method based on greedy heuristics [10] and *k*-means model [6] applied at present for splitting electronic components into the homogeneous production batches does not allow us to define the quantity of mixture components (quantity of clusters). This size has to be set before algorithm starts operating. Otherwise a series of problems has to be solved with various estimated number of clusters.

It is possible to carry to advantages of the EM algorithm [8]:

- selection of metrics is not required;
- a possibility of use it in combination with algorithms of data processing;
- it works on small volumes of data;
- it allows to allocate noise emissions and a rarefied hum noise [7];

IOP Conf. Series: Materials Science and Engineering **155** (2016) 012026 doi:10.1088/1757-899X/155/1/012026

- a number of the main algorithm modifications is developed [9]: so, a number of modifications of the main algorithm (median modifications, a SEM algorithm, a CEM algorithm, MCEM and SAEM algorithms) are developed for partial elimination of the disadvantages listed above [9].

Thus at application of more widespread *k*-means model to a problem of selection of the homogeneous production batches of EEE products the choice of a metrics or measure of distance [6, 11] is the complex problem which does not have the unique decision: the choice of the Euclidean distance square gives more compact clusters, the choice of a rectangular metrics makes the model less sensitive to the presence of outliers.

In practice, it is necessary to use the special methods of data normalization [6, 12].

The problem of the outliers selection is also very complicated. In [6], at application of the *k*-means model it is solved by introducing a special criterion. The threshold value of this criterion is rather difficult to define experimentally.

As well as the classical *k*-means algorithm, the EM algorithm is the prime two-step procedure. It contains a step of determination clusters parameters and a step of splitting a set of objects into clusters that allows us to use in the future the same methods of increase in accuracy and stability of solutions as those successfully applied for solving problems of the automatic grouping of EEE components applying *k*-means, *k*-medoid and *k*-medians models [6].

The methods based on the specified models remarkably proved on rather larger selections [6] from several hundred copies of EEE.

At the same time some the most expensive types of EEE components bearing the greatest functional loading in spacecrafts are shipped in lots from several pieces that makes the application of the EM algorithm and algorithms on its basis very perspective in relation to our problem.

The known versions of the EM algorithm (for example, in the R software environment [13]) include EMCluster [14] package and a mclust [15] package. In these packages, versions of the EM algorithm and various ways of initialization for a normal distribution of mixes are realized. The possibility of visualization of the work results is provided.

We applied a realization of the EM algorithm in the R software environment for automatic grouping of EEE parties from 50 to 620 pieces (see Fig. 1). In addition to the definition of belonging each EEE component (a point on the chart shown in Fig.1) to a cluster (which are highlighted in the color), the algorithm gives the table with probabilistic characteristics of belonging to a cluster and shows a cluster form, visualizing possible correlations of the EEE parameters.

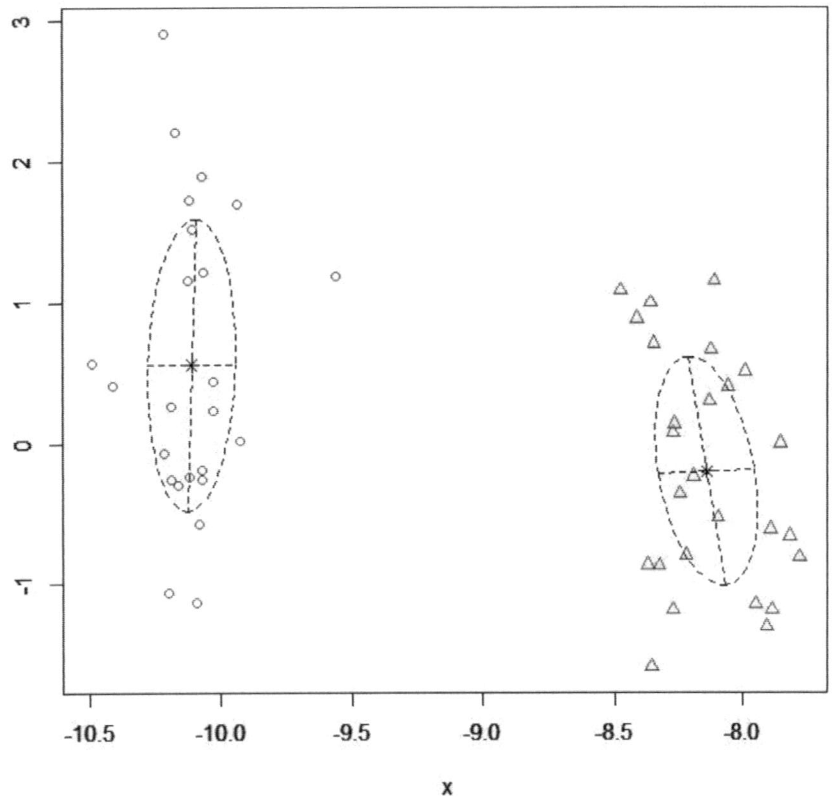

Figure 1. A projection (2 parameters from 32 are shown) of result of automatic group of amplifiers 140UD25AS1VK.

Conclusion

The application of the EM algorithm and algorithms on its basis opens new prospects in the solution of a problem of automatic group of EEE components on production batches, in particular, at the small volume of input data. At the same time, it is required to solve a problem of increase in stability of result of an algorithm.

Thus the application of the EM algorithm and algorithms on its basis opens the considerable prospects in the solution of a problem of automatic group of EEE on production parties, in particular, at the small volume of input data. However the problem of instable results of this algorithm must be solved for successful implementation in the space industry. The equivalence of the main steps of the EM algorithms and steps of the *k*-means algorithm allows to use the same methods for increasing the stability and preciseness of its results.

References

[1] Kazakovtsev L A, Antamoshkin A N, Fedosov V V 2016 Greedy heuristic algorithm for solving series of EEE components classification problems *IOP Conference Series: Materials Science and Engineering* Vol. **122** Article ID 012011. 7 P. DOI: 10.1088/1757-899X/122/1/012011.

[2] Fedosov V V, Orlov V I Minimal necessary extent of examination of microelectronic products at inspection test stage *Izvestiya Vuzov. Priborostroenie*, Volume **54(4)** pp. 62-68.

[3] Koplyarova N V, Orlov V I 2014 About a research of computer system of diagnostics of EEE on the basis of these tests *Vestnik SibGAU* Volume **1(53)**, pp.24-30.
[4] Kazakovtsev L A, Orlov V I, Stupina A A and Masich I S 2014 Problem of electronic components classifying *Vestnik SibGAU* **4(56)** pp 55-61
[5] Xindong Wu, Vipin Kumar, J. Ross Quinlan, Joydeep Ghosh, Qiang Yang, Hiroshi Motoda, Geoffrey J. McLachlan, Angus Ng, Bing Liu, Philip S. Yu, Zhi-Hua Zhou, Michael Steinbach, David J. Hand, Dan Steinberg. Top 10 algorithms in data mining. URL http://www.cs.uvm.edu/~icdm/algorithms/10Algorithms-08.pdf
[6] Kazakovtsev L A Greedy Heuristic Method for Systems of Automatic Object Groupping: Doctoral Dissertation. — Krasnoyarsk, SFU, 2016.
[7] Vorontsov K V Lectures on algorithms of a clustering and multidimensional scaling http://www.machinelearning.ru
[8] Cherezov D S, Tyukachev N A 2009 Review of the main methods of classification and clustering of data. *Voronezh State University. Vestnik VSU. Series: System analysis and information technologies* Vol. **2**
[9] Korolev V Y 2007 EM algorithm, his modifications and their application to a problem of division of mixes of probabilistic distributions. *Theoretical review*
[10] Kazakovtsev L A, Antamoshkin A N, Masich I S 2015 Fast deterministic algorithm for EEE components classification. *IOP Conference Series: Materials Science and Engineering* Vol. **94** Article ID 012015. 10 P. doi:10.1088/1757-899X/94/1/012015.
[11] Kazakovtsev L A, Orlov V I, Stupina A A 2015 [On Distance Metric for the System of Automatic Classification of the EEE Devives by Production Batches] // Programmnye produkty I sistemy Issue **2** P. 124-129. doi: 10.15827/0236-235X.110.124-129. In Russian.
[12] Fedosov V V, Kazakovtsev L A, Gudyma M N 2016 Problem of Normalization of Input Test Data of Space Industry EEE Components for Automatic Groupping Algorithm *Informatsionnye tekhnologii modelirovaniya I upravleniya* Issue **4** P. 263-268. In Russian.
[13] Software environment for statistical computing and graphics R. URL https://cran.r-project.org/
[14] Chen W.-C., Maitra R 2015 EMCluster: EM Algorithm for Model-Based Clustering of Finite Mixture Gaussian Distribution. R Package, URL http://cran.r-project.org/package=EMCluster Chen, W.-C., Maitra, R. (2015) A Quick Guid for the EMCluster Package (Ver. 0.2-5). R Vignette, URL http://cran.r-project.org/package=EMCluster
[15] Chris Fraley, Adrian E, Raftery T Brendan Murphy and Luca Scrucca (2012) mclust Version 4 for R: Normal Mixture Modeling for Model-Based Clustering, Classification, and Density Estimation Technical Report No. 597, Department of Statistics, University of Washington Chris Fraley and Adrian E. Raftery (2002) Model-based Clustering, Discriminant Analysis and Density Estimation Journal of the American Statistical Association 97:611-631

Measurements of high impedance two-terminal device with SMU NI PXIe-4139*

S V Bogdanov, E T Lelekov, I V Kovalev, P V Zelenkov and A T Lelekov

Reshetnev Siberian State Aerospace University,
31, Krasnoyarsky Rabochy av., Krasnoyarsk 660037, Russia

E-mail: a.t.lelekov@yandex.ru

Abstract. To measure high-frequency and low-frequency impedance of betavoltaic power sources (it can be represented as two-terminal device), measurement stand was created. To measure high-frequency part need to inject external test signal through the current transformer with waveform generator and need to use external high-frequency current sensor, because of SMU PXIe-4139 current channel limitations.

Introduction

Two-terminal device impedance measurements is necessary in many scientific fields, such as: electrical impedance spectroscopy in the chemistry, electrochemistry and corrosion, frequency characteristics measurements in control systems research, impedance measurements in semiconductor devices, biological objects, etc.

The principles of impedance measurement are deeply investigated, the results are published in books [1-3]. The measurement methods are usually based on insertion of current oscillations to the main current of two-pole circuit and then voltage response measurement. This method used if the device under test (DUT) has a low resistance; in case of high internal resistance of DUT the insertion of voltage oscillations performs and then the current response measured. Generally, oscillations produces with special source, able to directly generate waveform to power DUT.

There are a lot of oscillations insertion methods, single-frequency and multifrequency methods are the most common. This methods with direct power waveform generation allow to measure the impedance in the low frequency part of spectra, right down to very low frequencies, about few uHz. For using these methods for high frequencies, high quality ADC required. ADC accuracy requirements are increased with reducing the level of cur-rents or voltages, that is usual for high impedance devices.

To overcome this limitations, the second method of oscillations insertion can be used. It is to insert of oscillations with the current or voltage transformer, connected to the external generator. This method is usually applicable for input of high frequency oscillations.

*Research is performed with financial support from Russian Federal Targeting Program "Research and development for priority targets of Russian scientific-technological complex development years 2014-2020", in accordance with the agreement 14.577.21.0117, unique identifier RFMEFI57714X0117

IOP Conf. Series: Materials Science and Engineering **155** (2016) 012027 doi:10.1088/1757-899X/155/1/012027

The high-frequency part of impedance spectra is very important for the parameters identification of equivalent circuit diagrams, because when impedance grows up, the time constants are shifted to the high-frequency region.

Using of modular devices allows to develop measurement procedure more flexible than with stand-alone devices. The advantages is in the creation of different synchronized working devices, implementation of complex measurement procedures and mathematical processing of the results. Modular SMU, produced by National Instruments, have one of the best capabilities nowadays in this class of devices. To significant disadvantages one can include lack of Linux support, the rest are solvable in principle.

Experimental

To measure static and dynamic characteristics of betavoltaic power sources, measurement stand based on SMU PXIe-4139 was created. Betavoltaic power source can be represented as double-pole circuit [4]. Because the generation current of the small area prototypes has a fairly low value (about a tens of nA), it was necessary to develop an effective method of impedance measuring of such devices, with the aim of SPICE model building to use in further calculations.

Measurement of static current-voltage characteristic is not a problem for this SMU model, which have an effective on-board protection from AC 50Hz power supply interference and the instrumental means of preventing (guarding) current leakage. The advices in [5] allow to build an effective system of low current measurements. The designed wiring diagram represented on figure 1.

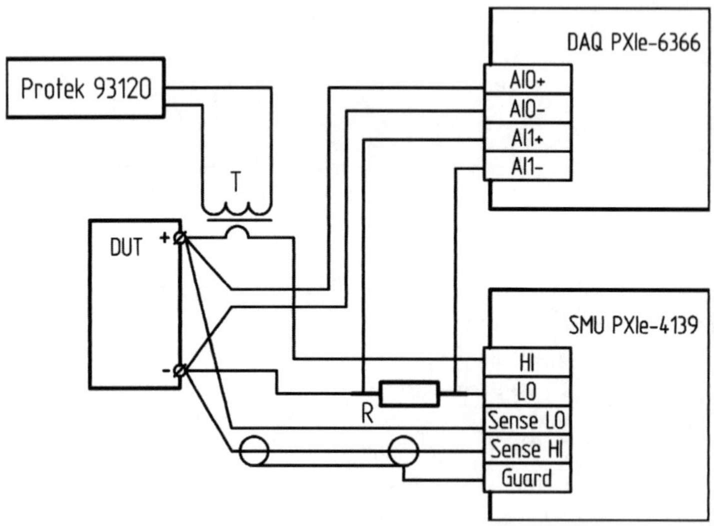

Figure 1. Wiring diagram of experiment.

While measuring low-frequence part of impedance, the insertion method of single-frequency oscillations was used, which realizes the following sequence for every point in a required frequency range:

1. Setting the DC operating point of the DUT.
2. Insertion of sinusoidal oscillations to the voltage of the DUT by the shape setting of SMU PXIe-4139 output signal. The oscillation amplitude shouldn't exceed the region of linearity, in order to the response signal must be also sinusoidal. This condition is satisfied if the nonlinear distortion coefficient of the signal response is not more than 5%.

3. Making current and voltage signals measurement.
4. Selection of the main harmonic amplitude of measured current and voltage, the impedance calculations for a given frequency point.

For the impedance measurement of the high frequency spectrum (10 kHz - 400 kHz) the above method can't be used, because the SMU's maximum sampling frequency is 100 kHz, and high-quality sine waveform requires at least 4 points on the period.

The desired signal is injected by an external generator with a current transformer with a variable transformation ratio. The rest of the measurement sequence is the same.

While impedance measuring as described way, the authors faced with feature of SMU PXIe-4139, not reflected in the documentation [6]. This was expressed as measured impedance overration, from the frequency of 100 kHz with a slope of +20 dB/decade, like as inductive component added. The influence of cables was checked on the test circuits, their effect was negligible at these frequencies.

To clarify the reasons of this effect, it was made an additional experiment. The meaning of experiment was in parallel current and voltage measuring with the second measuring system (multifunctional DAQ module PXIe-6366). The current sensor was a metallized heat-resistant lacquer resistors (russian MLT type), which do not have a reactive components up to frequencies of the 6 MHz (fig. 1).

The SMU configured to One-point voltage stabilization mode. After the source operation completes, the measurements of current and voltage channels were made. The AC 50Hz power supply frequency was filtered by second order filter. The SMU PXIe-4139 and DAQ PXIe-6366 measurements were synchronized. The start of the SMU measurement begins at the AI Start trigger from the DAQ board.

Results and discussion

The results revealed that the voltage channel measurements of SMU and DAQ are identical up to noise amplitude (see figure 2, for DAQ PXIe-6366 Fs = 2MS/s, for SMU PXIe-4139 Fs = 900kS/s).

Measurements of current with SMU showed, that sine amplitude was decreases from 100 kHz with the slope of −20 dB/decade, compared with DAQ current measurements.

The measured ratio of signal from the internal current sensor of the SMU to the external current sensor for different sampling frequency are shown at figure 3.

A detailed SMU documentation study did not give any information on the detected effect and means of extending this effect. Perhaps there is internal digital filtering (as the shape of Bode plot looks like) in the control loop of the SMU PXIe-4139.

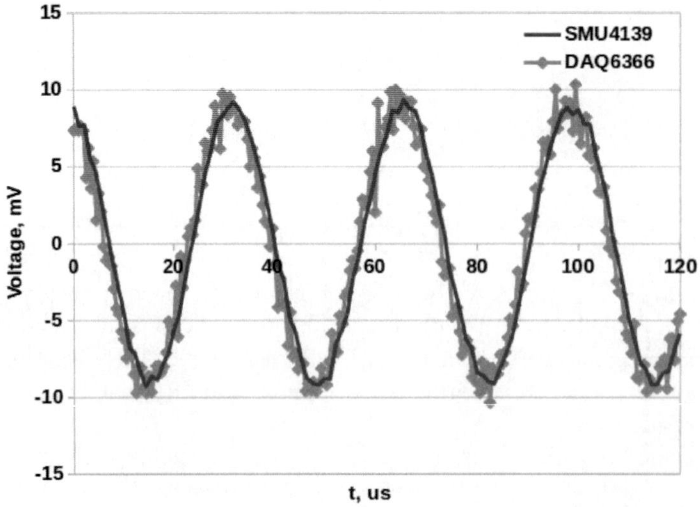

Figure 2. Voltage channel measurements of SMU PXIe-4139 and DAQ PXIe-6366.

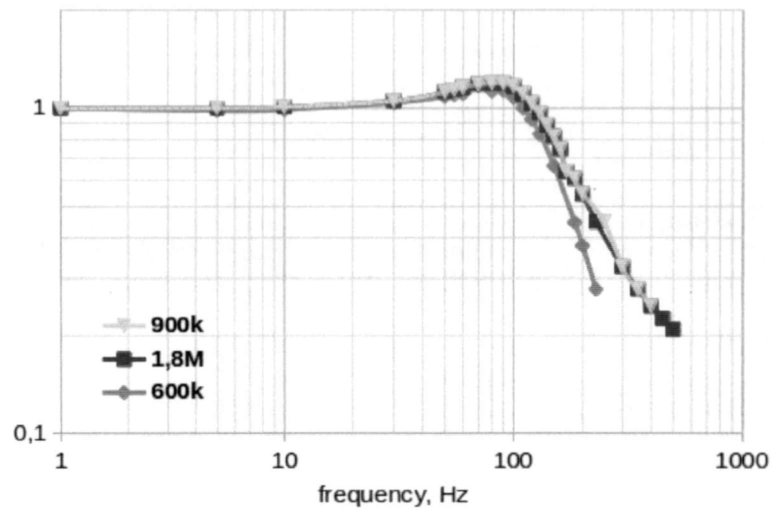

Figure 3. Ratio of signal from SMU internal current sensor to external current sensor for different sampling frequency.

Conclusion

Using of modular SMU gives wide possibilities for creation of measuring systems for specific needs, in particular for measuring the impedance characteristics in the range DC-400kHz. The low-frequency part of impedance spectra can be measured with the method of direct power waveform generation of the SMU PXIe-4139 output. High-frequency part can be measured with injection of external test signal through the current transformer with waveform generator. In the last case, for adequate current measurement we should use an external HF current sensor (resistive or transformer type). Also in this case SMU can be used only to set the operating point of the DUT.

References

[1] Mark E. Orasem and B. Tribollet, «Electrochemical Impedance Spectroscopy», Wiley, 2008, ISBN 978-0-470-04140-6.

[2] Agilent Impedance Measurements Handbook. A guide to measurement technology and techniques. 4th edition. Agilent Technologies, Inc. 2009.

[3] E. Barsoukov and J. Ross Macdonald «Impedance Spectroscopy. Theory, Experiment, and Applications» 2nd edition. Wiley, 2005, ISBN 0-471-64749-7.

[4] I.V Kovalev, V.G. Sidorov, P.V. Zelenkov, A.Y. Khoroshko, A.T. Lelekov. Electron-hole pairs generation rate estimation irradiated by isotope Nickel-63 in silicon using GEANT4. IOP Conference Series: Materials Science and Engineering, Volume 94, Issue **1**, article id. 012024 (2015), doi:10.1088/1757-899X/94/1/012024.

[5] P V Zelenkov, V G Sidorov, E T Lelekov, A Y Khoroshko, S V Bogdanov and A T Lelekov. Modeling of microporous silicon betaelectric converter with ^{63}Ni plating in GEANT4 toolkit. IOP Conference Series: Materials Science and Engineering, Volume 122, Issue **1**, article id. 012036 (2016), doi:10.1088/1757-899X/122/1/012036.

[6] KEITHLEY Nanotechnology Measurement Handbook. A Guide to Electrical Measurements for Nanoscience Applications. 1st edition.
`www.keithley.com`

[7] NI PXIe-4139 Specifications.
`http://sine.ni.com/nips/cds/view/p/lang/ru/nid/212038`

IOP Conf. Series: Materials Science and Engineering **155** (2016) 012028 doi:10.1088/1757-899X/155/1/012028

Ultrahigh vacuum holder-positioner for in situ studies of conductive nanostructures in a wide temperature range

D V Shevtsov [1,2], S A Lyaschenko[1,2], S N Varnakov[1,2]
[1] Reshetnev Siberian State Aerospace University, Krasnoyarsk, 660037, Russia
[2] Kirensky Institute of Physics, Federal Research Center KSC SB RAS, Krasnoyarsk, 660036, Russia

E-mail: snowman-82@mail.ru

Abstract. A holder-positioner was developed in order to be used in high vacuum systems designed for synthesis and in situ samples investigation by optical methods in 77÷1171 K temperature range. During tests of holder-positioner the designed system demonstrated independence from other process chamber components, compactness, a large number of the sample degrees of freedom and the stability of maintaining temperatures in the range from 77 to 1171 K with a maximum sample cooling rate about 0.1 K/s.

1. Introduction

Preparation and research of new functional materials for the aerospace industry necessitates the creation of equipment to carry out controlled synthesis of such materials and in situ analysis of their physical properties. In particular, epitaxial ferromagnetic nanostructures on the monocrystalline silicon surface [1] are important materials for the development of the element base for aerospace appliances. In situ analysis methods have a great practical value for obtaining quantitative information about the structure and morphology of the samples [2, 3]. For non-contact examination of magnetic properties and electronic structure of the absorbing ferromagnets, the method based on the magneto-optical Kerr effect has already become conventional [4, 5].

Trends in the development of cryogenic and high-temperature equipment for optical studies reveal that the current challenges in this area is the implementation of temperature inserts, which allow us to study nanomaterials physical properties in a wide temperature range (2-1300 K) [6, 7]. Detailed analysis of the world's scientific and technical literature has not revealed information on systems combining in situ spectroscopic magneto-optic ellipsometry with a range of the sample operating temperatures 77-1200 K. However, the this analysis made it possible to establish the most common approaches for heating and cooling of semiconductor substrates, in a special medium or vacuum.

Technical devices and methods that combine heating up to 1300 K and cooling down to several K units, with the possibility of spectral measurements in the optical wavelength range, can be divided into two classes. The first one is universal systems of the sample displacement and attachment, including temperature control option in a wide range. Such systems are usually built into the ultra-high vacuum sputtering research systems. The second one is systems for conducting optical investigations in a wide temperature range [5, 8-12].

In this regard, the modernization of technological equipment, which permits us to carry out the in situ spectral magneto-optical ellipsometric measurements over a wide temperature range, is the current trend in the development of the means for the production and in situ investigation of ferromagnetic nanostructures.

IOP Conf. Series: Materials Science and Engineering **155** (2016) 012028 doi:10.1088/1757-899X/155/1/012028

2. Results and discussion

The analysis of the existing solutions revealed that the most efficient cooling system is a heat-conducting copper rod with the sample fixed on one side and having a refrigerant on the other side [13-14]. We created a sample holder design that showed high stability in maintaining a constant temperature in the range from 77 to 470 K. The specimen heating system based on the direct current passage was also added to the cooling system developed. A key element of the holder is a sapphire dielectric plate placed between the specimen and the copper base. Sapphire has a fast-falling dependence of the thermal conductivity coefficient [15, 16] under the temperature rising from 80 to 1300 K. That allows us to use sapphire as a heat insulator at high temperatures and efficient heat-removal at the boiling point of nitrogen. Figure 1 is a schematic representation of the holder-positioner created.

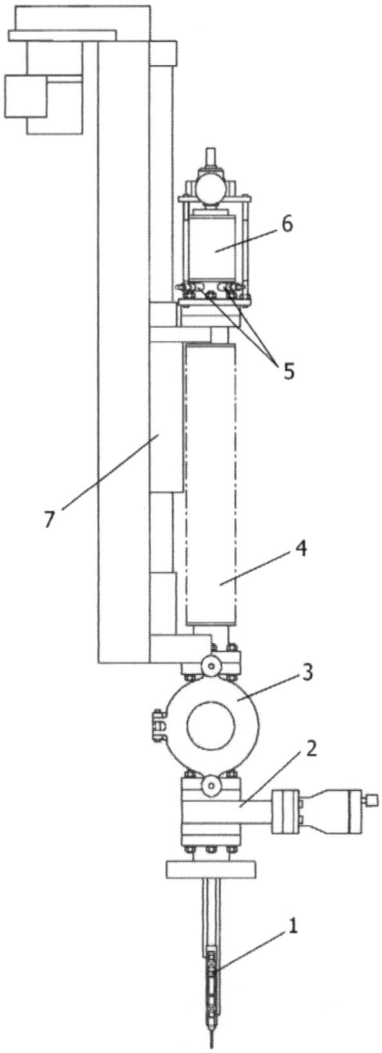

Figure 1. Schematic representation of the holder- positioner: 1) copper sample holder; 2) gate valve; 3) airlock; 4) bellows; 5) high vacuum current leads; 6) vacuum bellows inlet; 7) line translator

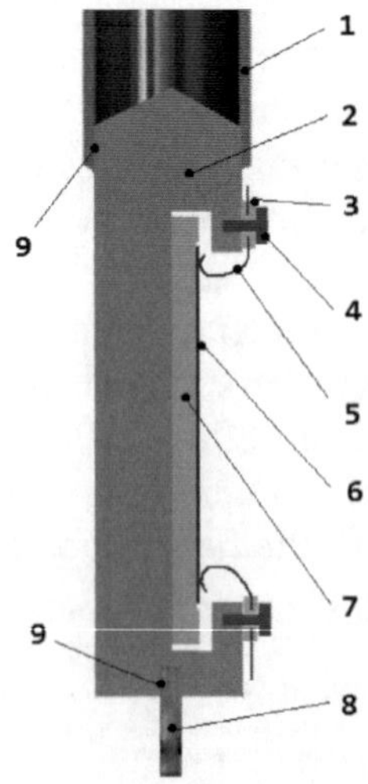

Figure 2. sample holder model profile with elements for annealing and cooling: 1 - stainless steel tube with refrigerant; 2 - solid copper sample holder; 3 – current leads electrical insulations; 4 - titanium fixing screws; 5 - tantalum leads in the form of hold-down clips; 6 - sample (silicon wafer); 7 - sapphire plate; 8 - steel latch; 9 - welded connections

The basis of holder-positioner is a vacuum-mechanical system (VMS). The VMS comprises the following elements: a vacuum shutter system, an airlock, the vacuum transport system, a nitrogen supply system, a sample heating system. The sample holder (Figure 2) is a coaxial design rod with a copper insert on which the specimen is mounted. The upper end of the rod is connected with the vacuum bellows inlet by means of tungsten-arc inert-gas (TIG). Catches mount with screws through the insulating sleeve on a copper insert. Catches provide a tight fixation of the specimen and serve as electrodes when the sample is heated by direct current passage.

Single-crystalline silicon substrate coated with a layer of SiO_2 was used for the holder-positioner temperature testing. Upon the sample holder nitrogen stream cooling temperature was being measured by a Pt100 resistance temperature detector (RTD). Figure 3 shows time dependences of the holder temperature and nitrogen pressure in the Dewar.

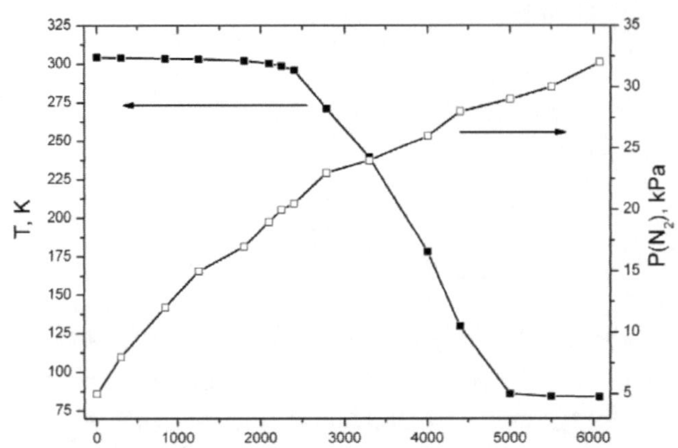

Figure 3. Time dependences of the holder temperature and nitrogen pressure in the Dewar

It can be clearly seen that the maximum sample cooling rate is approximately 0.1 K / s. After 2 hours of the cooling procedure the sample temperature has reached 84 K. To set an intermediate temperature in the range from 84 to 570 K a nitrogen flow heater is used. The flow heater power is smoothly given by a proportional-integral controller in the range of 0 to 1000 watts, which allows us to heat the nitrogen gas supplied to the sample holder from 76 to 800 K within a few minutes.

For the holder-positioner testing in the heating mode, voltage in current limiting mode was applied to the Si/SiO2 sample. The sample surface temperature was measured by the infrared pyrometer IMPAC IGA 5 Advanced (LumaSense Technologies). The current increase through the sample was discretely carried out from 0.05A up to 4 A in increments of 0.5 A. The maximum temperature of the sample was 1171 K. Figure 4 shows the electric current and sample temperature time dependences.

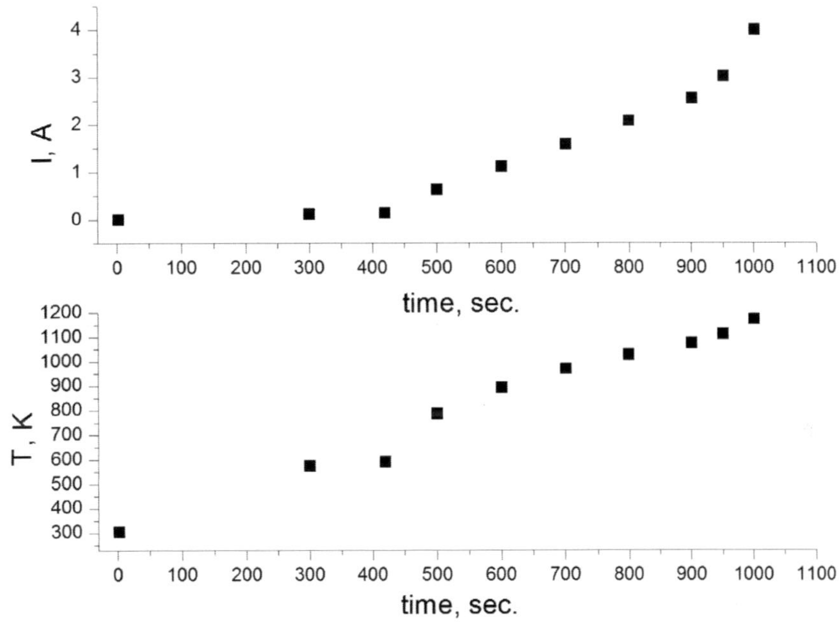

Figure 4. Time dependences of the electric current and sample temperature

As an example Figure 5 shows the results of the measurements of the ellipsometric parameter Ψ fulfilled on a silicon single crystal substrate with SiO2 film under temperature changing in the range of 85-800 K. In situ ellipsometric measurements were made at a wavelength of 470 nm (2.64 eV) and with the polarizer and analyzer positions 45 °, 0 ° relative to the incidence plane, respectively [17]. To subtract the background illumination during ellipsometry measurements we were using a light beam mechanical chopper at a frequency of interruptions of 10-20 s-1 [3].

The advantages of the holder-positioner construction are the ability to embed it in any vacuum system and two ways of the sample temperature setting: using the thermal contact with the heat-transfer and by the direct current passing.

The holder-positioner design is compact, the width of the copper holder with concomitant parts does not exceed 15 mm, which allows one to utilize standard ultra-high vacuum gate valves and flanges. The sample loading system is integrated with a goniometer to ensure the required degree of freedom when setting up the optical part. Thus, it is possible to fine-tune the sample holder space position for carrying out in situ spectral ellipsometry measurements directly in the coating chamber. The manipulator is integrated with a cryostat-heater to achieve the desired temperature range.

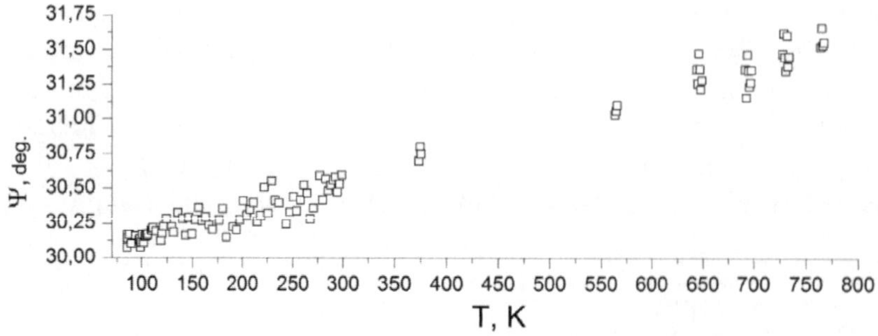

Figure 5. Temperature dependence of ellipsometric parameter Ψ for $SiO2/\ Si$ sample

This feature distinguishes the system from the proposed analogues in which such functionally different parts are implemented as various units.

3. Conclusion

Taking into account peculiar properties of the holder-positioner design developed, one can conclude it is versatile. Independence from the other chamber components, the compactness and the large number of freedom degrees permit the use of the holder-positioner in experimental devices equipped with both optical and diffraction, electron spectroscopic surface analysis methods.

4. Acknowledgments

The work was supported by the Ministry of Education and Science of the Russian Federation (State task no.16.663.2014K), RFBR (Grant no.16-42-243058) and Program of the President of the Russian Federation for the support of leading scientific schools (Scientific School 7559.2016.2).

References

[1] Fabian J, Matos-Abiague A, Ertler C, Stano P and Zutic I 2007 *Acta Phys. Slov.* 57 565-907
[2] Aspnes D E 1993 *Thin. Sol. Films* 233 1-8
[3] Lyaschenko S A, Varnakov S N, Tarasov I A, Shevtsov D V and Ovchinnikov S G 2012 *Vestnik SibGAU* (Krasnoyarsk) **4** (44) pp 162-167
[4] Vonsovskij S V *Magnetizm* 1971 (Moscow: Nauka) p 1032
[5] Rauer R, Neuber G, Kunze J, Bäckström J and Rübhausen M 2005 *Rev. Sci. Instrum.* **76** (2) 023910–023913
[6] Norrman S H and Andersson T G 1977 *Vacuum* **27** 97
[7] Belyaeva A I and Grebennik T G 1997 *Proc. SPIE, Optical Diagnostics of Materials and Devices for Opto-, Micro-, and Quantum Electronics* **3359** 401
[8] Oxford Instruments. Nitrogen Cryostats 77 K. URL: https://www.oxford-

IOP Conf. Series: Materials Science and Engineering **155** (2016) 012028 doi:10.1088/1757-899X/155/1/012028

instruments.com/products/cryogenic-environments/optical-cryostats-for-spectroscopy/nitrogen-cryostats-77k (date of the application: 20.09.2016)
[9] NIST Center for Neutron Research. Liquid Helium Cryostats. URL: http://www.ncnr.nist.gov/equipment/Cryostat.html (date of the application: 20.09.2016)
[10] LTLab, Inc. The Cryogen-free measurement system: optCRYO105. URL: http://www.ltlab.com/optcryo105.html (date of the application: 20.09.2016)
[11] Janis Research. Cryogenic Research Equipment. SuperTran Continuous Flow Cryostat Systems – Sample in Vacuum. URL: http://www.janis.com/Products/productsoverview/SuperTranContinuousFlowCryostatSystems.aspx (date of the application: 20.09.2016)
[12] Feng X H and Garfunkel E L 1989 *Vacuum* **39** 6
[13] Hernandez L M and Goldman A M 2002 *Rev. Sci. Instrum.* **73** 162
[14] Ekinci K L, Valles J M, Jr. 1997 *Rev. Sci. Instrum.* **68** 4152
[15] Berman R 1951 *Proc. R. Soc. Lond. A.* **208** 90-108
[16] Touloukian Y S and Ho C Y 1971 *Thermophysical Properties of Matter Volume 2: Thermal Conductivity: Nonmetallic* (NY: Macmillan) p 1172
[17] Shvets V A, Spesivtsev E V and Rykhlitskii S V 2004 *Optics and Spectroscopy* **97** 483

IOP Conf. Series: Materials Science and Engineering **155** (2016) 012029 doi:10.1088/1757-899X/155/1/012029

Anisotropic crystal of the δ-BiB_3O_6 investigated by vibrational spectroscopy

E A Strikina[1,2], A S Krylov[2,3], A S Oreshonkov[2,3], A N Vtyurin[2,3] and O A Maximova[1,2]
[1] Reshetnev Siberian State Aerospace University, Krasnoyarsk, 660037, Russia
[2] Siberian Federal University, Krasnoyarsk, 660041, Russia
[3] Kirensky Institute of Physics, Federal Research Center KSC SB RAS, Krasnoyarsk, 660036, Russia

E-mail: nas-nas@iph.krasn.ru

Abstract. The vibrational spectroscopy has been applied to investigate the structure the BiB3O6 (BIBO) crystal. Based on the experimental results, the total set of phonons mode of the polarized Raman spectra was proposed. To verify the obtained experimental data have been performed theoretical calculation in software package LADY.

1. Introduction

BiB_3O_6 (BIBO) has excellent optical nonlinear properties for to conversion fervency in the solid state lasers [1]. It is most effective in the optical range UV and visible light which have widely been used in medicine, signal progressing. BIBO have large nonlinear optical coefficient due to its structure and non-hydroscopic nature [2].

Long time thought that BiB_3O_6 have only alpha phase with space group symmetry C2. [3] Recently have been found new six phase of this compound. More extensively known and studied phase is α-BiB_3O_6, which build from chain bounded triangles [BO_3] and tetrahedrons [BO_4] in range 1:2. β- BiB_3O_6 is metastable phase and it can undergo the phase transition in other phase at the high temperature. As α - BiB_3O_6, β - BiB_3O_6 consist of chain bounded triangles [BO_3] and tetrahedrons [BO_4] in range 2:1. The first principles simulation showed that the main contribution in nonlinearity comes from [BO_4] tetrahedral units. Therefore, γ and δ – BiB_3O_6 attract the great interest due to the features of their structures (Figure 1) which are exclusively consisting of the chains of bounded tetrahedrons [BO_4].

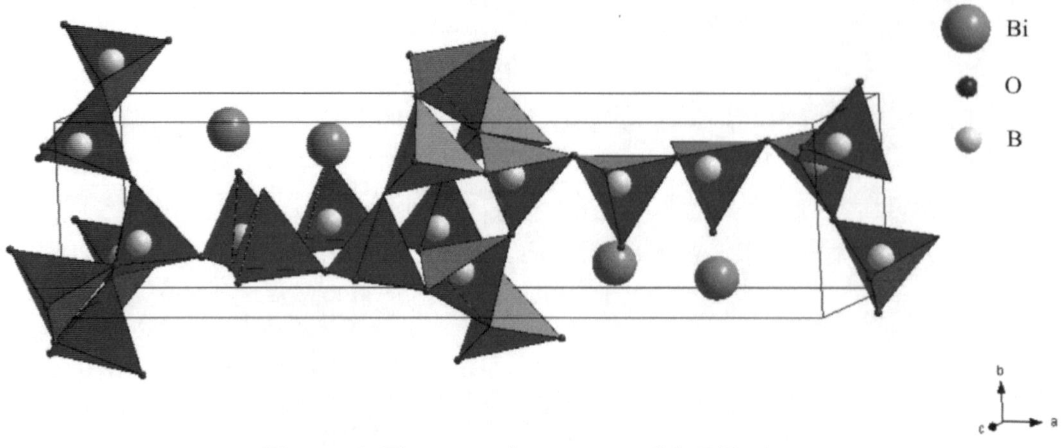

Figure 1. The crystal structure of δ–BiB_3O_6.

Since $\gamma - BiB_3O_6$ have inversion center, subject of the present research work is δ – BiB3O6. The symmetry space group of orthorhombic δ - phase is $Pca2_1$ (Z=4). The expansion of the total vibrational representation in the Brillouin zone center for the δ phase has the following view: $\Gamma V_{ibr} = 30A_1 + 30A_2 + 30B_1 + 30B_2$, acoustic and optic modes: $\Gamma_{acoustic} = A_1 + B_1 + B_2$, $\Gamma_{optic} = 29A_1 + 30A_2 + 29B_1 + 29B_2$.

2. Experiment

The samples have been obtained same as described in the paper [4]. The assembling is spectrometer Horiba Jobin Yvon T64000 equipped with a liquid nitrogen cooled charge coupled device detection system in subtractive dispersion mode in 10 to 1600 cm^{-1} range. The spectra were recorded in the backscattering geometry. It was used notation. [4]

3. Results and discussion

To simulate the $\delta - BiB_3O_6$ the vibrational spectrum package LADY was used. The program allows to obtain the full Raman spectrum using the model of 'rigid-ion'. The interatomic potential is considered as a sum of the short-range interaction potentials was taken in the Born–Mayer form:

$$V^{RIM}(r_{ij}) = \frac{1}{2}\sum_{ij}\frac{Z_i Z_j}{r_{ij}} + U(r_{ij}), \quad (1)$$

and of the of long range Coulomb electrostatic potentials:

$$U(r_{ij}) = \lambda \exp(-r_{ij}/\rho), \quad (2)$$

where r_{ij} is the interatomic distance and λ and ρ are the parameters characterizing of the short-range pair interionic interaction. Resulting model parameters were obtained by minimization of residual values of the simulated and experimental Raman frequencies using the Fletcher – Reeves method [5–7]. The values of λ, ρ and Z_{ij} are listed in Table 1.

Table 1. Parameters of the interatomic interaction potential:

Interactions	Radii of interaction, Å	λ, aJ/Å^2	ρ, Å
Bi – O	0–3.00	350.00	0.300
B1 – O	0–1.50	321.60	0.210
B2 – O	0–1.60	400.60	0.220
B3 – O	0–1.60	345.30	0.199
O – O	0–3.00	242.80	0.245
Ion	Bi	B	O
Z(e)	2.00	1.80	-1.2333

The Raman active frequency is shown in Table 2 in comparison with experiment

Table 2. The calculated Raman active frequencies are shown in Table 2 along with the experimental dates

A_1			A_2		B_1		B_2	
exp.		calc.	exp.	calc.	exp.	calc.	exp.	calc.
TO	LO							
1116	1097	1121	1076	1153		1212		1278
1081	1035	1093	1037	1119	1082	1132	1037	1132
1031	1007	979	1010	1090	1035	1023	1010	1029
996	975	959	978	981	997	949	997	965
959	935	908	944	945	955	938	956	929
919	907	894	905	912	919	910	936	905
864	865	789	864	805	897	854	880	892
807	783	752	784	755	863	803	784	786
783	727	721	728	751	813	749	727	737
752	650	697	709	730	783	732	665	689
726	607	634	607	648	753	648	607	647
639	530	598	558	606	727	618	515	623
606	515	552	534	553	607	575	499	563
530	502	510	514	529	578	533	468	518
499	467	490	501	511	531	504	415	483
467	416	447	466	490	515	497	393	459
415	394	428	417	428	500	432	327	442
393	329	384	393	416	467	421	312	400
328	310	345	330	382	415	392	304	383
303	256	320	313	332	394	347	249	338
272	244	274	255	310	327	319	226	307
228	235	255	246	262	304	299	204	280
186	225	221	226	227	245	267	185	240
175	194	159	196	191	229	229	177	185
122	165	128	126	170	184	156	166	135
97	120	100	110	94	123	116	123	119
71	88	69	89	89	116	89	112	94
67	69	46	68	70	97	76	98	66
46	49	18	51	54	71	55	69	35
			49	44	68		45	

Since the BO_4 bounded with B-O in crystal is not ideal we can observe a large number of vibrations in the range > 650 cm^{-1}. For this reason the whole spectrum is reached of lines. The simulations is shown that about 40 active modes appear in the range >650 cm^{-1} for the different position of the polarization.

The bands below 150 cm^{-1} are related with Bi translation vibrations. The lower wavenumber range of 150–350 cm^{-1} contains translational, rotational and mixed vibrations of BO_4 tetrahedra. Bands in the 350–650 cm^{-1} range are related with the distorted ν_2 and ν_4 BO_4 bending modes. Generally, frequency of ν_4 vibration should be above that of ν_2 vibration [8].

In the Figure 2 a,b is presented LO-TO splitting in depending on the different position of the polarization is presented. The lines corresponding to B_1 and B_2 modes are presented on Figure 3 a,b.

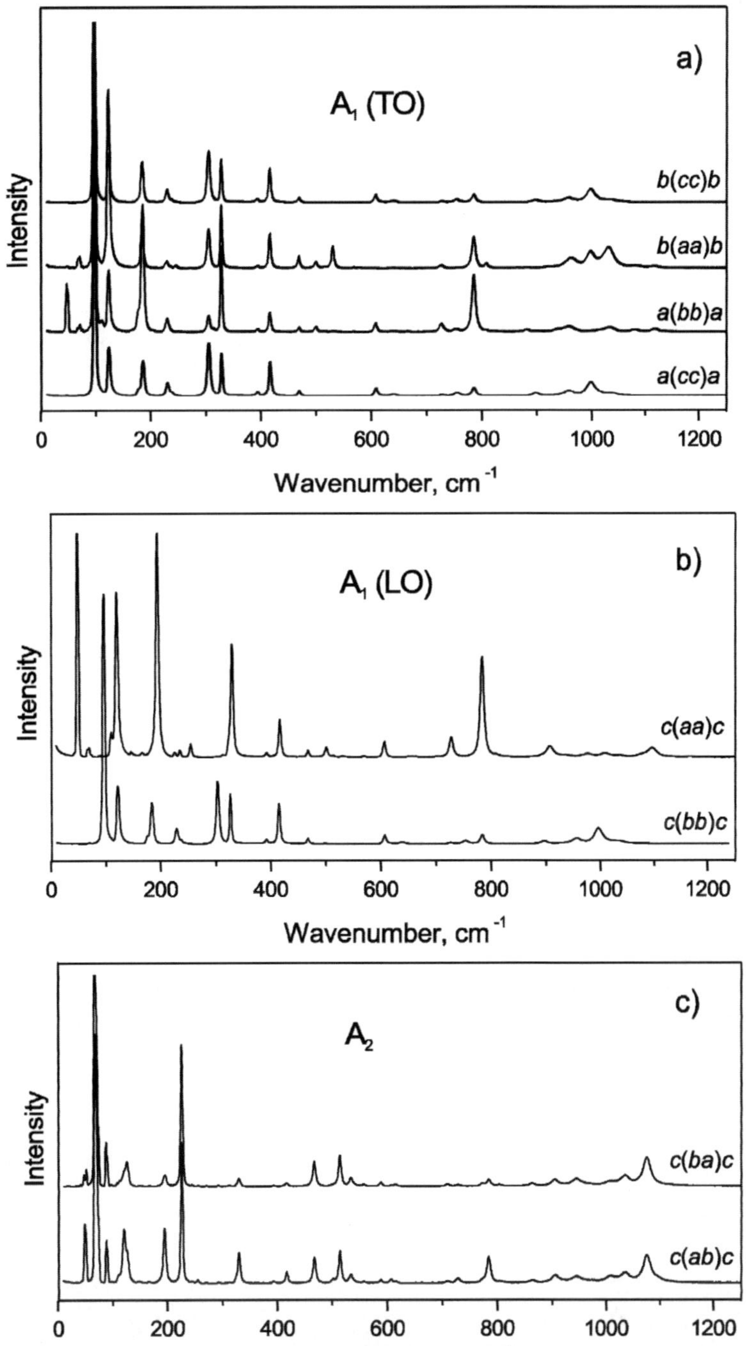

Figure 2. Raman spectra of the δ–BiB_3O_6 crystal at: a) b(cc)b, b(aa)b, a(bb)a, a(cc)a b) c(aa)c, c(bb)c, c) c(ba)c, c(ab)c geometries

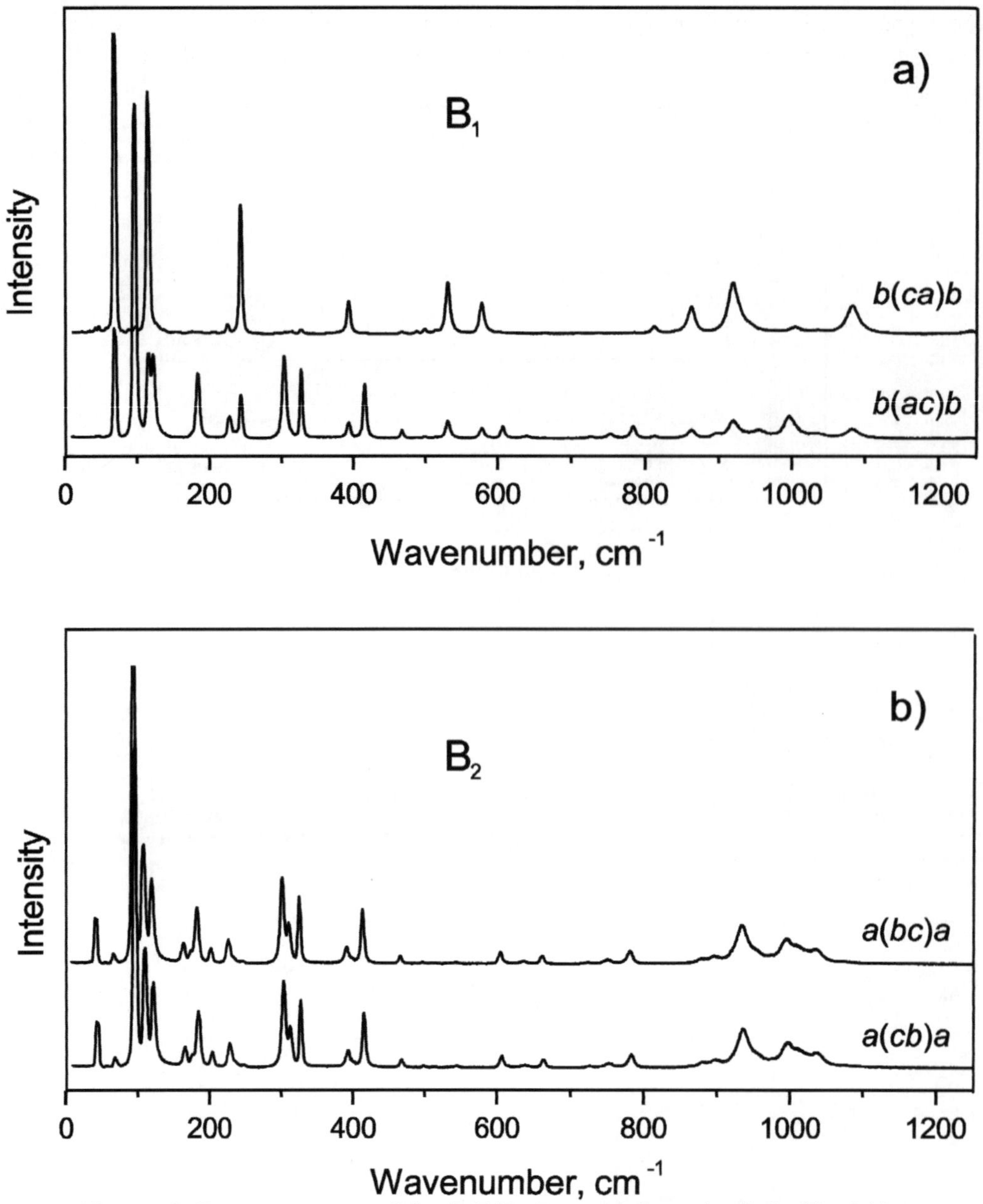

Figure 3. Raman spectra of the δ–BiB_3O_6 crystal at: a) a(bc)a b) a(cb)a

4. Conclusion

The polarization selection rules, lattice dynamic simulations and number of the spectral lines are in the good agreement with experiment. The full set of experimental Raman spectra polarizations is presented. The vibration modes are classified to corresponding Raman bands.

Acknowledgments

This study was partially supported by the Ministry of Education and Science of the Russian Federation, the “Krasnoyarsky regional fund of scientific support and scientific-technical activity” and the Russian Foundation for Basic Research Grant “15-42-04347 r_siberia_a".

References

[1] E I Algazin, A P Kovalevsky, V B Malinkin 2009 Signals transmission by invariant method with further non-linear processing under weak correlation *Vestnik SibGAU*. **5 (26)** pp. 4-7

[2] E M Levin, C L McDaniel 1962 The system Bi_2O_3 - B_2O_3. *J. Am. Ceram. Soc.* **45** pp 355–360

[3] Rihong Cong, Jinlong Zhu, Yingxia Wang, Tao Yang, Fuhui Liao, Changqing Jin and Jianhua Lin 2009 Phase transitions among four BiB3O6 polymorphs: a detailed investigation *J. Cryst. Eng. Comm.* **11** pp 1971–1978

[4] A S Aleksandrovsky, A D Vasiliev, A I Zaitsev, A V Zamkov 2008 Growth, optical and electromechanical properties of single-crystalline orthorhombic bismuth triborate. *J. of Crystal Growth* pp 4027–4030

[5] A S Krylov, A N Vtyurin, A S Oreshonkov, V N Voronov, S N Krylova 2013 Structural transformations in a single-crystal Rb2NaYF6: Raman scattering study *J. Raman Spectr.* **44** pp 763–770

[6] Y V Gerasimova, A S Oreshonkov, A N Vtyurin, A A Ivanenko, L I Isaenko, A A Ershov, E I Pogoreltsev 2013 Infrared Absorption Investigation of the Role of Octahedral Groups upon the Phase Transition in the Rb2KMoO3F3 *J.Crystal. Phys. Solid State* **55** pp 2331–2334

[7] Z Xia, M S Molokeev, A S Oreshonkov, V V Atuchin, Ru-Shi Liu, Cheng Dong 2014 Crystal and local structure refinement in Ca2Al3O6F explored by X-ray diffraction and Raman spectroscopy *J.Phys. Chem. Chem. Phys.* **16** pp 5952–5957

[8] K Nakomoto, Infrared and Raman Spectra of Inorganic and Coordination Compounds 2009 (Wiley, New York etc.)

IOP Conf. Series: Materials Science and Engineering **155** (2016) 012030 doi:10.1088/1757-899X/155/1/012030

Single-layer model of reflective nanostructures for magneto-ellipsometry data analysis

O A Maximova[1,2], N N Kosyrev[3], S N Varnakov[1,3], S A Lyashchenko[1,3] and S G Ovchinnikov[1,2,3]
[1] Reshetnev Siberian State Aerospace University, Krasnoyarsk, 660037, Russia
[2] Siberian Federal University, Krasnoyarsk, 660041, Russia
[3] Kirensky Institute of Physics, Federal Research Center KSC SB RAS, Krasnoyarsk, 660036, Russia

E-mail: maximo.a@mail.ru

Abstract. In this work we present the method of magneto-ellipsometry data analysis. Magneto-ellipsometry measurements are conducted in situ during nanostructures synthesis. Magnetic field is applied in configuration of magneto-optical transverse Kerr effect. Single-layer model of reflective nanostructures is in focus.

1. Introduction

Magneto-ellipsometry is considered as one of powerful reliable nondestructive methods for nanostructures synthesis control that is highly important for spintronics, electronics and nanotechnology. This technique combines ellipsometry and magneto-optical Kerr effect measurements. Magneto-ellipsometry has to be developed and in this work we report on magneto-ellipsometry measurements analysis for the case of single-layer nanostructures study. We have developed the approach that can be applied to investigation of reflective ferromagnetic/non-ferromagnetic nanostructures that are a subject of interest due to observed spin transport phenomena. We offer an algorithm that yields information about dielectric permittivity tensor of ferromagnetic layer [1], where diagonal tensor elements are responsible for refractive index and extinction coefficient, off-diagonal tensor elements are related to magneto-optical effects:

$$[\varepsilon]=\begin{bmatrix}\varepsilon_{11} & \varepsilon_{12} & 0\\ \varepsilon_{21} & \varepsilon_{22} & 0\\ 0 & 0 & \varepsilon_{33}\end{bmatrix}=\begin{bmatrix}\varepsilon'_{11}-i\varepsilon''_{11} & -i(\varepsilon'_{12}-i\varepsilon''_{12})(Q_1-iQ_2) & 0\\ i(\varepsilon'_{12}-i\varepsilon''_{12})(Q_1-iQ_2) & \varepsilon'_{11}-i\varepsilon''_{11} & 0\\ 0 & 0 & \varepsilon'_{11}-i\varepsilon''_{11}\end{bmatrix}, \tag{1}$$

where ε is a complex permittivity of a medium ($\varepsilon_{11}=\varepsilon_{22}\approx\varepsilon_{33}$, $\varepsilon_{12}=-\varepsilon_{21}$), real parts are marked by ', imaginary by ", $Q=Q_1-iQ_2$ is a proportional to magnetization magneto-optical parameter. In the non-magnetic condition ($Q=0$) the off-diagonal tensor elements vanish.

In the following, we describe the method of interpretation of the ellipsometric and magneto-ellipsometric measurements data from the in situ setup of a magneto-optical generalized ellipsometer, which is integrated into an ultra-high vacuum chamber with the electromagnet for magnetization reversal of the sample. The key idea of this approach and the case of the model of a homogeneous semi-infinite medium have been reported in [2] and at the 8th Joint European Magnetic Symposia (JEMS-2016) [3]. Here we repeat some of our basic ideas and present developed expressions for experimental data processing for a single-layer model of reflective nanostructures in order to study their optical and magneto-optical properties. We consider the case of electromagnetic wave incidence from non-magnetic dielectric medium (characterized by the refraction index N_0) onto ferromagnetic metal (the refraction index N_1) on substrate (the refraction index N_2). We set the magnetization vector to be z-axis directed, so that YX plane is a plane of incidence, YZ plane is a boundary plane. The transverse configuration is in focus because of the design features of high-vacuum chamber and electromagnet that are commonly used for magnetization reversal. In this paper, we consider the visible light range, as a great amount of modern ellipsometers work in this range.

2. Ellipsometry and magneto-ellipsometry measurements data

Let us denote the ellipsometric parameters in the non-magnetic condition ($Q=0$) as ψ_0 and Δ_0 [1]. In the case of magneto-ellipsomertic characterization of the sample ($Q=Q_1-iQ_2\neq 0$) the surface transverse magneto-optical Kerr effect results in the ellipsometric angles corrections $\delta\psi$ and $\delta\Delta$. Thus, the ellipsometric parameters become $\psi_0+\delta\psi$, $\Delta_0+\delta\Delta$. It means that four independent real-valued quantities (ψ_0, $\delta\psi$, Δ_0, $\delta\Delta$) are measured and, as a result, four real-valued quantities (ε'_{11}, ε''_{11}, ε'_{12}, ε''_{12}) can be derived.

To start analysis of magneto-ellipsometry experimental data (ψ and Δ) we have to write the real and imaginary parts of complex reflection coefficients in the basic equation of ellipsometry [4, 5]:

$$\rho = tg(\psi_0 + \delta\psi)\exp(i(\Delta_0 + \delta\Delta)) = R_p R_S^{-1} = (R'_p - iR''_p)(R'_S - iR''_S)^{-1}, \tag{2}$$

where ρ is the complex ellipsometric parameter, R_p and R_s are complex reflection coefficients corresponding to in-plane p-polarization and out-of-plane s-polarization respectively, real parts again are marked by ', imaginary by ". According to mode conversion from the p to the s polarized channel we can write that

$$R_p = R_{pp} + R_{ps} = R'_{p0} + R'_{p1} - i(R''_{p0} + R''_{p1}), \tag{3}$$

$$R_S = R_{SS} + R_{sp} = R_{S0} = R'_{S0} - iR''_{S0}, \tag{4}$$

where we have distinguished the magnetic field contribution and marked it by subscript 1, non-magnetic summands – by subscript 0. One can see that transverse Kerr effect yields to $R''_{s1}=0$, $R'_{s1}=0$.

By substituting equations (3-4) into (2) we obtain for non-magnetic condition:

$$tg\psi_0 = \sqrt{\frac{(R'_{p0}R'_{S0} + R''_{S0}R''_{p0})^2 + (R''_{S0}R'_{p0} - R''_{p0}R'_{S0})^2}{R'^2_{S0} + R''^2_{S0}}}, \tag{5}$$

$$\Delta_0 = arctg\frac{R''_{S0}R'_{p0} - R''_{p0}R'_{S0}}{R'_{p0}R'_{S0} + R''_{p0}R''_{S0}}, \tag{6}$$

while the influence of an external magnetic field leads to ellipsometric parameters $\delta\psi$ and $\delta\Delta$:

$$\delta\psi = \psi - \psi_0 = arc\,tg\left(F\,tg(\psi_0)\right) - \psi_0, \tag{7}$$

$$\delta\Delta = \Delta - \Delta_0 = arctg\frac{R''_{S0}(R'_{p0} + R'_{p1}) - (R''_{p0} + R''_{p1})R'_{S0}}{(R'_{p0} + R'_{p1})R'_{S0} + (R''_{p0} + R''_{p1})R''_{S0}} - arctg\frac{R''_{S0}R'_{p0} - R''_{p0}R'_{S0}}{R'_{p0}R'_{S0} + R''_{p0}R''_{S0}}, \tag{8}$$

where F is a helpful notation:

$$tg\left(\psi_0 + \delta\psi\right) = F\,tg\left(\psi_0\right) = tg\left(\psi_0\right)\times$$

$$\times\sqrt{1 + \frac{(R''_{S0}R''_{p1})^2 + (R''_{p1}R'_{S0})^2 + 2R''_{p0}R''_{p1}(R'^2_{S0} + R''^2_{S0})}{(R'_{p0}R'_{S0} + R''_{p0}R''_{S0})^2 + \left(R''_{S0}R'_{p0} - R''_{p0}R'_{S0}\right)^2} + \frac{(R'_{p1}R'_{S0})^2 + (R'_{p1}R''_{S0})^2 + 2R'_{p0}R'_{p1}(R'^2_{S0} + R''^2_{S0})}{(R'_{p0}R'_{S0} + R''_{p0}R''_{S0})^2 + \left(R''_{S0}R'_{p0} - R''_{p0}R'_{S0}\right)^2}}. \tag{9}$$

3. Data analysis

In Figure 1 one can see a diagram of single-layer model, where 0 – ambient medium, 1 – ferromagnetic metal (d – thickness), 2 – substrate, φ_0, φ_1 and φ_2 are the angles of incidence and refraction and related to each other by Snell's law.

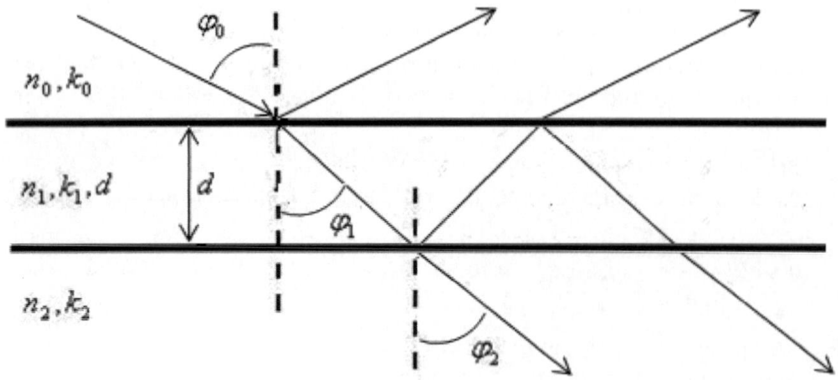

Figure 1. Single-layer model of reflective nanostructures

For a single-layer model complex refractive indices of the material under study ($N_1=n_1$-ik_1) are calculated from ψ_0, Δ_0 measurements by the Nelder-Mead method. The values of $N_0=n_0$-ik_0, $N_1=n_1$-ik_1, $N_2=n_2$-ik_2 are necessary for the following ellipsometric angles calculation.

Analytical expressions for the Fresnel coefficients that take into account the magneto-optical parameter in the off-diagonal permittivity tensor elements were presented in [2, 4]. It was shown that the following expressions should be used for a single-layer model:

$$R_p = r_{01p} + \frac{t_{01p} t_{10p} r_{12p} \exp(-i2\beta)}{1 - r_{10p} r_{12p} \exp(-i2\beta)}, \tag{10}$$

$$R_S = \frac{r_{01S} + r_{12S} \exp(-i2\beta)}{1 + r_{01S} r_{12S} \exp(-i2\beta)}, \tag{11}$$

$$r_{01p} = \frac{N_1 \cos\varphi_0 - N_0 \cos\varphi_1}{N_1 \cos\varphi_0 + N_0 \cos\varphi_1} - i\frac{2QN^2{}_0 \sin\varphi_0 \cos\varphi_0}{(N_1 \cos\varphi_0 + N_0 \cos\varphi_1)^2}, \tag{12}$$

$$r_{12p} = \frac{N_2 \cos\varphi_1 - N_1 \cos\varphi_2}{N_2 \cos\varphi_1 + N_1 \cos\varphi_2} - i\frac{2QN^2{}_1 \sin\varphi_1 \cos\varphi_1}{(N_2 \cos\varphi_1 + N_1 \cos\varphi_2)^2}, \tag{13}$$

$$r_{10p} = \frac{N_0 \cos\varphi_1 - N_1 \cos\varphi_0}{N_0 \cos\varphi_1 + N_1 \cos\varphi_0} + i\frac{2QN^2{}_1 \sin\varphi_1 \cos\varphi_1}{(N_0 \cos\varphi_1 + N_1 \cos\varphi_0)^2}, \tag{14}$$

$$r_{01S} = \frac{N_0 \cos\varphi_0 - N_1 \cos\varphi_1}{N_0 \cos\varphi_0 + N_1 \cos\varphi_1}, \tag{15}$$

$$r_{12S} = \frac{N_1 \cos\varphi_1 - N_2 \cos\varphi_2}{N_1 \cos\varphi_1 + N_2 \cos\varphi_2}, \tag{16}$$

$$t_{01p} = \frac{2N_0 \cos\varphi_0}{N_1 \cos\varphi_0 + N_0 \cos\varphi_1} + i\frac{2QN^3{}_0 \sin\varphi_0 \cos\varphi_0}{N_1(N_1 \cos\varphi_0 + N_0 \cos\varphi_1)^2}, \tag{17}$$

$$t_{10p} = \frac{2N_1 \cos\varphi_1}{N_1 \cos\varphi_0 + N_0 \cos\varphi_1} - i\frac{2QN^3{}_1 \sin\varphi_1 \cos\varphi_1}{N_0(N_1 \cos\varphi_0 + N_0 \cos\varphi_1)^2}, \tag{18}$$

$$\beta_1 = \frac{2\pi}{\lambda} N_1 \cos\varphi_1 d_1, \tag{19}$$

where β_1 is phase thickness of the film. Indices r_{01p}, r_{01s} and r_{12p}, r_{12s} in expressions (12, 15) and (13, 16) are the refractive indices for interfaces 0-1 and 1-2, respectively. Indices t_{01p} and t_{10p} in expressions (17, 18) are transmission coefficients. Indices r_{01p} and t_{01p} correspond to the wave propagation from medium 0 to medium 1, while r_{10p} and t_{10p} – to the backward propagation. Taking into account expressions (3, 4) let us write r_{01p}, r_{12p}, r_{01s}, r_{12s}, r_{10p}, t_{01p}, t_{10p} in the same manner as the refractive indices and transmission coefficients for the model of a homogeneous semi-infinite medium:

$$r_{01S} = (R'_{S0})_{01} - i(R''_{S0})_{01}\,. \tag{20}$$

$$r_{12S} = (R'_{S0})_{12} - i(R''_{S0})_{12}\,. \tag{21}$$

$$r_{01p} = (R'_{p0})_{01} + (R'_{p1})_{01} - i((R''_{p0})_{01} + (R''_{p1})_{01}) = rr_{01} - i\, ri_{01}, \tag{22}$$

$$r_{12p} = (R'_{p0})_{12} + (R'_{p1})_{12} - i((R''_{p0})_{12} + (R''_{p1})_{12}) = rr_{12} - i\, ri_{12}, \tag{23}$$

$$r_{10p} = (R'_{p0})_{10} + (R'_{p1})_{10} - i((R''_{p0})_{10} + (R''_{p1})_{10}) = rr_{10} - i\, ri_{10}, \tag{24}$$

$$t_{01p} = (T'_{p0})_{01} + (T'_{p1})_{01} - i((T''_{p0})_{01} + (T''_{p1})_{01}) = tr_{01} - i\, ti_{01}, \tag{25}$$

$$t_{10p} = (T'_{p0})_{10} + (T'_{p1})_{10} - i((T''_{p0})_{10} + (T''_{p1})_{10}) = tr_{10} - i\, ti_{10}, \tag{26}$$

where $(R'_{s0})_{01}$, $(R''_{s0})_{01}$, $(R'_{p0})_{01}$, $(R''_{p0})_{01}$, $(R'_{p1})_{01}$, $(R''_{p1})_{01}$ are R'_{s0} , R''_{s0}, R'_{p0}, R''_{p0}, R'_{p1}, R''_{p1} in the model of a homogeneous semi-infinite medium, respectively. Subscript 01 denotes the electromagnetic wave incidence from ambient medium 0 onto layer 1. Indices $(R'_{s0})_{12}$, $(R''_{s0})_{12}$, $(R'_{p0})_{12}$, $(R''_{p0})_{12}$, $(R'_{p1})_{12}$, $(R''_{p1})_{12}$ are also calculated by formulae for the model of a homogeneous semi-infinite medium, the only difference is that subscript 12 denotes the electromagnetic wave incidence from layer 1 onto substrate 2 that leads to the following changes: $\cos\varphi_0 \rightarrow \cos\varphi_1$, $\cos\varphi_1 \rightarrow \cos\varphi_2$, $\sin\varphi_0 \rightarrow \sin\varphi_1$, $n_1 \rightarrow n_2$, $n_0 \rightarrow n_1$, $k_1 \rightarrow k_2$, $k_0 \rightarrow k_1$. Likewise, indices $(R'_{p0})_{10}$, $(R''_{p0})_{10}$, $(R'_{p1})_{10}$, $(R''_{p1})_{10}$ describe the electromagnetic wave propagation from layer 1 to medium 0: $\cos\varphi_0 \leftrightarrow \cos\varphi_1$, $\sin\varphi_0 \leftrightarrow \sin\varphi_1$, $n_0 \leftrightarrow n_1$, $k_0 \leftrightarrow k_1$.

Transmission coefficients were not involved into algorithm of data processing for the model of a homogeneous semi-infinite medium. Therefore, we report on $(T'_{p0})_{01}$, $(T''_{p0})_{01}$, $(T'_{p1})_{01}$, $(T''_{p1})_{01}$ here:

$$(T'_{p0})_{01} = 2\frac{(n_0 n_1 + k_0 k_1)(a^2 + c^2) + (n_0^2 + k_0^2)(ab + cd)}{A_3^2 + B_3^2}, \tag{27}$$

$$(T''_{p0})_{01} = 2\frac{(n_0^2 + k_0^2)(ad - bc) + (n_1 k_0 - n_0 k_1)(a^2 + c^2)}{A_3^2 + B_3^2}, \tag{28}$$

$$(T'_{p1})_{01} = 2\frac{Q_1(pq + rs) - Q_2(pr - sq)}{(n_1^2 + k_1^2)(A_3^2 + B_3^2)^2}, \tag{29}$$

$$(T''_{p1})_{01} = 2\frac{Q_1(pr - sq) + Q_2(pq + rs)}{(n_1^2 + k_1^2)(A_3^2 + B_3^2)^2}, \tag{30}$$

where

$$A_3 = n_1 a + k_1 c + n_0 b + k_0 d\,, \tag{31}$$

$$B_3 = k_1 a - n_1 c + k_0 b - n_0 d\,, \tag{32}$$

$$p = N(3n_0^2 k_0 - k_0^3) + P(n_0^3 - 3n_0 k_0^2), \tag{33}$$

$$q = n_1(A_3^2 - B_3^2) - 2A_3 B_3 k_1, \tag{34}$$

$$r = k_1(B_3^2 - A_3^2) - 2A_3 B_3 n_1, \tag{35}$$

$$s = N(n_0^3 - 3n_0k_0^2) - P(3n_0^2k_0 - k_0^3), \tag{36}$$

$$a = \mathrm{Re}(\cos\varphi_0), \tag{37}$$

$$b = \mathrm{Re}(\cos\varphi_1), \tag{38}$$

$$c = \mathrm{Im}(\cos\varphi_0), \tag{39}$$

$$d = \mathrm{Im}(\cos\varphi_1), \tag{40}$$

$$N = \mathrm{Re}(\sin\varphi_0)a - \mathrm{Im}(\sin\varphi_0)\mathrm{c} \tag{41}$$

$$P = -\mathrm{Re}(\sin\varphi_0)c - \mathrm{Im}(\sin\varphi_0)a \tag{42}$$

Coefficients $(T'_{p0})_{10}$, $(T''_{p0})_{10}$, $(T'_{p1})_{10}$, $(T''_{p1})_{10}$ correspond to the electromagnetic wave propagation from layer 1 to medium 0, that leads to the changes: $\cos\varphi_0 \leftrightarrow \cos\varphi_1$, $\sin\varphi_0 \leftrightarrow \sin\varphi_1$, $n_0 \leftrightarrow n_1$, $k_0 \leftrightarrow k_1$.

Let us take into account $N_0=n_0-ik_0$, $N_1=n_1-ik_1$, $N_2=n_2-ik_2$, $Q=Q_1-iQ_2$ and compare expressions (10, 11) with (2, 3). Thus we obtain expressions for R'_{p0}, R''_{p0}, R'_{p1}, R''_{p1}, R'_{s0} and R''_{s0}.

$$R'_{p0} = (((R'_{p0})_{01} + \xi_1(R'_{p0})_{12} - \eta_1(R''_{p0})_{12})(1+\xi_1 L_{0112} - \eta_1 M_{0112}) + \\ +((R''_{p0})_{01} + \eta_1(R'_{p0})_{12} + \xi_1(R''_{p0})_{12})(\xi_1 M_{0112} + \eta_1 L_{0112}))((1+\xi_1 L_{0112} - \eta_1 M_{0112})^2 + (\xi_1 M_{0112} + \eta_1 L_{0112})^2)^{-1}, \tag{43}$$

$$R''_{p0} = (((R''_{p0})_{01} + \eta_1(R'_{p0})_{12} + \xi_1(R''_{p0})_{12})(1+\xi_1 L_{0112} - \eta_1 M_{0112}) - \\ -((R'_{p0})_{01} + \xi_1(R'_{p0})_{12} - \eta_1(R''_{p0})_{12})(\xi_1 M_{0112} + \eta_1 L_{0112}))((1+\xi_1 L_{0112} - \eta_1 M_{0112})^2 + (\xi_1 M_{0112} + \eta_1 L_{0112})^2)^{-1}, \tag{44}$$

$$R'_{p1} = \frac{\Omega\chi - \Gamma\varpi}{\Omega^2 + \Gamma^2} - R'_{p0}, \tag{45}$$

$$R''_{p1} = \frac{\Omega\varpi + \Gamma\chi}{\Omega^2 + \Gamma^2} - R''_{p0}, \tag{46}$$

$$R'_{s0} = (((R'_{s0})_{01} + \xi_1(R'_{s0})_{12} - \eta_1(R''_{s0})_{12})(1+\xi_1 H_{0112} - \eta_1 J_{0112}) + \\ +((R''_{s0})_{01} + \eta_1(R'_{s0})_{12} + \xi_1(R''_{s0})_{12})(\xi_1 J_{0112} + \eta_1 H_{0112}))((1+\xi_1 H_{0112} - \eta_1 J_{0112})^2 + (\xi_1 J_{0112} + \eta_1 H_{0112})^2)^{-1}, \tag{47}$$

$$R''_{s0} = (((R''_{s0})_{01} + \eta_1(R'_{s0})_{12} + \xi_1(R''_{s0})_{12})(1+\xi_1 H_{0112} - \eta_1 J_{0112}) - \\ -((R'_{s0})_{01} + \xi_1(R'_{s0})_{12} - \eta_1(R''_{s0})_{12})(\xi_1 J_{0112} + \eta_1 H_{0112}))((1+\xi_1 H_{0112} - \eta_1 J_{0112})^2 + (\xi_1 J_{0112} + \eta_1 H_{0112})^2)^{-1}, \tag{48}$$

where the following notations are used:

$$L_{0112} = (R'_{p0})_{12}(R'_{p0})_{01} - (R''_{p0})_{12}(R''_{p0})_{01}, \tag{49}$$

$$M_{0112} = (R'_{p0})_{01}(R''_{p0})_{12} + (R'_{p0})_{12}(R''_{p0})_{01}, \tag{50}$$

$$\xi_1 = \mathrm{Re}(\exp(-i2\beta_1)), \tag{51}$$

$$\eta_1 = -\mathrm{Im}(\exp(-i2\beta_1)), \tag{52}$$

$$J_{0112} = (R'_{s0})_{01}(R''_{s0})_{12} + (R'_{s0})_{12}(R''_{s0})_{01}, \tag{53}$$

$$H_{0112} = (R'_{s0})_{01}(R'_{s0})_{12} - (R''_{s0})_{12}(R''_{s0})_{01}, \tag{54}$$

$$\Omega = 1 - \xi_1(rr_{10}rr_{12} - ri_{10}ri_{12}) + \eta_1(ri_{10}rr_{12} + rr_{10}ri_{12}), \tag{55}$$

$$\varpi = ri_{01} - (\xi_1 rr_{12} - \eta_1 ri_{12})(ri_{01}rr_{10} + ri_{10}rr_{01} - \tau) - (\xi_1 ri_{12} + \eta_1 rr_{12})(rr_{01}rr_{10} - ri_{01}ri_{10} - \theta), \tag{56}$$

$$\Gamma = \xi_1(ri_{10}rr_{12} + rr_{10}ri_{12}) + \eta_1(rr_{10}rr_{12} - ri_{10}ri_{12}), \tag{57}$$

$$\chi = rr_{01} - (\xi_1 rr_{12} - \eta_1 ri_{12})(rr_{01}rr_{10} - ri_{01}ri_{10} - \theta) + (\xi_1 ri_{12} + \eta_1 rr_{12})(ri_{01}rr_{10} + ri_{10}rr_{01} - \tau), \tag{58}$$

$$\theta = tr_{01}tr_{10} - ti_{01}ti_{10}, \tag{59}$$

$$\tau = ti_{01}tr_{10} + ti_{10}tr_{01}. \tag{60}$$

Thus, we have all formulae that are necessary for theoretical calculation of the ellipsometric angles (2-9) in case of a single-layer model. Final step is giving the best fit to the experimental data by the use of the wavelength-to-wavelength Nelder–Mead minimization of the ellipsometric angles. It yields the spectral dependences of real (Q_1) and imaginary parts (Q_2) of magneto-optical parameter Q. So, we have information about all elements of the dielectric permittivity tensor.

4. Conclusion

To conclude, we have proposed an approach to studying single-layer nanomaterials by means of magneto-ellipsometry. The algorithm of experimental data analysis (ψ_0, Δ_0, $\psi_0+\delta\psi$, $\Delta_0+\delta\Delta$) is presented. As a result, optical and magneto-optical properties can be easily and reliably characterized during synthesis.

5. Acknowledgements

The work was supported by the Russian Foundation for Basic Research, Grant No. 16-32-00209 mol_a, Grant No. 16-42-243058, Grant No. 14-02-01211. The work was supported partly by The Complex program of SB RAS № II.2P, project 0358-2015-0004, the Ministry of Education and Science of the RF (State task No. 16.663.2014K), grant Scientific School 7559.2016.2.

References

[1] Sokolov A V 1961 *Optical Properties of Metals* (Moscow: GIFML) p.322
[2] Maksimova O A et al. 2014 *Journal Of Structural Chemistry* **55**. No. 6 pp 1134–41
[3] Maximova O, Kosyrev N, Varnakov S et al. *Proc of JEMS-2016* (Glasgow: IOP) p 117-118
[4] Maximova O A, Ovchinnikov S G, Hartmann U et al. 2013 *J. SibSAU Vestnik* (Krasnoyarsk) **49**. No. 3 pp121–127
[5] Azzam R M A and Bashara N M 1977 *Ellipsometry and Polarized Light* (Amsterdam: North-Holland) chapter 4

IOP Conf. Series: Materials Science and Engineering **155** (2016) 012031 doi:10.1088/1757-899X/155/1/012031

Neural-net modeling for direct and inverse problems of shell theory

O M Maximova[1,2], **O A Maximova**[1,2,3]
[1] Reshetnev Siberian State Aerospace University, Krasnoyarsk, 660037, Russia
[2] Siberian Federal University, Krasnoyarsk, 660041, Russia
[3] Kirensky Institute of Physics, Federal Research Center KSC SB RAS, Krasnoyarsk, 660036, Russia

E-mail: maximom_7@mail.ru

Abstract. Effectiveness of the use of neural-net technology for the solving of shell theory problems is shown. Some results of neural-net interpolation and extrapolation for direct and inverse problems are discussed. Exact accuracy of neural-net solving opens wide latitude for shell constructions engineering design and optimization.

1. Introduction

Neural network technologies as a modern universal approximation tool are widely applied in both classic and new areas of science and technology [1, 2]. Their use in the theory of shells, shells designing and research seems relevant and useful.

The joint use of solutions of direct and inverse problems can provide substantial assistance in the process of design engineering, which can be considered as finding the optimal form in the selection of certain parameters and characteristics.

The solution of the inverse problem is usually fraught with difficulties related to the ambiguity of the decision. We have to use different classical and non-classical mathematical methods for solving. For example, the definition of the inverse function or matrix inversion is required for solutions in linear problems without restrictions. For problems with constraints and nonlinear problems it is possible to apply the methods of mathematical programming.

In cases where the problem is poorly formalized or not formalized, we have to look for special innovative ways of solving. They include neural network approaches.

In terms of a neural network, inverse problem is an identification of dependencies between the input and output data, which are interchanged with respect to the direct problem. In this regard, the solution of direct and inverse problems can be carried out by a single algorithm. The neural network solution of the inverse problem will be simplified if we can build such a neural network, for which it would be possible to get a set of ambiguous solutions.

In this case the use of neural network technology, specifically neural network modeling, is convenient and efficient, it allows high-speed seek adoption of rational and optimal solutions.

Note that in this study, we have used standard neural network simulators, that were unfortunately not assumed to be applied to the problems of mechanics while being developed. It seems promising to develop specialized neural simulators, as well as the individual neural networks, which take into account the properties of the structure and, apparently, can significantly improve the results. However, it should be noted that this approach can restrict the class of problems.

As a neural simulator for most of the problems we used the program "Modely" written on the basis of interpolation of the solutions using Fourier series and minimization of the functional of squared deflection error using the Lagrange multipliers [3]. We consider the examples of solving direct and inverse problems [4] using neural networks.

2. Thin smooth shallow spherical shell on a square plan

Area of moments is shown for a smooth spherical shell under the load uniformly distributed over the band of width λ along the Y axis. It shows a change of the bending moment in the center of band-pass load, depending on the load bandwidths (Figure 1).

Material for training the neural network: the value of the bending moment M_X in the local center of the site, depending on the width of the band-pass load λ.

M_X, kg·cm	11.5	10.3	10	9.4	8	6.2	4.2
λ, mm	16.67	27	33.3 3	37	50	66.67	100

Direct problem: to investigate the influence of bandwidth λ of uniformly distributed load on the value of the bending moment.

The input parameters of the network – the bandwidth of load λ.

The output network parameters – values of the moment M_X.

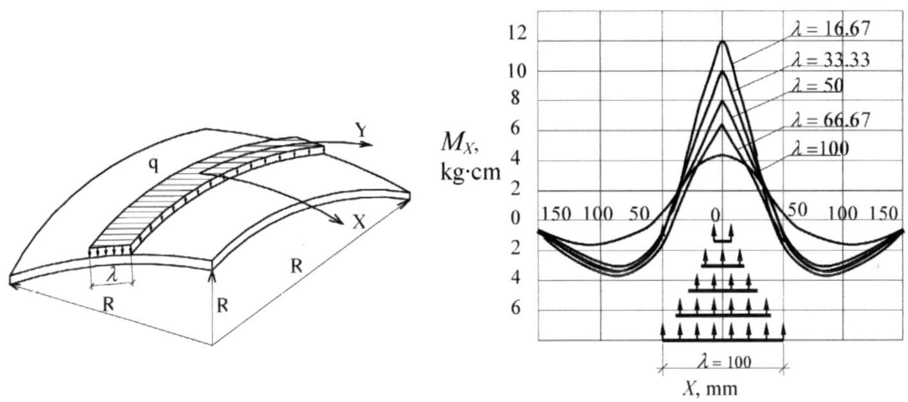

Figure 1. The dependence of the bending moment in the center of band-pass load, depending on the load bandwidths for a spherical shell

Features of the neural network solving the problem. In the case where M_X is determined for λ, located within the training pattern interval (interpolation problem – training pattern: moments for λ = 16.67; 33.33; 50; 66.67; 100; tests: λ = 27; 37), the neural network training takes instantaneously. In the training pattern, consisting of just five examples with the minimum number of neurons, equal to 2, and unit spectral density, training error is 1.4%. If λ extends beyond training pattern interval, for example to decrease the bandwidth load (extrapolation problem – training pattern: moments for λ = 33.33; 37; 50; 66.67; 100, tests: λ = 16.67; 27), to achieve a similar accuracy of the solution the number of neutrons should be increased to 9 (with unchanged other network parameters). Adding just one neuron reduces training error to 0.5%. The process of training a neural network in the case of interpolation and in the case of extrapolation takes seconds.

This problem is neural forecasting problem of design type. With the help of the graph of dependence of the moment M_X of the load bandwidth λ, it is possible to determine the maximum possible bending moment with the minimum allowable load bandwidth (Figure 2).

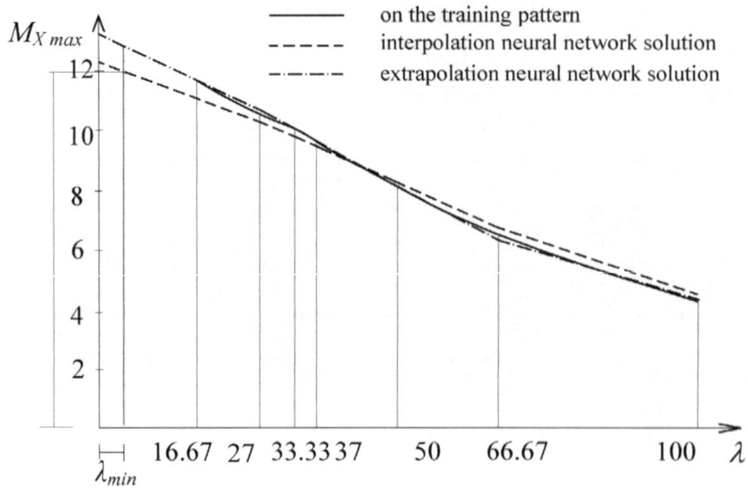

Figure 2. Changing the maximum bending moment in the shell depending on the width of the band-pass load

Inverse problem: to find λ (width of band-pass load), corresponding to the desired moment (M_X) in the center of the local area.

This problem can be formulated differently: "what should the bandwidth of the current distributed load be to produce a bending moment that does not exceed a predetermined value?" In this case, we are dealing with the search for optimal solutions.

The input parameters of the network – values of the moment M_X.

The output network parameters – the bandwidth of load λ.

Similarly to the direct problem, we solved the problem of interpolation and extrapolation type. Interpolation problem was converted to determining load bandwidth λ for a given value of the bending moment M_X within the training pattern interval (training pattern $M_X = 4.2$; 6.2; 8; 10; 11.5; tests $M_X = 9.4$; 10.4). The training pattern consisted of 5 problems. Compared with the direct problem the neural network training process for the inverse problem (due to the complexity of the reversal function) is more time-consuming and complicated.

Features of the neural network solving the problem. The best results that have been achieved are within 4% error and obtained by means of a neural network with 15 neurons and spectral density equal to 2.6. The process of network training, the selection of its architecture has a singularity. In the process of training the spectral density should be increased gradually, starting with 1.0 in increments of 0.1. Otherwise, training error increases greatly to 15% or higher. Changing the size of the network (number of neurons) and spectrum as downwards and upwards does not give positive results (Fig. 3).

If given moment M_X, for which you want to determine the appropriate load bandwidth λ, is outside the training pattern interval (extrapolation problem – training pattern: $M_X = 4.2$; 6.2; 8; 9.4; 10; tests: $M_X = 10.3$; 11.5), the network training is very long and painstaking. It requires the use of step method neural network forecasting [3, 5]. At each step of training the magnitude of the spectral density must be increased from 1, not more than 0.01 to 1.3. In this case, two-neuron network allows to achieve error on the test cases within 6%.

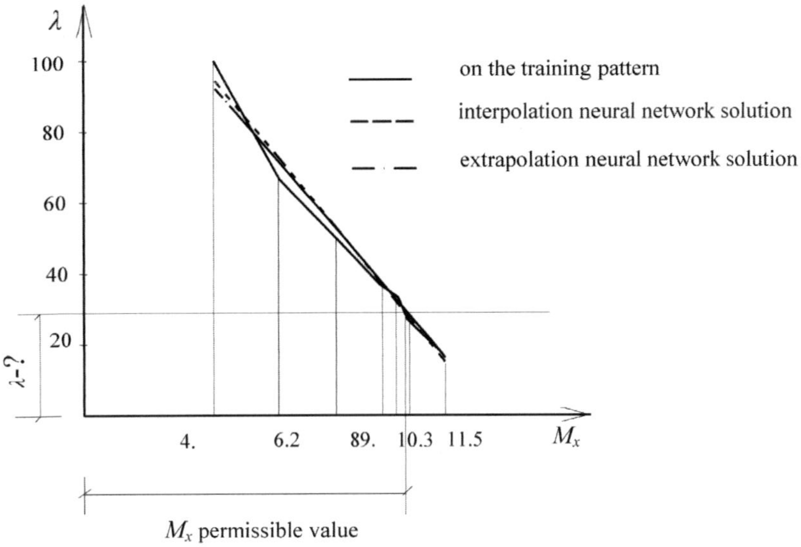

Figure 3. Dependence of λ on M_X for determining the lowest possible band width of the load at which the bending moment does not exceed a predetermined value

3. The thin shallow spherical shell on a square plan, backed by edges on a contour

In this example, the load is distributed evenly over the entire surface of the shell.

Material for training the neural network: values of deflections w/q and bending moments M_X/q in grid nodes on the coordinate axes for some pliability parameter values of contour arches ν (Figures 4, 5).

Direct problem: to determine the impact of pliability parameter $\nu=h/t$ of contour arch (h – section height of contour arch, t – shells thickness) on the deflections and bending moments.

The input parameters of the network – the value of pliability parameter ν.

The output network parameters – values of deflections w/q and values of moments M_X/q in shell grid nodes.

If deflections and moments are determined for pliability parameter ν, without departing from the training pattern interval (interpolation problem), learning is fast, does not cause difficulties.

Features of the neural network solving the problem. In determining the deflections (training pattern: data for $\nu = 2, 3, 4, 6, 8$; test: $\nu = 5$) he best result of training have been achieved for 6- neuron network with the spectral density 1.2. Errors of deflections calculating δ are in the interval $0.05\ \% \le \delta \le 5.72\ \%$. The maximum error is noted in the contour arch. In determining the moments (training pattern: data for $\nu = 2, 3, 4, 6, 9$, test: $\nu = 5$) the most effective is an 8-neuron network with the spectral density 0.8. Here network training error does not exceed 2%. The maximum errors are in the area of moments values close to zero.

When deflections and moments are determined for pliability parameter extending beyond the training pattern interval (extrapolation problem – training pattern: data for $\nu=2, 3, 4, 5, 6$, test: $\nu=8$) the problem is complicated. The number of examples in the training pattern has a significant impact on the results of the network functioning.

For instance, for the training pattern of five examples (4- neuron network with the spectral density 2.5) the maximum error in the tests for deflections is 5.5 %. Reducing the pattern by a single example leads to an increase in errors to 8%.

The accuracy of neural network solutions can be significantly improved if instead of one-step (traditional) neural network extrapolation the step method of neural network forecasting is used [3, 5].

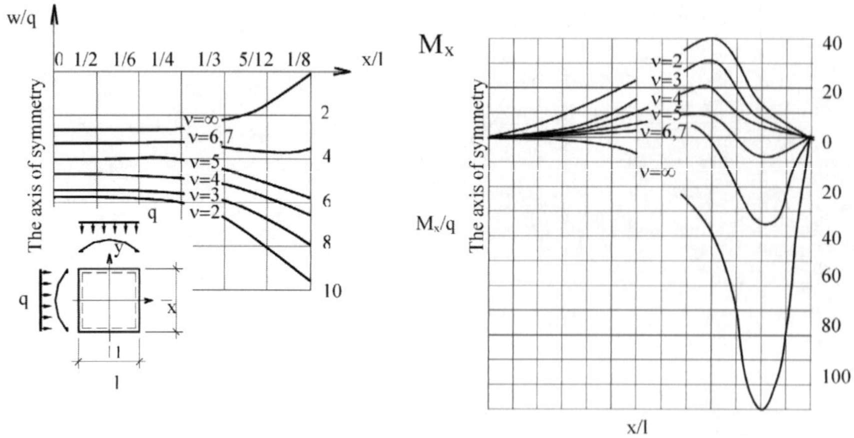

Figure 4. Epures of deflections and bending moments in the cross section *y=0* (ν*=h/t*, where h – section height of contour arch, t – shells thickness)

Figures 4 and 5 show that with increasing pliability, quality change (even in sign) of moments epures near the contour and significant growth of deflections throughout the area occurs. The value $\nu = 5$ can be taken as conventional limit, below which a reinforcement shell contour can be called a very pliable. It may be recommended for design purposes to take $\nu \geq 5$.

***Inverse problem**:* what should be the parameter ν (pliability of contour rib), to the deflections w/q and moments M_X/q do not exceed the specified values?

The input parameters of the network – values of deflections w/q and the values of the bending moments M_X/q at the nodes of the shell mesh.

The output network parameters – pliability ribs parameter ν.

Feature of the neural network solving the problem is the fact, that the inverse problem is solved much more accurate than direct both for the interpolation problem when the network functioning is carried out within the training pattern interval, and for extrapolation problem, when network functioning is carried out outside the training pattern interval.

For example, in the problem of extrapolation (training pattern: five variants of deflection for $\nu = 2, 3, 4, 5, 6$, test: deflection for $\nu = 8$) with the network of the two neurons with spectral density 0.8 deflection values are determined with an error not exceeding 1.7%.

To determine the moments (training pattern: five variants of moments for $\nu = 2, 3, 4, 5, 6$, test: moments for $\nu = 9$) the most effective is the network of three neurons with spectral density 0.8. Maximum network training error on the test examples is 1.95%.

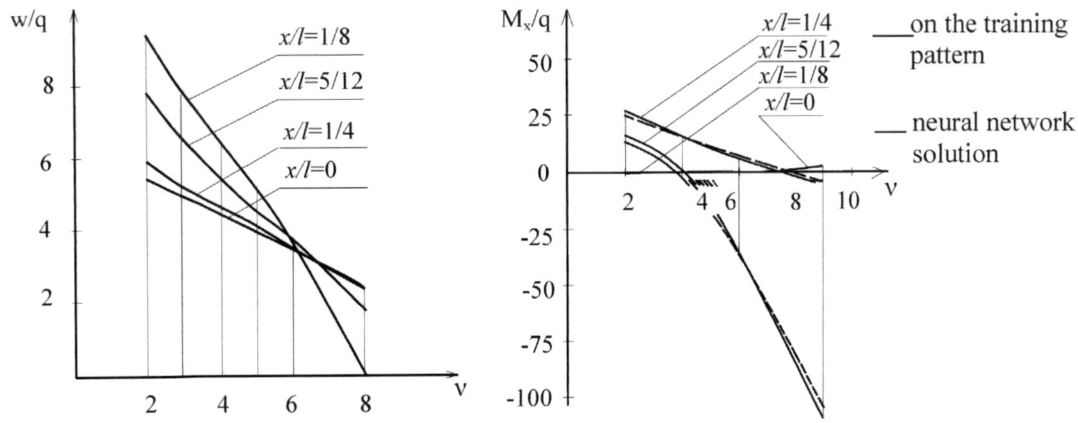

Figure 5. Dependence of deflections functions and moments on pliability of contour ribs $\nu = h/t$ for the various points of the shell region

In both cases, the training pattern consists of only 5 examples. But even with such a small number of training examples the neural network gives a fairly accurate results. The error of the neural network solution does not exceed 2%. However, reducing the pattern by even one example in the problem of deflections forecast causes a significant increase in error.

It confirms the need for reasonably select training examples, their number, as well as network architecture and its characteristics for each specific problem in view of its features.

In this problem, there is no need to use step method of neural forecasting as "traditional" (one-step) neural forecasting provides acceptable accuracy of the solution. However, the use of this method could significantly improve the results

This problem can be considered as an optimization problem. Plotting of dependence of pliability parameter ν on the value of the deflections and moments and their analysis can help in the selection of such pliability parameter of contour ribs, which would provide the desired (required) shell behavior, its stress-strain state (SSS).

4. Conclusion

Thus, the results of solving the problems under consideration show, that the use of neural network models allows to solve with sufficient accuracy (within 5% error) as a direct problem of calculating shells (to receive SSS parameters according to known geometric characteristics of shell and its load) and inverse problems (to select the design dimensions according to known moments and deflections). This opens up opportunities for the design of shell structures and their optimization. Note that the neural network solution of the inverse problem will be simplified, if we can build such neural network model, by which it would be possible to get a set of ambiguous solutions.

The examples discussed herein are illustrative a greater extent for the traditional method of using neural network technology (interpolation, extrapolation). It should be noted that the development and application of stepwise technology to these problems, set out in [3, 5], using the completion of the training pattern by interpolation, additional training the neural network model and the step process, can significantly improve the results.

Neural network technologies have much more applications, and can include not only the static problem, but also the dynamics, to take into account the experimental data in order to bring the calculated results to the actual design.

References

[1] Sidorov M Yu, Zablotskiy S G, Semenkin E S and Minker W 2012 Evolutionary design of networks for forecasting of financial time series *Vestnik SibGAU* (Krasnoyarsk) **4 (44)** pp 106-110

[2] Popov E A and Semenkina M E 2013 Self-configuring genetic programming algorithm for medical diagnostic problems *Vestnik SibGAU* (Krasnoyarsk) **4 (50)** pp 116-123

[3] Abovskiy N P, Deruga A P, Maximova O M and Svetashkov P A 2003 *Neuro-Controlling Structures and Systems (Scientist series "Neuro-computers and their application", book 13)* ed A I Galushkin (Moskva: Radiotechnica) p 368

[4] Endzhievskiy L V 1982 *Nonlinear Deformation of Ribbed Shells* (Krasnoyarsk: Krasnoyarsk State University) p 296

[5] Abovskiy N P and Maximova O M 2007 Neuro-prognosis based on step model with teaching for natural tests results of building structures *J.Optical Memory & Neural Networks (Information Optics), SPRINGER* **16** chapter 1 pp 40-46

IOP Conf. Series: Materials Science and Engineering **155** (2016) 012032 doi:10.1088/1757-899X/155/1/012032

Processing information system for highly specialized information in corporate networks

M O Petrosyan, I V Kovalev, P V Zelenkov, VV Brezitskaya, G A Prohorovich
Siberian State Aerospace University named after Academician MF Reshetnev
31 “KrasnoyarskiyRabochiy” prospect, Krasnoyarsk, 660014, Russia.

E-mail: mopetrosyan@gmail.com

Abstract. The new structure for formation system and management system for highly specialized information in corporate systems is offered. The main distinguishing feature of this structure is that it involves the processing of multilingual information in a single user request.

Currently, there is an active development of information technologies. One of the most interesting questions while using information technologies is question about how to collect, process and control information. [1] Computer information retrieval systems hold all the more important place in science and education, especially the Internet, which is an extensive reference tool. Development of the Internet technology ensures that the daily increasing number of information resources made available for public use, increasing the volume of thematic and focused information on the various subject areas.

The expansion of the global network and connectivity to it resulted in a significant increase in the number of its users. The increase of users in recent years, are most pronounced in Russia. Since 2006, the number of Internet users has increased by 10 times. A large part of the Russian-speaking segment of the Internet users in information gathering use existing search services of general purpose. By data for June 2015, most popular information retrieval services are: Yandex, Google, Mail, Rambler. (fig.1)

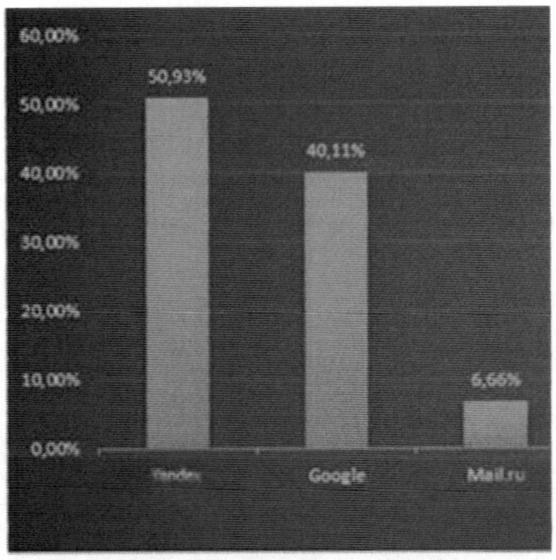

Figure 1. The share of search queries in the Russian-language segment of the Internet

However, it should be noted that these services allow good results only when working with general topics, and there are difficulties in carrying out a highly specialized retrieval.

Moreover, the problem of these systems lay in multilingual presentation of information in the Internet. [1,] Search general-purpose services operate only in the language set on which an a request has been set, but when looking for a highly specialized personalized information you can immediately arrange of multilingual search procedure. [1, 2]

To solve the problems mentioned above could be used existing technologies and approaches, but with emphasis on the processing of multilingual thematic information. In this regard, technology of implementation of information management systems based on multi-agent approach is well proven. Creating a search systems based on agent-based principle allows modification of individual agents, without significant impact on system performance as a whole. Search multi-agent systems are a kind of metasearch systems. metasearch system architecture provides a single point of access to multiple search engines. In other words serves user requests by polling the other user systems, which are completely independent and do not provide any specific information about the contents of their indices or used search methods. This leads to the fact that the construction of metasearch systems have to solve a series of new problems. For example, query languages, often differ greatly and therefore a simplified search language for metasearch system or reformulation of requests for a particular search system is required.

Another problem is the fusion of the responses from the various search systems. Authors propose the following structure of the organization of interaction between agents of multi-agent corporate systems. (fig.2)

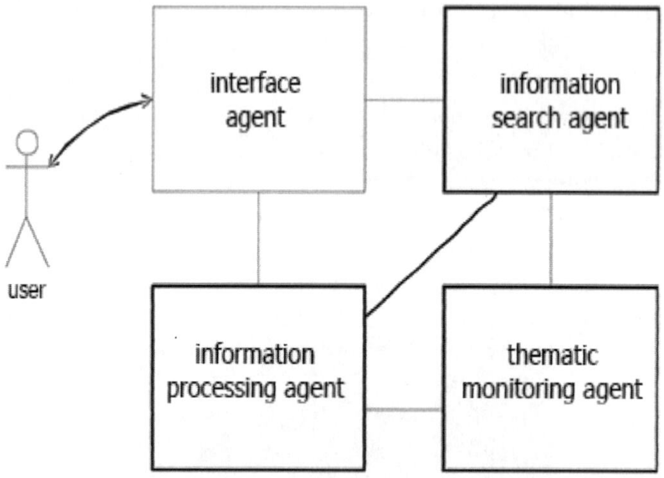

Figure 2. Scheme of multi-agent system

This system consists of four logically related software modules (agents), the purpose and structure of each of which will be shown below. It is connected with two other agents - search agent and processing agent. This agent is simple in structure and performance and performs all operations while interacting with user. It receives a search query, transmits a request to search agents, outputs the search results to a user, and performs a search process of adaptation to the used search systems. Each search agent interacts with a particular search system. (fig.3) This agent requires a more detailed description, as its implementation is proposed to make in metasearch multilingual performance.

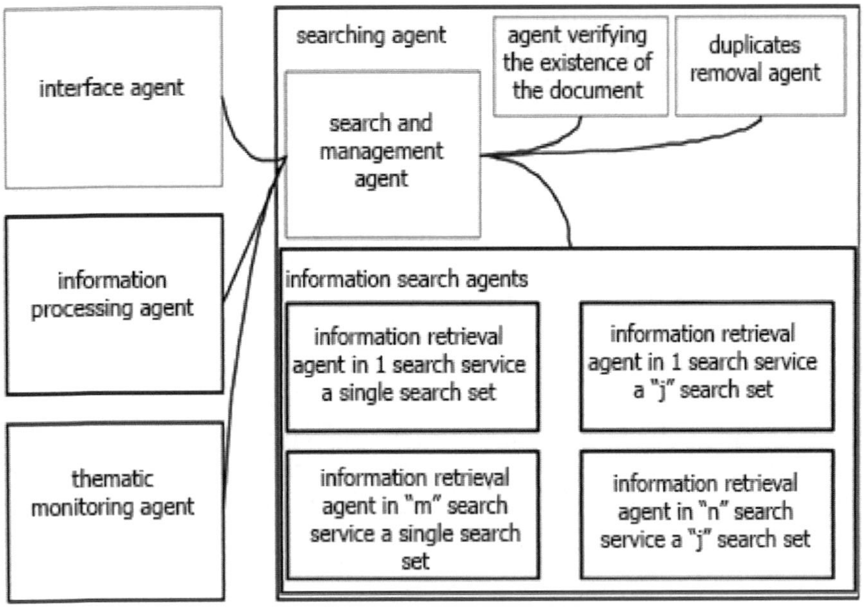

Figure 3. Scheme of the search agent

Search agent is primary. Its main task - is to process user search line that the agent receives from the interface agent. Once the line has been processed, you must initialize the process of multilingual metasearch in the corporate network and the Internet. Next, a process of verification of the existence of documents and removal of duplicate documents is performed. Then, all received data sample is transferred to the information processing agent. (fig.4) This agent is responsible for the information management of the thematic collection obtained in the search stage (from the point of view of the user of corporate system).

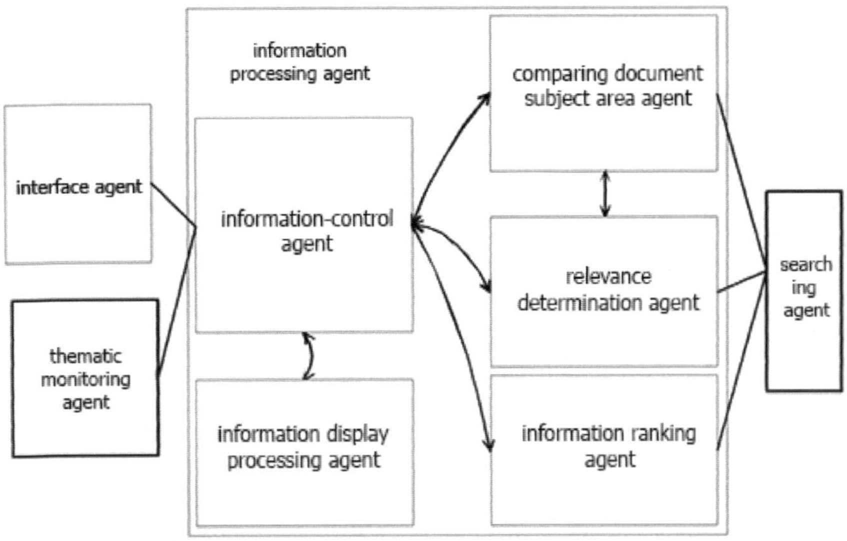

Figure 4. Scheme of information processing agent

Information processing agent consists of:

- information-control agent (this is functionally the main agent of the procedure);
- two agents, which are rigidly connected to each other: relevance determination agent and comparing document subject area agent;
- information ranking agent;
- information display processing agent.

Let us consider in more detail relevance determination agent and comparing document subject area agent. The first agent holds definition of relevance of the documents from the proposed sample. Using algorithms to determine relevance, it can be shown that some documents are more relevant to the request, and some - less. Thus, the problem of processing conditionally relevant documents arise, in other words documents from related subject areas. Therefore, when searching is necessary to determine the possibility of falling into the resulting data sample from such areas. And here we must take into account system user preferences and solve the problem of the inclusion of documents from related subject areas in the result set (or exclusion from it). That is the problem and solves relevance determination agent and comparing document subject area agent. In addition, in the chosen text of the selected subject area relevant can be not the entire document, but only a part, such as separate sections of the general purpose books, selected articles from collections of articles, sections of reports, and organizations and so on. Taking account of this limitation will help in deciding whether to grant the user only the necessary pieces of information.

The next agent - a ranking agent. It is no less important in the processing of information, since when issuing to the user several thousands of documents in the first place on the display list must be the most valuable documents. Thematic monitoring agent is responsible for the analysis of corporate system user information preferences within the information and thematic collections and providing him personalized navigation support and personalized data. We reduce time needed to find the right information, and user traffic in both corporate and in the external network to see only high-quality information by providing users with information collections of personalized navigation menu links to pages that are close to their thematic preferences. Thus, the proposed solutions should enhance the user experience with the corporate system of information resources and serve as an additional incentive for them to more frequent visit these resources. This approach would also greatly reduce the burden on both the inner (corporate) and the external traffic.

References

[1] Karaseva M V, Bachurina E P, Zelenkov P V, Brezitskaya V V 2015 Information search module based on multilinguistic thesauruses *IOP Conference Series: Materials Science and Engineering* Volume **70** Issue 1 Article number 012011

[2] Kovalev I V, Polyanskiy K V, Zelenkov P V, Brezitskaya V V, Sidorova GA 2013 System of search, analysis and processing of multilinguistic texts, integrated with data retrieval systems *Vestnik SibGAU* Volume: **1 (47)** P. 48-52.

Methodological approach for identification the function of the object for automatic regulation system of the continuous technological processes

Kovalev I V[1], Zelenkov P V[2], Losev V V[3], Kovalev D I[4], Perantseva A V[5], Burdina E V[6]

[1]Dr. Sc., Professor of System analysis department, Rector, Siberian State Aerospace University named after Academician M. F. Reshetnev 31 "Krasnoyarskiy Rabochiy" prospect., Krasnoyarsk, 660014, Russia.

[2]Cand. Sc., Vice-Rector for IKNR, Siberian State Aerospace University named after Academician M. F. Reshetnev 31 "Krasnoyarskiy Rabochiy" prospect., Krasnoyarsk, 660014, Russia.

[3]Cand. Sc., Deputy Vice-Rector for Informatization, Siberian State Aerospace University named after Academician M. F. Reshetnev 31 "Krasnoyarskiy Rabochiy" prospect., Krasnoyarsk, 660014, Russia.

[4] Graduate student, Siberian State Aerospace University named after Academician M. F. Reshetnev 31 "Krasnoyarskiy Rabochiy" prospect., Krasnoyarsk, 660014, Russia.

[5]Scientific worker, Siberian State Aerospace University named after Academician M. F. Reshetnev 31 "Krasnoyarskiy Rabochiy" prospect., Krasnoyarsk, 660014, Russia.

[6] Engineer, Siberian State Aerospace University named after Academician M. F. Reshetnev 31 "Krasnoyarskiy Rabochiy" prospect., Krasnoyarsk, 660014, Russia.

E-mail: basilos@mail.com

Abstract. The article describes one of the approaches to the implementation of the functional approximation problem of nonlinear characteristics of the object of automatic regulation system based on the identification the function of the object, the analysis of input and output characteristics of the object, and numerical characteristics of random values.

In conditions of the continuous technological processes of the complex dynamic systems, modes which involve a constant change of adjustable parameters it is not possible to obtain the experimental static characteristic of the automatic regulation system volume.

Though the task to raise the regulation quality, to improve transient characteristics including the regulation decrease and totally the reduce of the excessive energy consumption for the own needs requires to build a dynamic model able to show quite fully the technological control object [1]. The presence of a priori information, in this case, the input and output of nonlinear characteristics of the object allows accelerating the process of object identifying, more precisely, obtaining its abstraction expressed with transfer function with a high degree of reliability. Turning to the application software, in particular, System Identification Toolbox Matlab package, we analyze the available a priori information for the object (Fig. 1) performing surface heat transfer function.

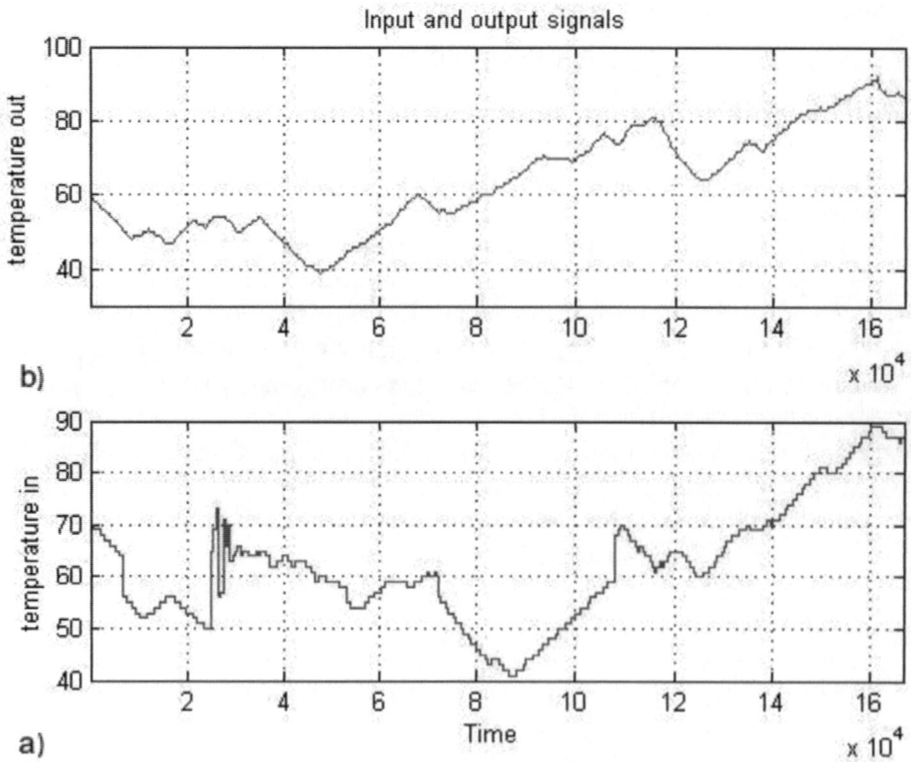

Figure 1. Nonlinear object characteristics a) input b) output

A priori information is presented as the input object characteristic - change of the thermal agent temperature of the supply circuit (C^0) over time, and the output characteristic - change of the thermal agent temperature of the output circuit (C^0) over time.

Time series analysis is based on the alternation of transfer functions with a variable number of poles and subsequent finding of the optimal conversion coefficient values Kp and the time constant Tpi by Gauss-Newton method, iterative numerical method of finding a solution to the problem of least squares, by means of System Identification Toolbox, in which the posteriori output characteristic will be brought most closer to a priori output characteristic by means of functional approximation method [2, 3].

Time series analysis of the full length at transfer of the alternation functions with a variable number of poles has shown that with number increase of transfer function poles the degree of reliability curves - F increases, though the achieved maximum is F = 38,05% (Table 1.).

Table 1. Results of the analysis of time series of full length (N = 1)

Segment number, n	Quantity of Poles, i	Range of authenticity F, %	View of Transfer Function	Conversion Factor (Kp), Time Constant (Tpi)
1000	0	0,8598	$G(s) = Kp$	Kp=0.86003

	1	3,436	$G(s)=\dfrac{Kp}{1+Tp_1*s}$	Kp=9.7581e5 Tp_1=3.8838e12
	2	-0,7378	$G(s)=\dfrac{Kp}{(1+Tp_1*s)(1+Tp_2*s)}$	Kp=1.1301 Tp_1=9.9231 Tp_2=0.002
	3	37,05	$G(s)=\dfrac{Kp}{(1+Tp_1*s)(1+Tp_2*s)(1+Tp_3*s)}$	Kp=-1.7007e6 Tp_1=5.176e11 Tp_2=3.3712e5 Tp_3=0.0072701

Initial knowledge about the object can judge that the surface heat transfer process is very inertial in nature; therefore, the transition process from one stable state to another, within the static characteristic, is also a long time [4]. Thus, the task of enhancing the credibility of curves is based on the allocation of the static characteristic from the time series of full length.

Not taking into account the structural features of the object, the outdoor temperature and the room temperature, we carry out site investigation of time series with a variable number of sampling points by separation and subsequent analysis of the original time series of full length. This approach will also allows to analyze the numerical characteristics of the obtained random variables, provided a sufficient number of time series sections [5].

The first step is to separate three sections (N = 3) from the starting time series of full length. We introduce the concept of conditional probability of Pi - the weight of a particular structure of the transfer function in the task of curves approximation - N sections of time series, where i - the number of poles of the transfer function. The obtained results (Table 2), including the conditional probability P_1 = 0,68 and P_3 = 0,32, do not allow to make an unambiguous conclusion about the functional relationship between parameters such as of N, Pi, and F, and also of the character pattern between Pi and F at a variable N. We note that at N = 3, the transfer function of the lower weight (P_3) corresponds to a high degree of curves reliability F = 64,35 %. For answers to these questions we continue the analysis of time series of full length sections at N = 10, 20, 40.

Table 2. Results of the time series analysis with separation of three sections (N = 3)

Segment number, n	Quantity of Poles, i	Range of authenticity F, %	View of Transfer Function	Conversion Factor (Kp), Time Constant (Tpi)
1	0	16,08	$G(s)=Kp$	Kp=1.0157
	1	54,05	$G(s)=\dfrac{Kp}{1+Tp_1*s}$	Kp=2.9334e9 Tp1=6.7235e 14
	2	65,23	$G(s)=\dfrac{Kp}{(1+Tp_1*s)(1+Tp_2*s)}$	Kp= 1.4082e7 Tp1= 2.3103e12 Tp2= 21035

	3	66,03	$G(s)=\dfrac{Kp}{(1+Tp_1*s)(1+Tp_2*s)(1+Tp_3*s)}$	Kp=1.7031 Tp1= 417.13 Tp2= 49031 Tp3= 49115
	0	-16,5	$G(s)=Kp$	Kp= 0.86751
	1	0,2535	$G(s)=\dfrac{Kp}{1+Tp_1*s}$	Kp=3772.3 Tp1=3.3409e10
2	2	-15,85	$G(s)=\dfrac{Kp}{(1+Tp_1*s)(1+Tp_2*s)}$	Kp=0.86755 Tp1=134.52 Tp2= 134.64
	3	-15,45	$G(s)=\dfrac{Kp}{(1+Tp_1*s)(1+Tp_2*s)(1+Tp_3*s)}$	Kp=0.86442 Tp1=0.001 Tp2=730.04 Tp3=0.001
	0	12,05	$G(s)=Kp$	Kp= 0.79075
	1	40,08	$G(s)=\dfrac{Kp}{1+Tp_1*s}$	Kp= 0.76735 Tp1= 13701
3	2	40,14	$G(s)=\dfrac{Kp}{(1+Tp_1*s)(1+Tp_2*s)}$	Kp= 0.7757 Tp1= 0.001 Tp2= 8890.7
	3	30,03	$G(s)=\dfrac{Kp}{(1+Tp_1*s)(1+Tp_2*s)(1+Tp_3*s)}$	Kp=0.72007 Tp1= 1595 Tp2=39384 Tp3= 39375

As the number of sections (N = 10, 20, 40) increases at separation of full length initial time series and their subsequent analysis the following results were obtained - Table 3, 4, 5, as well as numerical characteristic of F reliability degree as a discrete random variable.

Table 3. Results of the time series analysis with separation of 10 sections (N = 10), the number of sampling points - 100

Quantity of Poles, i	Conditional Probability, Pi	Range of authenticity F, % (average)	Mathematical Expectation $M_N(F)$, %	View of Transfer Function
0	0,12	-5,891		$G(s)=Kp$
1	0	0		$G(s)=\dfrac{Kp}{1+Tp_1*s}$
2	0,18	30,01	38,75	$G(s)=\dfrac{Kp}{(1+Tp_1*s)(1+Tp_2*s)}$
3	0,7	44,63		$G(s)=\dfrac{Kp}{(1+Tp_1*s)(1+Tp_2*s)(1+Tp_3*s)}$

Table 4. Results of the time series analysis with separation of 20 sections (N = 20), the number of sampling points - 50

Quantity of Poles, i	Conditional Probability, Pi	Range of authenticity F, % (average)	Mathematical Expectation $M_N(F)$, %	View of Transfer Function
0	0	0	61,45	$G(s) = Kp$
1	0,13	69,37		$G(s) = \frac{Kp}{1 + Tp_1 * s}$
2	0,37	53,77		$G(s) = \frac{Kp}{(1 + Tp_1 * s)(1 + Tp_2 * s)}$
3	0,5	65,1		$G(s) = \frac{Kp}{(1 + Tp_1 * s)(1 + Tp_2 * s)(1 + Tp_3 * s)}$

Table 5. Results of the time series analysis with separation of 40 sections (N = 40), the number of sampling points - 25

Quantity of Poles, i	Conditional Probability, Pi	Range of authenticity F, % (average)	Mathematical Expectation $M_N(F)$, %	View of Transfer Function
0	0	0	75,76	$G(s) = Kp$
1	0,06	68,15		$G(s) = \frac{Kp}{1 + Tp_1 * s}$
2	0,22	77,97		$G(s) = \frac{Kp}{(1 + Tp_1 * s)(1 + Tp_2 * s)}$
3	0,72	73,54		$G(s) = \frac{Kp}{(1 + Tp_1 * s)(1 + Tp_2 * s)(1 + Tp_3 * s)}$

These results allow observing the effect of the following pattern: in the problem of curves sections approximation prevails the weight of transfer function with a large number of poles and as the number of sections (N = 10, 20, 40) increases the curves reliability rises. The character of this pattern in this study is not specified, though the obtained characteristics can be shown in the following functional relationship:

$$M_N(F) = \sum_{j=1}^{i} F_i \cdot P_i, \tag{1}$$

Expression (1) is mathematical expectation of F reliability degree in solving the problem of N sections approximation by means of transfer functions with different numbers of i poles and Pi conditional probabilities.

An interesting result of the time series analysis is the transfer function (2) with number of poles i = 3, reliability degree F = 94,65 % and mathematical expectation $M_N(F)$=75,76 %:

$$G(s) = \frac{133{,}45}{(1+5045{,}3*s)(1+2789{,}3*s)(1+1.7955\cdot e\cdot 7*s)} \tag{2}$$

The obtained expression (2) with a maximum degree of reliability determines the surface heat transfer function in one of N = 40 sections (n = 32) and does not allow to describe other 39 sections with the similar reliability, however, the conditional probability for this type of transfer function - P_3 is 0.725 that says of the maximum weight of this type transfer function in the solution of the problem of N sections curves approximation.

Thus, it is possible to increase the curves reliability degree by following steps:
- The separation of time series sections with a variable number of sampling points;
- The analysis of time series sections, alternating the transfer functions with a variable number of i poles and subsequent finding of the optimal values of Kp transformation coefficients and Tpi time constant by Gauss-Newton;
- The acquisition and subsequent analysis of the numerical characteristics of F reliability degree.

It is acceptable to assume that the final result of this approach is the value of mathematical expectation of reliability degree - $M_N(F)$, which physical meaning, in this study, can be interpreted as a display indicator of the static characteristics from the full length time series, expressed in percentage, in solving the problem of curves approximation by alternating of four transfer functions with a variable N.

The results obtained in this study, namely methodological aspects of ASR object identification in conditions of continuous TP, may be applied as one of the approaches for assessing the reliability degree of the transfer function in solving the problem of curves approximation [6, 7].

References

[1] Kovalev I V, Zelenkov P V, Ognerubov S, Bahmareva K, Denisova E 2015 The efficiency analysis of the automated plants *IOP Conf. Series: Materials Science and Engineering* vol. **70** pp. 012007.
[2] Engel E A, Kovalev I V 2011 Information processing using intelligent algorithms by solving WCCI 2010 tasks *Vestnik SibGAU* vol. **3**, no. 36 pp. 4–8.
[3] Raspopin N A, Karaseva M V, Zelenkov P V, Kayukov E V, Kovalev I V 2012 Models and methods of optimization of information acquisition and processing *Vestnik SibGAU* vol. **2**, no. 42 pp. 69–72.
[4] Kuznetsov P A, Kovalev D I, Losev V V, Kalinin A O 2015 To the question about the states of workability for automatic control systems with complicated structure *Vestnik SibGAU* vol. **16**, no. 4 pp. 941–945.
[5] Kuznetsov P A, Kovalev I V, Losev V V, Kalinin A O, Murygin A V 2016 To the question about the states of workability for automatic control systems with complicated structure *IOP Conf. Series: Materials Science and Engineering* vol. 122.
[6] Kovalev I V, Zelenkov P V, Tsarev M Y 2015 The control of developing a structure of a catastrophe-resistant system of information processing and control *IOP Conf. Series: Materials Science and Engineering* vol. **70**. pp. 012008.

[7] Kovalev I V, Kuznetsov P A, Zelenkov P V, Shaydurov V V, Bahmareva K K 2013 Assessing the reliability of ACS with a blocking protection module 2013 *Pribory* no. **6** pp. 20–23.

IOP Conf. Series: Materials Science and Engineering **155** (2016) 012034 doi:10.1088/1757-899X/155/1/012034

Testing of hypothesis of two-dimensional random variables independence on the basis of algorithm of pattern recognition

A V Lapko[1,2], V A Lapko[1,2], E A Yuronen[1]
[1]Reshetnev Siberian State Aerospace University
31, Krasnoyarsky Rabochy Av., Krasnoyarsk, 660037, Russian Federation
[2]Institute of Computer Modeling Siberian Branch of the RAS
50, Akademgorodok, Krasnoyarsk, 660036, Russian Federation

E-mail: lapko@icm.krasn.ru

Abstract. The new technique of testing of hypothesis of random variables independence is offered. Its basis is made by nonparametric algorithm of pattern recognition. The considered technique doesn't demand sampling of area of values of random variables.

1. Introduction

The testing of hypothesis about random distributions with use of nonparametric algorithms of a pattern recognition is considered in works [1, 2]. Possibility of transition from a problem of comparison of distribution laws of random values to check of a hypothesis of equality of probability of an error of pattern recognition to threshold value is proved. The offered approach, for example, when checking a hypothesis of uniformity of distribution laws of two sequences of random values consists in realization of the following actions:

- according to the compared random series to create the training selection for the solution of the two-alternate problem of a pattern recognition;
- to carry out synthesis of nonparametric algorithm of the pattern recognition corresponding to criterion of maximum likelihood (for estimation of probability densities of random values in classes are used statistics like Rosenblatt-Parzen);
- in the mode of "the sliding examination" to calculate an assessment of probability of an error of pattern recognition;
- to check a hypothesis of equality of probability of an error of pattern recognition to threshold value.

This approach allows bypassing a problem of decomposition of a range of values of random values which is peculiar to Pearson's criterion.

In this work nonparametric algorithms of a pattern recognition are used at the solution of a problem of testing of a hypothesis of independence of random values.

2. Properties of nonparametric estimates of the probability density dependent and independent of random values

Let's compare asymptotic properties of nonparametric estimates of a probability density in the conditions of dependence and independence of random values. Statistical data are used at synthesis of nonparametric algorithms of pattern recognition at a test of hypothesis about random distributions.

Let there is a selection $V = \left(x^i, i = \overline{1, n}\right)$ of n statistically independent supervision of a two-dimensional random value $x = (x_1, x_2)$ with a priori unknown probability density $p(x)$. It is known, as x_1 and x_2 are independent.

In these conditions for estimation of a probability density $p(x)$ let's use nonparametric statistics

$$\bar{p}(x) = \bar{p}_1(x_1)\,\bar{p}_2(x_2), \tag{1}$$

Where

$$\bar{p}_v(x_v) = \frac{1}{n\,c_v}\sum_{i=1}^{n}\Phi\left(\frac{x_v - x_v^i}{c_v}\right),\ v=1, 2. \tag{2}$$

Nuclear functions $\Phi(u_v)$ obey H :

$$\Phi(u_v) = \Phi(-u_v),\quad 0 \le \Phi(u_v) < \infty,$$
$$\int \Phi(u_v)du_v = 1,\quad \int u_v^2\,\Phi(u_v)du_v = 1,$$
$$\int u_v^m\,\Phi(u_v)du_v < \infty,\quad 0 \le m < \infty,\ v=1, 2.$$

Parameters of nuclear functions $c_v = c_v(n), v = 1, 2$ decrease with body height n. Hereinafter the infinite limits are passed.

Fairly following statement:

Theorem. Let density of probabilities $p_v(x_v)$ of random values x_v, $v = 1, 2$ and their first derivative be limited and continuous; nuclear functions $\Phi(u_v)$ obey H; sequences $c_1 = c_1(n)$, $c_2 = c_2(n)$ of coefficients of a diffuseness of nuclear functions of a nonparametric assessment of a probability density $\bar{p}(x_1, x_2)$ are, that $n \to \infty$ values $c_1 \to 0$, $c_2 \to 0$, and $n\,c_1 \to \infty$ and $n\,c_2 \to \infty$.

Then for a nonparametric assessment $\bar{p}(x_1, x_2) = \bar{p}_1(x_1)\,\bar{p}_2(x_2)$ of a probability density $p(x_1, x_2) = p_1(x_1)\,p_2(x_2)$ asymptotic expression of a mean squared deviation will register in a look

$$M\iint\left(p_1(x_1)\,p_2(x_2) - \bar{p}_1(x_1)\,\bar{p}_2(x_2)\right)^2 dx_1\,dx_2 \sim$$
$$\sim \frac{\left(\|\Phi(u)\|^2\right)^2}{n^2\,c_1\,c_2} + \frac{\|\Phi(u)\|^2\,\|p_2(x_2)\|^2}{n c_1} + \frac{\|\Phi(u)\|^2\,\|p_1(x_1)\|^2}{n c_2} +$$
$$+\iint\left(\frac{p_2(x_2)\,p_1^{(2)}(x_1)\,c_1^2}{2} + \frac{p_1(x_1)\,p_2^{(2)}(x_2)\,c_2^2}{2}\right)dx_1\,dx_2\ . \tag{3}$$

Here the following designations are used:

$$\|\Phi(u)\|^2 = \int \Phi^2(u)\,du\,; \ \|p_\nu(x_\nu)\|^2 = \int p_\nu^2(x_\nu)\,du\,;$$

$p_\nu^{(2)}(x_\nu)$ - density flexon $p_\nu(x_\nu)$, $\nu = 1, 2$; M - is the sign of expected value.

Convergence in mean squared statistics follows from the analysis of expression (3) when performing conditions of the theorem (1).

At the proof of the theorem the technique offered in work [3] and developed in [4-8] is used.

Let's compare approximating properties of nonparametric estimates of a probability density (1) and

$$\tilde{p}(x) = \frac{1}{n\,c_1\,c_2} \sum_{i=1}^{n} \prod_{\nu=1}^{2} \Phi\left(\frac{x_\nu - x_\nu^i}{c_\nu}\right). \tag{4}$$

For this purpose we will define best values c_ν from a condition of a minimum of asymptotic expression of a mean squared deviation

$$M \int \left(\bar{p}_\nu(x_\nu) - p_\nu(x_\nu)\right) dx_\nu \sim \frac{1}{n\,c_\nu}\|\Phi(u)\|^2 + \frac{c^4}{4}\left\|p_\nu^{(2)}(x_\nu)\right\|^2.$$

Then it is easy to show that best value of diffuseness coefficient of nuclear function is defined by expression

$$c_\nu^* = \left[\frac{\|\Phi(u)\|^2}{n\left\|p_\nu^{(2)}(x_\nu)\right\|^2}\right]^{\frac{1}{5}}, \ \nu = 1, 2.$$

Substituting c_ν^*, $\nu = 1, 2$ in (3) we will receive

$$W_2 = \left(\frac{\|\Phi(u)\|^2}{n}\right)^{\frac{4}{5}} \left[\left(\frac{\|\Phi(u)\|^2}{n}\right)^{\frac{4}{5}} \left(\left\|p_1^{(2)}(x_1)\right\|^2 \left\|p_2^{(2)}(x_2)\right\|^2\right)^{\frac{1}{5}} + \right.$$

$$+ \frac{5}{4}\left(\left(\left\|p_1^{(2)}(x_1)\right\|^2\right)^{\frac{1}{5}} \|p_2(x_2)\|^2 + \left(\left\|p_2^{(2)}(x_2)\right\|^2\right)^{\frac{1}{5}} \|p_1(x_1)\|^2\right) +$$

$$+\frac{1}{2}\left(\left\|p_1^{(2)}(x_1)\right\|^2\left\|p_2^{(2)}(x_2)\right\|^{2^2}\right)^{-\frac{2}{5}}\prod_{v=1}^{2}\int p_v(x_v)\,p_v^{(2)}(x_v)\,dx_v\Bigg]\square$$

According to results of researches [4] the minimum mean squared deviation of a nonparametric assessment of a probability density $\tilde{p}(x_1, x_2)$ at $k=2$ and $c_1=c_2$ is defined by expression

$$W_2' = \frac{5}{2^{7/3}}\left[\left(\frac{\left(\left\|\Phi(u)\right\|^2\right)^2}{n}\right)^4\left(\int\int\left(p_1^{(2)}(x_1, x_2)+p_2^{(2)}(x_1, x_2)\right)^2 dx_1\,dx_2\right)^2\right]^{\frac{1}{6}},$$

where $p_v^{(2)}(x_1, x_2)$ - probability density flexon $p(x_1, x_2)$ on a variable x_v, $v=1,2$.

At terminating $p_v(x_v)$, $p_v^{(2)}(x_v)$, $v=1,2$ with body height of volume n of statistical data of value W_2 of a minimum mean squared deviation $\bar{p}(x_1, x_2)=\bar{p}_1(x_1)\bar{p}_2(x_2)$ aspire to zero in proportion to $r=n^{-4/5}$. And the order of similar convergence is higher, than for W_2', which values decrease in proportion to $n^{-4/6}$.

Distinction of approximating properties of nonparametric estimates of a probability density $\bar{p}(x)$, $\tilde{p}(x)$ is a basis of a technique of check of a hypothesis of independence of random values with use of nonparametric algorithms of pattern recognition.

3. Technique of the hypothesis testing of independence of random values

There is a selection $V=\left(x_1^i, x_2^i, i=\overline{1,n}\right)$ from n statistically independent supervision of a two-dimensional random value $x=(x_1, x_2)$. Random values are characterized by probability densities $p(x_1, x_2)$, $p_1(x_1)$, $p_2(x_2)$. It is necessary to confirm or disprove a hypothesis H_0 about independence of distribution laws of random values x_1, x_2.

Let's assume that there are two classes Ω_1, Ω_2. The first class Ω_1 is characterized by a probability density $p_1(x_1)p_2(x_2)$. It can be defined as a nonparametric assessment of a probability density $\bar{p}_1(x_1)\bar{p}_2(x_2)$ type (2) which is restored on selection V. The second class Ω_2 is defined by a probability density $p(x_1, x_2)$ and is estimated by statistics (4).

On this basis we will construct nonparametric algorithm of pattern recognition

$$\bar{m}(x):\begin{cases}x\in\Omega_1,\ if\ \ \bar{f}_{12}(x)\geq 0\\ x\in\Omega_2,\ if\ \ \bar{f}_{12}(x)<0,\end{cases} \qquad (5)$$

where

$$\bar{f}_{12}(x)=\bar{p}_1(x_1)\bar{p}_2(x_2)-\bar{p}(x_1, x_2).$$

The choice of best values c_ν^*, $\nu = 1, 2$ is carried out from a condition of a minimum of an assessment of probability of an error of pattern recognition

$$\bar{\rho} = \frac{1}{n} \sum_{j=1}^{n} 1(\sigma(j), \bar{\sigma}(j))$$

on the training selection $V' = \left(x^i, \sigma(i), i = \overline{1, n}\right)$. Here all $\sigma(i)$ - are instructions on situation accessory x^i to the first class Ω_1. Decisions $\bar{\sigma}(i)$ are defined by algorithm (5).

Indicator function

$$1(\sigma(j), \bar{\sigma}(j)) = \begin{cases} 0, if & \sigma(j) = \bar{\sigma}(j) \\ 1, if & \sigma(j) \neq \bar{\sigma}(j). \end{cases}$$

When forming "decision" $\bar{\sigma}(j)$ the situation x^j is excluded from process of calculation of statisticians $\bar{p}_1(x_1)\bar{p}_2(x_2)$, $\bar{p}(x_1, x_2)$.

To define a minimum error of pattern recognition $\bar{\rho}^*$, which correspond to values c_ν^*, $\nu = 1, 2$.

Using traditional criteria, to check a hypothesis H_1 about equality of probability of an error of pattern recognition to value $1/2$. The initial hypothesis H_0 is fair if the hypothesis H_1 is carried out, differently the hypothesis H_0 is rejected.

The offered technique can is generalized on a problem of hypothesis testing of independence of sets of random values $x(1) = \left(x_\nu, \nu = \overline{1, k1}\right)$, $x(2) = \left(x_\nu, \nu = \overline{k1+1, k}\right)$.

4. Acknowledgment

Difference of approximating properties of nonparametric estimates of probability densities of dependent and independent random values is established. On this basis possibility of application of nonparametric algorithm of pattern recognition in a problem of hypothesis testing of independence of random values is proved. The offered technique allows bypassing difficult formalizable procedure of decomposition of a range of values of random values.

Further development of the offered approach is bound to its generalization on a problem of hypothesis testing of independence of sets of random values.

This work was carried out as part of assignment No. 2.914.2014/K of the Ministry of Education and Science of the Russian Federation.

References

[1] Lapko A V, Lapko V A 2010 Nonparametric algorithms of pattern recognition in the problem of testing a statistical hypothesis on identity of two distribution laws of random variables *Opt. Instrum. Data Proc.*, **46** pp 545-550.

[2] Lapko A V , Lapko V A 2012 Comparison of empirical and theoretical distribution functions of a random variable on the basis of a nonparametric classifier *Opt. Instrum. Data Proc.*, **48** pp37-41.

[3] Epanechnikov V A 1969 Nonparametric estimator of multidimensional probability density *Teor. Veroyatn. Ee Primen.*, **14** pp 156–161 (In Russ).
[4] Lapko A V, Lapko V A 2012 Properties of nonparametric estimates of multidimensional probability density of independent random variables *Informatika i systemy upravleniya*, **31** (1) pp166-174 (In Russ).
[5] Lapko A V, Lapko V A 2013 Optimal selection of the number of sampling intervals in domain of variation of a one-dimensional random variable in estimation of the probability density *Measur. Techn.*, **56** pp 763 – 767.
[6] Lapko A V, Lapko V A 2012 Properties of the nonparametric decision function with a priori information on independence of attributes of classified objects *Opt. Instrum. Data Proc.*, **48** pp 416-422.
[7] Lapko A V, Lapko V A 2014 Effect of data incompleteness on the approximation properties of nonparametric estimation of the two-dimensional probability density of independent random variables *Opt. Instrum. Data Proc.*, **50** pp 68-74.
[8] Lapko A V, Lapko V A 2014 Regression estimate of the multidimensional probability density and its properties *Opt. Instrum. Data Proc.*, **50** pp148-153.

IOP Conf. Series: Materials Science and Engineering **155** (2016) 012035 doi:10.1088/1757-899X/155/1/012035

Testing of hypothesis of random variables independence on the basis of nonparametric method of confidence estimation of probability density

A V Lapko[1,2], V A Lapko[1,2], E A Yuronen[1]
[1]Reshetnev Siberian State Aerospace University
31, Krasnoyarsky Rabochy Av., Krasnoyarsk, 660037, Russian Federation
[2]Institute of Computer Modeling Siberian Branch of the RAS
50, Akademgorodok, Krasnoyarsk, 660036, Russian Federation

E-mail: lapko@icm.krasn.ru

Abstract. Algorithmic approach of hypothesis testing of identity of distribution laws of random values is offered. Its basis is made by technique of creation of confidence borders for a probability density.

1. Introduction

Problems of hypothesis testing about random distributions are one of the main in the theory of mathematical statistics. For their decision the Pearson's criterion which does not depend on random distributions and their dimension is widely used [1]. However the formation technique of this criterion contains difficult formalizable stage of splitting a range of values of random values into intervals. This problem is absent in criterion of Kolmogorov-Smirnov which is used at a hypothesis testing about distributions of one-dimensional random values [2]. The technique based on replacement of a problem of hypothesis testing about random distributions by a problem of hypothesis testing about equality of probabilities of an error of pattern recognition to threshold value [3, 4].

Problems of statistical hypotheses testing of random distributions are intimately bound to creation of confidence regions for their probability densities or cumulative distribution functions [1].

In this thesis the algorithmic approach of hypothesis testing of random distributions using confidence borders of their probability densities is offered and investigated. The basis of approach is made by a regression assessment of a probability density.

2. Regression assessment of the probability density and its property

Let there is a selection $V=\left(x^i,\, i=\overline{1,n}\right)$ from n independent values of a one-dimensional random value x with unknown probability density $p(x)$.

For estimation of a probability density $p(x)$ in the conditions of selections of large volume we will use a technique of "compression" of input statistical data V [5]. Let's break definition range $p(x)$ into N not crossed intervals 2β long and we will create sets of random values $X^j,\, j=\overline{1,N}$. As characteristics X^j we will accept frequency $\overline{P}^j$ hits of a random value x in the j-th interval and its center z^j. On the basis of the received information we will define a data array $V_1 = \left(z^j,\ y^j = \overline{P}^j/(2\beta),\ j = \overline{1, N}\right)$□ made of the centers z^j injected intervals and values of estimates of probability densities y^j corresponding to them.

For a quantity choice N intervals of sampling of a range of values of a random value it is possible to use recommendations of publications [6-10].

As an assessment of a required probability density $p(x)$ we will accept statistics

$$\tilde{p}(x) = c^{-1} \sum_{j=1}^{N} \overline{P}^j \Phi\left(\frac{x - z^j}{c}\right) \tag{1}$$

in which nuclear functions $\Phi(u)$ obey H:

$$\Phi(u) = \Phi(-u), \quad 0 \le \Phi(u) < \infty,$$

$$\int_{-\infty}^{+\infty} \Phi(u)\,du = 1, \quad \int_{-\infty}^{+\infty} u^2 \Phi(u)\,du = 1.$$

Values of diffuseness coefficients of nuclear functions $c(N)$ in expression (1) decrease with body height N.

Asymptotic properties of statistics (1) are defined by the following statement.

Theorem. Let: 1) $p(x)$ is limited and continuous with all the derivatives to the second order inclusive; 2) nuclear functions $\Phi(u)$ obey H; 3) at $N \to \infty$, c and $\beta \to 0$; $\beta/c \to 0$, and $nc \to \infty$. Then the regression assessment of a probability density possesses properties of an asymptotic unbiasedness and convergence in the mean squared:

Asymptotic unbiasedness

$$M(\overline{p}(x) - p(x)) \sim \frac{p^{(2)}(x)\left(\frac{\beta^2}{3} + c^2\right)}{2}.$$

Convergence in the mean squared

$$M(\overline{p}(x) - p(x))^2 \sim \frac{\|\Phi(x)\|^2}{nc} + \frac{\left\|p^{(2)}(x)\right\|^2\left(\frac{\beta^2}{3} + c^2\right)}{4} +$$

$$+ \frac{\|\Phi(x)\|^2 \beta}{c}\left(2\|p(x)\|^2 + \frac{\beta^4\left\|p^{(2)}(x)\right\|^2}{18}\right) + \beta c\|u\Phi(u)\|^2\left(2\left\|p^{(1)}(x)\right\|^2 + \frac{\beta^2\left\|p^{(2)}(x)\right\|^2}{3}\right) +$$

$$+ \frac{\beta c^3\left\|p^{(2)}(x)\right\|^2\left\|u^2\Phi(u)\right\|^2}{2}.$$

Here the following designations are accepted:

$$\|\Phi(u)\|^2 = \int_{-\infty}^{+\infty} \Phi^2(u)\,du,\ \|p(x)\|^2 = \int_{-\infty}^{+\infty} p^2(x)\,dx,\ \left\|p^{(1)}(x)\right\|^2 = \int_{-\infty}^{+\infty} \left(p^{(1)}(x)\right)^2 du,$$

$$\left\|p^{(2)}(x)\right\|^2 = \int_{-\infty}^{+\infty} \left(p^{(2)}(x)\right)^2 du,\ \|u\Phi(u)\|^2 = \int_{-\infty}^{+\infty} u^2\,\Phi^2(u)\,du;$$

$p^{(1)}(x)$, $p^{(2)}(x)$ are the first and flexon of probability density $p(x)$; M - is the sign of expected value.

3. Creation of confidence borders for the probability density

The assessment of a probability density (1) allows to construct confidence borders for $p(x)$ on the basis of confidence estimation of probabilities P^j of events $x \in X^j$, $j = \overline{1, N}$.

Top P_h^j and bottom P_d^j borders of an interval assessment of probability P^j of event $x \in X^j$ with a confidence coefficient γ are defined by expressions

$$P_h^j = \overline{P}^j + \frac{u_{1-\alpha/2}}{\sqrt{n}} \sqrt{\overline{P}^j\left(1-\overline{P}^j\right)}, \tag{2}$$

$$P_d^j = \overline{P}^j - \frac{u_{1-\alpha/2}}{\sqrt{n}} \sqrt{\overline{P}^j\left(1-\overline{P}^j\right)}, \tag{3}$$

where $u_{1-\alpha/2}$ - quantile of level $1-\alpha/2$ of a reference normal distribution. Values $u_{1-\alpha/2}$ are determined by tables of quantiles of a normal distribution at $\alpha = 1-\gamma$.

Let's make computing experiment and determine by $\left(\overline{P}^j,\ j = \overline{1, N}\right)$ according to (2), (3), values $P_h^j,\ j = \overline{1, N}$ and $P_d^j,\ j = \overline{1, N}$.

On this basis let's carry out synthesis top and bottom confidence borders

$$\tilde{p}_h(x) = c_h^{-1} \sum_{j=1}^{N} P_h^j \Phi\left(\frac{x - z^j}{c_h}\right), \tag{4}$$

$$\tilde{p}_d(x) = c_d^{-1} \sum_{j=1}^{N} P_d^j \Phi\left(\frac{x - z^j}{c_d}\right) \tag{5}$$

for probability density $p(x)$.

Let's define optimum coefficients of a diffuseness of nuclear functions in expressions (4), (5) with use of a method of "the sliding examination". For example, for an upper bound (4) choice of optimum diffuseness coefficient c_h of nuclear functions is carried out as a result of expression minimization

$$\sum_{i=1}^{N}\left(\frac{P_h^i}{2\beta}-\tilde{\tilde{p}}_h\left(z^i\right)\right)^2 .$$

Here $\frac{P_h^i}{2\beta}$ - us value of the upper confidence bound $p(x)$ within the i -th interval of sampling of a random value x, and

$$\tilde{\tilde{p}}_h\left(z^i\right)=\frac{1}{c_h}\sum_{\substack{j=1\\ j\neq i}}^{N} P_h^j \Phi\left(\frac{z^i-z^j}{c_h}\right)$$

represents its assessment on this interval.

4. Technique of hypothesis testing of identity of distribution laws of random values

There are two independent selections $V_1=\left(x^i, i=\overline{1,n_1}\right)$ and $V_2=\left(x^i, i=\overline{1,n_2}\right)$ the one-dimensional random values taken from universes X_1, X_2, which are characterized by unknown probability densities $p_1(x)$, $p_2(x)$. It is necessary to confirm or disprove a hypothesis H_0 of identity of their distribution laws.

Let Ω_t, $t=1,2$ confidence regions for probability densities $p_t(x)$, $t=1,2$ at the given confidence coefficient γ.

Then realization of a hypothesis H_0 is defined by the following rule

$$m_{12}(\gamma):\begin{cases} \textit{hypothesis } H_0 \textit{ is fair if} \\ \Omega_1\subseteq\Omega_2 \quad or \quad \Omega_2\subseteq\Omega_1 , \\ \textit{otherwise the hypothesis } H_0 \textit{ is rejected} . \end{cases}$$

For implementation of this rule according to the technique offered above let construct for probability densities $p_1(x)$, $p_2(x)$ confidence borders

$$\tilde{p}_h^t(x)=\frac{1}{c_h^t}\sum_{j=1}^{N_t} P_h^j(t)\Phi\left(\frac{x-z^j}{c_h^t}\right),\ \tilde{p}_d^t(x)=\frac{1}{c_d^t}\sum_{j=1}^{N_t} P_d^j(t)\Phi\left(\frac{x-z^j}{c_d^t}\right),\ t=1,2,$$

where $P_h^j(t)$, $P_d^j(t)$ - interval estimates $P^j(t)$ of an event $x\in X_t^j$ with a confidence coefficient γ; N_t - quantity of not crossed intervals of areas X_t; c_h^t, c_d^t - diffuseness coefficients of nuclear functions in statisticians $\tilde{p}_h^t(x)$, $\tilde{p}_d^t(x)$.

Let's designate through a V_t^γ - a range of a random value of an $x^j\in V_t$, $j=\overline{1,N_t}$ getting to a confidence region of a probability density of a $p_t(x)$, $t=1,2$ at the given confidence coefficient γ. Formation of a set V_t^γ is carried out on the basis of the following decisive rule

$$m_t\left(x^j\right): x^j \in V_t^\gamma \text{ if } \tilde{p}_h^t\left(x^j\right) \geq \tilde{p}^t\left(x^j\right) \geq \tilde{p}_d^t\left(x^j\right),\ j=\overline{1,n_t},\ t=1,2.$$

It is easy to notice that the set V_t^γ represents an assessment of a confidence region Ω_t, $t=1,2$. Therefore there is an opportunity instead of $m_{12}(\gamma)$ use the rule

$$\overline{m}_{12}(\gamma): \begin{cases} \textit{hypothesis } H_0 \textit{ is fair if} \\ \tilde{p}_d^2\left(x^j\right) \geq \tilde{p}_d^1\left(x^j\right) \ \textit{and} \ \ \tilde{p}_h^2\left(x^j\right) \leq \tilde{p}_h^1\left(x^j\right) \forall\, x^j \in V_2^\gamma \\ \textit{or} \\ \tilde{p}_d^1\left(x^j\right) \geq \tilde{p}_d^2\left(x^j\right) \ \textit{and} \ \ \tilde{p}_h^1\left(x^j\right) \leq \tilde{p}_h^2\left(x^j\right) \forall\, x^j \in V_1^\gamma . \end{cases}$$

This rule is based on replacement of operations with confidence regions Ω_t, $t=1,2$ set by sets V_t^γ, $t=1,2$ on check of ratios between their borders $\tilde{p}_h^t(x)$, $\tilde{p}_d^t(x)$, $t=1,2$.

5. Acknowledgment

The structure of a regression assessment of a probability density allows to solve a problem of confidence estimation of a probability density on its basis. The idea of the offered approach consists in decomposition of input statistical data and the subsequent analysis of probabilistic characteristics of the received sets of random values. On this basis, using a nonparametric assessment of a regression curve, synthesis of confidence borders of a probability density is carried out. The area sizes determined by confidence borders depend on quantity of intervals of sampling of random values, their probabilistic characteristics and volume of input statistical data.

If the confidence region of a probability density includes a confidence region of other compared density, they are identical. Otherwise the hypothesis of identity of distribution laws of random values is rejected.

The offered approach opens possibility of generalization of the received results on a hypothesis testing about distributions of many-dimensional random values.

This work was supported within the basic part of the State Task of the Ministry of Education and Sciences of the Russian Federation for higher educational institutions in 2014–2016 (SibGAU No. B121/14) and Program of Siberian Branch of the Russian Academy of Sciences IV.35.1 «Theoretical bases and technologies of creation and use of the integrated information systems for problem solving of support of a decision making».

References

[1] Pugachev V S Probability Theory and Mathematical Statistics (Nauka, Moscow, 1979).
[2] Smirnov N V 1930 Estimation of the Difference between the Distribution Curves in Two Independent Samples *Bull. Mosk. Univ.*, **2** pp3–14.
[3] Lapko A V , Lapko V A 2010 Nonparametric algorithms of pattern recognition in the problem of testing a statistical hypothesis on identity of two distribution laws of random variables *Opt. Instrum. Data Proc.*, **46** pp 545-550.
[4] Lapko A V, Lapko V A 2012 Comparison of empirical and theoretical distribution functions of a random variable on the basis of a nonparametric classifier *Opt. Instrum. Data Proc.*, **48** pp37-41.

[5] Lapko A V , Lapko V A 2014 Regression estimate of the multidimensional probability density and its properties *Opt. Instrum. Data Proc.*, **50** pp148-153.
[6] Heinhold I, Gaede K W *Ingenieur Statistik* (Springer-Verlag, Munich, Vienna, 1964).
[7] Freedman D, Diaconis P 1981 On the histogram as a density estimator: L2 theory *Probability Theory and Related Fields*, **57** pp 453–476.
[8] Scott D 1979 On optimal and data-based histograms *Biometrika*, **66** pp 605–610.
[9] Sturges H A 1926 The Choice of a Class Interval *J. Amer. Stat. Association*, **21** pp 65–66.
[10] Lapko A V, Lapko V A 2013 Optimal selection of the number of sampling intervals in domain of variation of a one-dimensional random variable in estimation of the probability density *Measur. Techn.*, **56** pp 763 – 767.

IOP Conf. Series: Materials Science and Engineering **155** (2016) 012036 doi:10.1088/1757-899X/155/1/012036

Synthesis of hybrid systems of pattern recognition on the basis of procedure of consecutive correction of decision functions

V A Lapko[1,2], A V Lapko[1,2], Yu P Yuronen[1]
[1]Reshetnev Siberian State Aerospace University
31, Krasnoyarsky Rabochy Av., Krasnoyarsk, 660037, Russian Federation
[2]Institute of Computer Modeling Siberian Branch of the RAS
50, Akademgorodok, Krasnoyarsk, 660036, Russian Federation

E-mail: lapko@icm.krasn.ru

Abstract. Hybrid systems of pattern recognition in the conditions of large volumes of the training selections and not stationarity of classification objects are offered. Asymptotic properties of their decision function are investigated.

1. Introduction

Perspective nature of sharing of parametrical and local methods of approximation was for the first time noted in work [1]. Its idea was used at restitution of stochastic dependences in the conditions of existence of partial data on their look [2, 3]. The developed models were called hybrid. This concept is natural and is based on procedure of correction of initial parametrical approximation. The next analogs of the used approach are semi-parametrical models of stochastic dependences [4].

The offered technique was developed at the solution of the two-alternate problem of pattern recognition in the conditions of existence of partial data on a type of the dividing surface equation [5, 6].

In this work the technique of synthesis of hybrid systems of pattern recognition in the conditions of large volumes of the training selections and not stationarity of objects of classification is offered. The basis of a technique is made by serial procedure of correction of decision functions.

2. Synthesis of hybrid system of pattern recognition

Let there is a training selection $V=\left(x^i,\sigma\left(x^i\right), i=\overline{1,n}\right)$ of large volume n, where $\sigma(x^i)$ - instructions on situation x^i belonging to one of two classes Ω_1, Ω_2. The objects of classification which are characterized by a feature set $x=\left(x_v, v=\overline{1,k}\right)$, can be nonstationary. These conditions are characteristic, for example, at data processing of remote sensing. Let's apply the principles of hybrid model operation at synthesis of system of pattern recognition.

The idea of the offered approach consists in realization of the following actions:

1. Let's carry out decomposition of the initial training selection V on N parts $V_j=\left(x^i,\sigma\left(x^i\right), i\in I_j\right)$, $j=\overline{1,N}$. The basis for realization of this operation is formation of selections V_j according to their accessory to various time frames of overseeing by signs of the classified objects.

2. By data V_1 construct the decisive rule

$$\overline{m}_{12}^{1}(x):\begin{cases} x\in\Omega_1,\ if\ \ \bar{f}_{12}^{1}(x)\geq 0,\\ x\in\Omega_2,\ if\ \ \bar{f}_{12}^{1}(x)<0, \end{cases} \qquad (1)$$

where the nonparametric assessment of the equation of the dividing surface is

$$\bar{f}_{12}^{1}(x)=\frac{1}{n_1\prod\limits_{v=1}^{k}c_v(n_1)}\sum_{i\in I_1}\sigma(x^i)\prod_{v=1}^{k}\Phi\left(\frac{x_v-x_v{}^i}{c_v(n_1)}\right);$$

n_1 - quantity of elements of set I_1.

Here values are $\sigma(x^i)$=1, when $x^i\in\Omega_1$ and $\sigma(x^i)$=-1, when $x^i\in\Omega_2$. Nuclear functions $\Phi(u_v)$ in nonparametric assessment $\bar{f}_{12}^{1}(x)$ obey H: positive, the symmetric and the normalized [7]. Diffuseness coefficients c_v of nuclear functions decrease with body height n_1.

In this case the decisive rule (1) corresponds to criterion of a maximum of a posterior probability.

3. By results of computing experiment to create data $V_2=\left(x^i,q_1(x^i),i\in I_2\right)$. They reflect difference of "decisions" $\overline{\sigma}_1(x^i)$ of rule $\overline{m}_{12}^{1}(x)$ (1) from "instructions" $\sigma(x^i)$ from the training selection V_1. Here values of the correcting function are

$$q_1(x^i)=\begin{cases} 0\ \forall\ \sigma(x^i)=\overline{\sigma}_1(x^i),\\ \left(\bar{f}_{12}^{1}(x^i)+\Delta\right)\ \forall\ \overline{\sigma}_1(x^i)=-1\ \ and\ \ \sigma(x^i)=1,\\ -\left(\bar{f}_{12}^{1}(x^i)+\Delta\right)\ \forall\ \overline{\sigma}_1(x^i)=1\ \ and\ \ \sigma(x^i)=-1. \end{cases}$$

In the presence of a pattern recognition error the correcting function accepts value inverse on a sign of a nonparametric assessment of the equation of the dividing surface $\bar{f}_{12}^{1}(x)$ also exceeds it on a small Δ.

4. To estimate the correcting function $q_1(x)$ on selection V_2 on the basis of nonparametric regression

$$\overline{q}_1(x)=\frac{\sum\limits_{i\in I_2}q_1(x^i)\beta_i(x)}{\sum\limits_{i\in I_2}\beta_i(x)},\ \beta_i(x)=\prod_{v=1}^{k}\Phi\left(\frac{x_v-x_v{}^i}{c_v(n_2)}\right).$$

5. Then the intermediate hybrid algorithm of pattern recognition will register as

$$\overline{m}_{12}^{2}(x):\begin{cases} x\in\Omega_1,\ if\ \ \bar{f}_{12}^{2}(x)\geq 0,\\ x\in\Omega_2,\ if\ \ \bar{f}_{12}^{2}(x)<0,\end{cases}$$

where

$$\bar{f}_{12}^{2}(x)=\bar{f}_{12}^{1}(x)+\bar{q}_1(x).$$

The choice of best values $c_v(n_2),\ v=\overline{1,k}$ is carried out from a condition of a minimum of a statistical assessment of probability of an error of pattern recognition in the mode of "the sliding examination".

6. Generally serial correction procedure of the dividing surface equation has an appearance as

$$\bar{f}_{12}^{t}(x)=\bar{f}_{12}^{t-1}(x)+\bar{q}_{t-1}(x),\ t=\overline{2,N}. \tag{2}$$

Nonparametric assessment of the correcting function

$$\bar{q}_{t-1}(x)=\frac{\sum\limits_{i\in I_t} q_{t-1}(x^i)\beta_i(x)}{\sum\limits_{i\in I_t}\beta_i(x)},\ \ \beta_i(x)=\prod_{v=1}^{k}\Phi\left(\frac{x_v-x_v{}^{i}}{c_v(n_t)}\right)$$

is formed according to computing experiment

$$q_{t-1}(x^i)=\begin{cases}0\ \forall\ \sigma(x^i)=\overline{\sigma}_{t-1}(x^i),\\ \left(\bar{f}_{12}^{t-1}(x^i)+\Delta\right)\ \forall\ \overline{\sigma}_{t-1}(x^i)=-1\ \ and\ \ \sigma(x^i)=1,\\ -\left(\bar{f}_{12}^{t-1}(x^i)+\Delta\right)\ \forall\ \overline{\sigma}_{t-1}(x^i)=1\ \ and\ \ \sigma(x^i)=-1.\end{cases}$$

Here $\overline{\sigma}_{t-1}(x^i)$ - is "solution" of algorithm $\overline{m}_{12}^{t-1}(x)$ in conditions x^i, and $\sigma(x^i)$ - are "instructions" from selection $V_t=\left(x^i,\sigma(x^i),i\in I_t\right)$.

On the basis of decision functions $\bar{f}_{12}^{t}(x),\ t=\overline{2,N}$ serial synthesis of algorithms of pattern recognition is carried out

$$\overline{m}_{12}^{t}(x):\begin{cases} x\in\Omega_1,\ if\ \ \bar{f}_{12}^{t}(x)\geq 0,\\ x\in\Omega_2,\ if\ \ \bar{f}_{12}^{t}(x)<0,\ t=\overline{2,N}.\end{cases} \tag{3}$$

Expression (2) is a nonparametric assessment of decision function

$$f_{12}^{t}(x)=f_{12}^{t-1}(x)+q_{t-1}^{0}(x). \tag{4}$$

For simplification of the subsequent transformations we will consider that

$$f_{12}^{t-1}(x) = p_1^{t-1}(x) - p_2^{t-1}(x).$$

In expression (4)

$$q_{t-1}^{0}(x) = \int q_{t-1}^{0}\; p\left(q_{t-1}^{0} / x \right) d\, q_{t-1}^{0} .$$

Hereinafter the infinite limits are passed.

3. Asymptotic properties of hybrid decision function

Let's consider asymptotic properties of one-dimensional hybrid model (2) type at a known collateral probability density $p(x)$ of distribution x in classes Ω_1, Ω_2. Let the algorithm of classification (3) correspond to criterion of maximum likelihood. The correcting function is optimum in sense of a minimum of a mean squared deviation between $q_{t-1}^{0}(x)$ and $\bar{q}_{t-1}(x)$. In these conditions nonparametric estimates of the equation of the dividing surface $f_{12}^{t-1}(x)$ and the correcting function $q_{t-1}^{0}(x)$ in (4) are presented in the form

$$\bar{f}_{12}^{t-1}(x) = \bar{p}_1^{t-1}(x) - \bar{p}_2^{t-1}(x),$$
$$\bar{q}_{t-1}(x) = \left(n_t c_t\, p_t(x)\right)^{-1} \sum_{i \in I_t} q_{t-1}\left(x^i\right) \Phi\left(\frac{x - x^i}{c_t} \right). \quad (5)$$

Here $\bar{p}_j^{t-1}(x)$ - is nonparametric assessment of a probability density of distribution x in class Ω_j.

Let $f_{12}^{t-1}(x)$, $p_t(x)$ are limited and continuous with all the derivatives to the second order inclusive. We will designate these conditions through G_{t-1}.

Theorem. Let: 1) the equation of the dividing surface $f_{12}^{t-1}(x)$ and a collateral probability density $p_t(x)$ of distribution x in classes obey G_{t-1}; 2) nuclear functions $\Phi(u) \geq 0$ in nonparametric statistics $\bar{q}_{t-1}(x)$, $\bar{f}_{12}^{t-1}(x)$ obey H; 3) sequences $c_t = c(n_t) \to 0$, $\Delta = \Delta(n_t) \to 0$, $\dfrac{\left(n_{1,t-1} + n_{2,t-1}\right)}{n_{1,t-1}\, n_{2,t-1}\, c_{t-1}} \to 0$, $n_t c_t \to \infty$ at $n_t \to \infty$, $n_{1,t-1} \to \infty$, $n_{2,t-1} \to \infty$.

Then the hybrid model $\bar{f}_{12}^{t}(x)$ possesses properties of an asymptotic unbiasedness and solvency of rather decision function $f_{12}^{t}(x)$.

Proof.

1) Let's consider expression
2)

$$M\left(\bar{f}_{12}^{t}(x)\right) = M\left(\bar{f}_{12}^{t-1}(x) + \bar{q}_{t-1}(x)\right) = M\left(\bar{f}_{12}^{t-1}(x)\right) + M\left(\bar{q}_{t-1}(x)\right), \quad (6)$$

where M - is the sign of expected value.

By definition of expected value, if $q_{t-1}\left(x^i\right)=q_{t-1}^0\left(x^i\right)+\Delta$, we have

$$M\left(\bar{q}_{t-1}(x)\right)=\left(c_t\, p_t(x)\right)^{-1}\int q_{t-1}^0(\tau)\,\Phi\left(\frac{x-\tau}{c_t}\right)p_t(\tau)\,d\tau+$$

$$+\Delta\left(c_t\, p_t(x)\right)^{-1}\int\Phi\left(\frac{x-\tau}{c_t}\right)p_t(\tau)\,d\tau\,.$$

Let's carry out replacement of variables in integrals of the received expression $(x-\tau)c_t^{-1}=u$. Let's spread out functions $q_{t-1}^0(x-c_t u)$, $p_t(x-c_t u)$ in Taylor series in a point x. Taking into account properties of nuclear functions at rather great values n_t we will receive

$$M\left(\bar{q}_{t-1}(x)\right)=q_{t-1}^0(x)+\Delta+\frac{c_t^2}{2p_t(x)}\left(\left(q_{t-1}^0(x)\,p_t(x)\right)^{(2)}+\Delta\, p_t^{(2)}(x)\right)+0\left(c_t^4\right). \quad (7)$$

Here $\left(q_{t-1}^0(x)\,p_t(x)\right)^{(2)}$, $p_t^{(2)}(x)$ - are flexons of the corresponding functions. By symbol $0\left(c_t^4\right)$ are designated the composed order trifles c_t^4.

From the analysis of expression (7) when performing conditions $c_t\to 0$, $\Delta\to 0$ $\forall$ $n_t\to\infty$ property of asymptotic not shift of the correcting function $\bar{q}_{t-1}(x)$ follows.

Asymptotic expression first composed in (6) is determined by analogy

$$M\left(\bar{f}_{12}^{t-1}(x)\right)=M\left(\bar{p}_{1,t-1}(x)-\bar{p}_{2,t-1}(x)\right)=$$

$$=\frac{1}{c_{t-1}}\int\Phi\left(\frac{x-\tau}{c_{t-1}}\right)p_{1,t-1}(\tau)d\tau-\frac{1}{c_{t-1}}\int\Phi\left(\frac{x-\tau}{c_{t-1}}\right)p_{2,t-1}(\tau)d\tau\,.$$

After simple transformations we will receive

$$M\left(\bar{f}_{12}^{t-1}(x)-f_{12}^{t-1}(x)\right)\sim\frac{c_{t-1}^2}{2}\left(p_{1,t-1}^{(2)}(x)-p_{2,t-1}^{(2)}(x)\right). \quad (8)$$

In this expression $p_{j,t-1}^{(2)}(x)$ - density flexon of probability $p_{j,t-1}(x)$ on x, $j=1,2$.

From condition $c_{t-1}\to 0$ at $n_{1,t-1}\to\infty$, $n_{2,t-1}\to\infty$, property of asymptotic not shift of nonparametric statistics $\bar{f}_{12}^{t-1}(x)$ follows.

Taking into account expression (6) and the received results (7), (8) asymptotic not shift of hybrid decision function $\bar{f}_{12}^{t}(x)$ follows.

2) Let's investigate asymptotic properties of a mean square deviation

$$M\left(f_{12}^{t}(x)-\bar{f}_{12}^{t}(x)\right)^2 = M\left(f_{12}^{t-1}(x)+q_{t-1}^{0}(x)-\bar{f}_{12}^{t-1}(x)-\bar{q}_{t-1}(x)\right)^2 =$$
$$= M\left(f_{12}^{t-1}(x)-\bar{f}_{12}^{t-1}(x)\right)^2 + M\left(q_{t-1}^{0}(x)-\bar{q}_{t-1}(x)\right)^2 +$$
$$+2M\left(\left(f_{12}^{t-1}(x)-\bar{f}_{12}^{t-1}(x)\right)\left(q_{t-1}^{0}(x)-\bar{q}_{t-1}(x)\right)\right). \qquad (9)$$

Let’s consider expression

$$M\left(q_{t-1}^{0}(x)-\bar{q}_{t-1}(x)\right)^2 = \left(q_{t-1}^{0}(x)\right)^2 - 2\,q_{t-1}^{0}(x)\,\bar{q}_{t-1}(x) + M\left(\bar{q}_{t-1}(x)\right)^2. \qquad (10)$$

Carrying out similar transformations, we will calculate

$$M\left(\bar{q}_{t-1}(x)\right)^2 = \left(n_t c_t\, p_t(x)\right)^{-2}\left[\sum_{i\in I_t} M\left(\left(q_{t-1}^{0}(i)+\Delta\right)^2 \Phi^2\left(\frac{x-x^i}{c_t}\right)\right)+\right.$$
$$\left.+\sum_{i\in I_t}\sum_{\substack{j\in I_t\\ j\neq i}} M\left(\left(q_{t-1}^{0}(i)+\Delta\right)\Phi\left(\frac{x-x^i}{c_t}\right)\right) M\left(\left(q_{t-1}^{0}(j)+\Delta\right)\Phi\left(\frac{x-x^j}{c}\right)\right)\right] =$$
$$= \left(n_t c_t\, p_t(x)\right)^{-2}\left[n_t\left(\int\left(q_{t-1}^{0}(\tau)\right)^2 \Phi^2\left(\frac{x-\tau}{c_t}\right) p_t(\tau)\,d\tau+\right.\right.$$
$$\left.+2\Delta\int q_{t-1}^{0}(\tau)\,\Phi^2\left(\frac{x-\tau}{c_t}\right) p_t(\tau)\,d\tau + \Delta^2\int \Phi^2\left(\frac{x-\tau}{c_t}\right) p_t(\tau)\,d\tau\right)+$$
$$\left.+n_t\left(n_t-1\right)\left(\int\left(q_{t-1}^{0}(\tau)+\Delta\right)\Phi\left(\frac{x-\tau}{c_t}\right) p_t(\tau)\,d\tau\right)^2\right] \sim$$
$$\sim \frac{\left(q_{t-1}^{0}(x)\right)^2 \int \Phi^2(u)\,du}{n_t c_t\, p_t(x)} + \left(q_{t-1}^{0}(x)+\Delta\right)^2 + \frac{q_{t-1}^{0}(x)\,\Delta\, c_t^2}{p_t(x)}\left(\left(q_{t-1}^{0}(x)\, p_t(x)\right)^{(2)} + \Delta\, p_t^{(2)}(x)\right). \qquad (11)$$

Substituting expressions (7), (11) in (10), at $n_t \to \infty$ we will receive

$$M\left(q_{t-1}^{0}(x)-\bar{q}_{t-1}(x)\right)^2 \sim \frac{\left(q_{t-1}^{0}(x)\right)^2 \int \Phi^2(u)\,du}{n_t c_t\, p_t(x)} +$$
$$+\frac{q_{t-1}^{0}(x)\,\Delta\, c_t^2}{p_t(x)}\left(\left(q_{t-1}^{0}(x)\, p_t(x)\right)^{(2)} + \Delta\, p_t^{(2)}(x)\right)(1-\Delta)+\Delta^2.$$

When performing these transformations composed trifles $0(1/n_t)$, $0(c_t/n_t)$, $0(\Delta/(n_t c_t))$, $0\left(c_t^4\right)$ weren't considered.

Let’s define asymptotic expression for a mean square deviation

$$M\left(f_{12}^{t-1}(x)-\bar{f}_{12}^{t-1}(x)\right)^2=M\left(p_{1,t-1}(x)-\bar{p}_{1,t-1}(x)\right)^2-$$
$$-2M\left(\left(p_{1,t-1}(x)-\bar{p}_{1,t-1}(x)\right)\left(p_{2,t-1}(x)-\bar{p}_{2,t-1}(x)\right)\right)+M\left(p_{2,t-1}(x)-\bar{p}_{2,t-1}(x)\right)^2. \quad (12)$$

Asymptotic expression for a mean square deviation $\bar{p}_{j,t-1}(x)$ from $p_{j,t-1}(x)$, $j=1,2$ is received in [8]

$$M\left(p_{j,t-1}(x)-\bar{p}_{j,t-1}(x)\right)^2\sim\frac{\int\Phi^2(u)du}{n_{j,t-1}\,c_{t-1}}+\frac{c_{t-1}^4\int\left(p_{j,t-1}^{(2)}(x)\right)^2dx}{4}.$$

It is easy to show that

$$M\left(\left(p_{1,t-1}(x)-\bar{p}_{1,t-1}(x)\right)\left(p_{2,t-1}(x)-\bar{p}_{2,t-1}(x)\right)\right)\sim c_{t-1}^4\,p_{1,t-1}^{(2)}(x)\,p_{2,t-1}^{(2)}(x).$$

Then at rather great values of $n_{j,t-1}$, $j=1,2$ asymptotic expression for a mean square deviation (12) is presented in the form

$$M\left(f_{12}^{t-1}(x)-\bar{f}_{12}^{t-1}(x)\right)^2\sim\frac{\int\Phi^2(u)du\left(n_{1,t-1}+n_{2,t-1}\right)}{n_{1,t-1}\,n_{2,t-1}\,c_{t-1}}. \quad (13)$$

The composed trifles $0\left(c_{t-1}^4\right)$ are excluded. Substituting expression (11), (13) in (9) at $n_{1,t-1}\to\infty$, $n_{2,t-1}\to\infty$, $n_t\to\infty$, we will receive

$$M\left(f_{12}^{t}(x)-\bar{f}_{12}^{t}(x)\right)^2\sim\frac{\int\Phi^2(u)du\,\left(n_{1,t-1}+n_{2,t-1}\right)}{n_{1,t-1}\,n_{2,t-1}\,c_{t-1}}+$$
$$+\frac{\left(q_{t-1}^0(x)\right)^2\int\Phi^2(u)\,du}{n_tc_t\,p_t(x)}+\left(q_{t-1}^0(x)+\Delta\right)^2+\frac{q_{t-1}^0(x)\,\Delta\,c_t^2}{p_t(x)}\left(q_{t-1}^0(x)\,p_t(x)\right)^{(2)}+\Delta^2.$$

In the course of transformations the composed trifles $0\left(c_{t-1}^4\right)$, $0\left(c_t^2\,\Delta\right)$ weren't considered. From conditions $c_t\to0$, $\dfrac{\left(n_{1,t-1}+n_{2,t-1}\right)}{n_{1,t-1}\,n_{2,t-1}\,c_{t-1}}\to0$, $n_t\,c_t\to\infty$, $\Delta\to0$ at $n_t\to\infty$, $n_{1,t-1}\to\infty$, $n_{2,t-1}\to\infty$, convergence in mean square statistics $\bar{f}_{12}^{t}(x)$ follows. The last statement and asymptotic not shift of hybrid model $\bar{f}_{12}^{t}(x)$ defines property of its solvency.

4. Acknowledgment

Hybrid systems provide the effective solution of problems of pattern recognition in the conditions of non-uniform data of large volume. Sources of the non-uniform training selection are: not stationarity of objects of research, existence of the completed admissions of data and their polytypic character. The structure of hybrid system of pattern recognition in a two-alternative problem of classification is based on consecutive procedure of correction of the equation of the dividing surface. At each its stage the intermediate hybrid model of the equation of the dividing surface is specified by the correcting function. Estimation of the correcting function is carried out by results of computing experiment with the intermediate decisive rule and use of one of uniform parts of the training selection. As its model nonparametric regression is used. The hybrid equation of the dividing surface possesses properties of asymptotic convergence.

The type of the correcting function and feature of the training selection generate family of hybrid systems of pattern recognition . Use of consecutive procedure of formation of structure of hybrid system allows to organize the accounting of partial aprioristic data on a type of the equation of the dividing surface. Further development of a technique of synthesis of hybrid system of pattern recognition is connected with its generalization on a multialternative problem of classification and the analysis of its properties in a multidimensional case.

This work was carried out as part of assignment No. 2.914.2014/K of the Ministry of Education and Science of the Russian Federation.

References

[1] Katkovnik V Ya 1979 Linear and nonlinear methods of nonparametric regression analysis *Avtomatika*, **5** pp 35-46
[2] Lapko A V, Lapko V A 2002 Hybrid models of stochastic dependences *Opt. Instrum. Data Proc.*, **5** pp 38-48
[3] Lapko V A 2004 Synthesis and analysis of hybrid models of stochastic dependencies in the presence of their partial description *Opt. Instrum. Data Proc.*, **40** pp42-49
[4] Hardle W *Applied nonparametric regression* (Mir, Moscow, 1993)
[5] Lapko A V, Lapko V A, Yaroslavtsev S G 2006 Development and analysis of hybrid algorithms in pattern recognition problems *Avtometria*, **42** pp 32-39
[6] Lapko A V, Lapko V A 2008 Hybrid systems of pattern recognition *Pattern recognition and image analysis*, **18** pp 7-13
[7] Parzen E 1962 On estimation of a probability density function and mode *Ann. Math. Statistic.*, **33** pp 1065-1076
[8] Lapko A V, Lapko V A *Multi-level nonparametric data processing systems* (Reshetnev siberian state aerospace university, Krasnoyarsk, 2013)

Prospects for using sonar for underwater archeology on the Yenisei: surveying a 19th century shipwreck*

A E Goncharov[1], D M Mednikov[2], N M Karelin[3], I R Nasyrov[4]
[1] Associate professor, Reshetnev Siberian State Aerospace University, Krasnoyarsk, Russia
[2] Lecturer, Reshetnev Siberian State Aerospace University, Krasnoyarsk, Russia
[3] Senior lecturer, Reshetnev Siberian State Aerospace University, Krasnoyarsk, Russia
[4] Post-graduate student, Reshetnev Siberian State Aerospace University, Krasnoyarsk, Russia

E-mail: goncharovae@sibsau.ru

Abstract. Current progress in underwater archeology is based on a rich arsenal of high-tech appliances, among which sonar technology plays a key role; it enables scientists not only to detect submerged archeological objects, but to examine them in high definition without having to conduct diving operations or use expensive underwater unmanned vehicles. While the majority of sensational scientific discoveries using sonar have been made in saltwater environments, freshwater ones, rivers in particular, have seen limited activity. The river Yenisei in central Siberia contains an unrecorded number of shipwrecks that await being discovered and studied. In this article we focus on the peculiarities of using sonar for detecting archeological sites on the Yenisei. This article is based on the results of the 2016 expedition, which has determined the location of *Thames*, a 19th century British steam schooner which was wrecked on the Yenisei.

Keywords: Sonar, side-scan sonar, echo sounding, bathymetry, fishfinder, Yenisei, underwater archeology, shipwrecks, *Thames*, Northern Sea Route

1. Introduction

Underwater archeology like no other archeological discipline has been employing high-technological equipment and methods from its beginnings as an independent branch of science. While early underwater archeologists relied largely on insufficient and even dangerous methods, modern researchers have at their disposal the advantages of sophisticated electronics, including spin-offs from military and space programs. Perhaps the most indispensible tool in an underwater archeologist's stockpile of tools is the sonar. Active sonar technologies such as echo sounding and side-scan sonar have found the broadest use for detecting and studying submerged objects on a preliminary stage: this includes imaging and mapping the archeological site to determine the area of the proposed investigation. While sonar technology had found wide application for military needs since World War I, it wasn't

*This work was funded by a joint grant of the Russian Fund of Humanities Research and the Krasnoyarsk Regional Science Fund under Grant number 16-11-24010.

until the invention of the side-scanning sonar that this technology firmly established itself in underwater archeology [1, p. 22]. Traditional sonar limits the image to a two-dimensional projection, while side-scanning is capable of providing researchers with high definition 3D imagery of the archeological site and the object itself.

Employing side-view sonar for underwater archeology was first proposed by W.D. Chersterman in the late 1960s. In the Soviet Union this technology was first applied in 1978; this process was accelerated after a contract was signed between the Soviet Academy of Sciences and the Soviet Navy, which constructed the side-scan sonar apparatuses [2].

Recent significant finds in the Arctic, attributed to side-scanning sonar technology include both of Sir John Franklin's ships – *Erebus* and *Terror*, found by Parks Canada in 2014 and the Arctic Research Foundation in 2016, the Soviet Arctic Second World War BD-5 convoy ship *Marina Raskova* and the icebreaker *Aleksandr Sibiriakov*, sunken by the German cruiser *Admiral Scheer* in 1942. Both ships were found in 2014 by the Russian Geographical Society.

The first recorded use of side-view sonar on the Yenisei took place in 1993 for assessing the condition of submerged pipelines in the vicinity of the port of Dudinka. Later, the same technology was used to survey hydrotechnical structures of the Sayano-Shushenskaya Dam on the upper Yenisei [2]. However, neither of these operations included any archeological investigations. Needless to say, the Yenisei has been a major shipping lane since central Siberia was settled by the Russians in the 1600s. Intense navigation with oceangoing vessels began in the 1870s and continues to this day. The depth of the river allows certain oceangoing craft to reach areas as far south as Lesosibirsk and even Krasnoyarsk. Domestic and foreign shipping activity has produced unique shipwrecks on the river, in its delta and in the Eniseiskii zaliv – the estuary of the river which is an inlet of the Kara Sea. The location of a number of these wrecks has been determined recently. Further development of the oil and gas mining industry on the Taymyr Peninsula, resulting in the construction of potential pipelines crossing the Yenisei. This opens new perspectives for conducting sonar and riverbed surveys, thus providing opportunities for new archeological investigations.

2. Existing electronic methods applied in underwater archeology

Due to its nature underwater archeology tends to incorporate various technological methods and devices for detecting and studying objects. In fact, this branch of archeology is not at all possible without technology; even underwater diving, such as scuba diving, which is considered to be a traditional method of research, employs electronics.

Taskaev compares a number of technological methods employed for detecting and positioning submerged archeological objects (Fig. 1) [3, p. 92].

As we can see from the table, the most effective device in terms of range is the side scanning sonar. Optical methods for capturing images such as cameras offer limited usage and cannot be considered effective for detecting submerged objects and their preliminary investigation. Results of the 2016 expedition *Where Thames meets the Yenisei* of Reshetnev Siberian State Aerospace University with support from the Russian Geographical Society demonstrated that underwater cameras have limited application in the murky waters of the Yenisei and its tributaries. The search for the wreckage of Joseph Wiggins's polar steam schooner *Thames* in the mouth of the river Sal'naia Kur'ia (Figs. 2, 3) at 66°22′41.5″N and 87°35′04.4″E using an underwater camera showed that fine particulate deposits (silts) from the riverbed constantly obstructed visibility and caused the automatic focusing system to capture images of silt particles instead of the river bottom. Silt outs occurred not only due to motion around the riverbed; slow (<0.5 m/s) currents from the Yenisei and the current of the

Sal'naia Kur'ia (<0.2 m/s) colliding in the river mouth create continuous movement of sediment in this area significantly affecting the visibility of the water.

Device type	Swath range	Riverbed penetration depth	Distance from river bottom	Definition	Material of object
Side-scan sonar	100 m	-	5 m	0.1 m	All
Acoustic Doppler current profiler (ADCP)	-	10 m	5 m	1 m	All
Video camera	4–6 m	-	3–5 m	0.03–0.05 m	All
Photo camera	4–6 m	-	3–5 m	0.02–0.03 m	All
Magnetometer	-	0.5 m	1–2 m	-	Ferromagnetic metals
Metal detector	<1 m	0.5 m	0.1–0.3 m	0.05–0.1 m	Metals

Figure 1. Comparative table of specification of devices used in underwater archeology (source [3, p. 92]).

The acoustic Doppler current profiler may be a useful tool in the arsenal of an underwater archeologist working on the Yenisei as it can be used for conducting further investigation on submerged objects that have been covered by silt or sand. A penetration of 10 m should be considered enough for any river wreck, as ships rarely had this height from keel to deck.

Devices which react to large masses of metal such as metal detectors are relatively useless in cases when the studied object is wooden, as in case of *Thames*. They cannot be used remotely for archeological purposes, requiring the operator to be underwater as well.

More expensive tools include various remotely operated underwater vehicles (ROVs), pioneered for studying historical shipwrecks by Robert Ballard in the 1980s, and autonomous underwater vehicles (AUV) However these mobile devices will hardly find wide application in the relatively shallow waters of rivers such as the Yenisei.

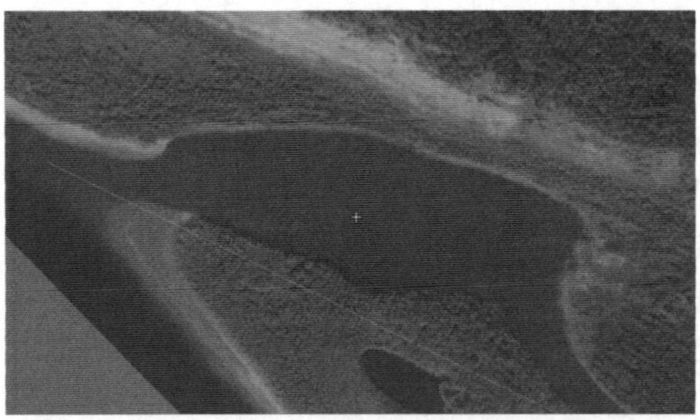

Figure 2. Satellite image of mouth of Sal'naia Kur'ia (image from http://wikimapia.org)

Figure 3. Sal'naya Kur'ia looking southwest. Photo from archive of 2016 *Where Thames meets Yenisei Expedition*.

To conclude on this passage it should be said that with the exception of side-scanning sonars, the remaining devices can be used only for supplementary or specialized purposes such as further investigation, producing high definition images and footage, determining the position on the object under a layer of sediment and detecting metal structures and artifacts. Thus, side-scan sonars are currently the most functional and productive tools for underwater archeology. They are, virtually, unlimited by depth, water clarity, (turbidity) and overhead cover [4].

3. Characteristics of side-scan sonars

Side-scanning sonars differ from other systems in having two arrays. Each array produces one sonar beam which follows a 90-degree arc (horizontally and vertically). The main advantage to using this technology is its ability to "see" almost any submerged object, using a sound spectrum. Side-scanning sonars are able to provide a resolution of one tenth of a degree, whereas in multi-beam sonars this resolution is equal to whole degrees. Thus, side-scanners provide the user with a sound "photograph" of the sea or riverbed tens of times better than multi-beam devices; it is more energy efficient, less cumbersome, has a simpler design and, as a result, a lower price [5, p. 66]. However, the image quality largely depends on the specifications of the concrete sonar.

The market offers a wide selection of various side-scan sonar apparatuses, making it difficult to select a specific model for using in specific environments. Let's look at some side-scanning sonars that have found successful application in marine archeology. In Fig. 4 we have presented the specifications of two devices: the first is an example of a highly-sophisticated expensive piece of equipment, designed for deep-water application in marine environments, the second sonar in a simpler design having limited application as a recreational or amateur piece of electronics.

The table shows that the frequency for the first sonar ranges from 100 to 500 kHz, whereas the second apparatus demonstrates a constant frequency of 1,000 kHz; this enables the researcher to see the smallest details of the submerged object. However, this frequency range limits the range of the sonar to dozens of meters. In terms of marine archeology this range may be inadequate, whereas in rivers it is more than enough, as the average depth rarely exceeds 10–15 meters. The second device is more accessible for smaller groups of researches working on aboard small craft and in shallower enclosed areas of the river, such as coves, bays, river inlets and island channels. As aforementioned, we are focusing on using this technology on the Yenisei, which is a freshwater northern river with current speeds ranging from 1.5 m/s in the tailwater of the Krasnoyarsk Dam to 0.3 m/s below Dudinka. The lower Yenisei, specifically its delta and estuary are subjected to tidal currents and rises in water levels, caused by northerly winds. The riverbed in this area consists primarily out of sand and silt. This inevitably affects the clarity of the water, which as aforementioned, has negative impact on the usage of various underwater optical instruments. Dense particle suspension is known to be a source of acoustic interference [8, p. 79]. Further development in side-scanning technology includes advances in hydroacoustic antennas with controlled characteristics through phase response [9]. This might provide solutions for using sonars both in deep and shallow waters with silt outs. Another direction for the development of the side-scanner includes improvements in positioning accuracy using global navigation satellite systems (GPS/GLONASS). Recent development in phased array technology [10], which has been successfully applied since the 1990 [11] might provide further improvement in positioning accuracy of side-scanning sonars, some of which already boast an accuracy of 15 cm [12].

	System 3000 Towfish™	Starfish 990F™
Model of operation	Simultaneous dual frequency	Single frequency
Location of digitation	Towfish	Topside box
Data processing and imaging capabilities	SDF or XTF or both selectable	Real time waterfall display
Standard system interfacing capabilities	Interfaces to all major Sonar Data Processors	USB Connection to PC/Laptop
Generation of firing trigger	Transceiver Processing Unit (TPU)	Topbox
Power consumption [W]	120	6
Length [m]	1.22	0.378 (0.166 for the surface stack)
Width [m]	0.089	0.097 (0.034 for the surface stack)
Height [m]	0.089	0.11 (0.106 for the surface stack)
Weight in air [kg]	29	2 (0.4 for the surface stack)
Max. depth rating [m]	1500	50
Transducer type and material	Single Beam-Piezo-Electric Ceramic Array	Monolithic Ceramic
Materials used	Stainless Steel	Reinforced red polyurethane rubber
Fish height from seabed determination method	Altitude Algorithm	First Return Algorithm
Deployment methods	Winch with coaxial; hand deployed; Kevlar reinforced lightweight cables & AUV platforms	By hand and towed from boat
Min. frequency [kHz]	100	1000
Max. frequency [kHz]	500	1000
Min. pulse length [μs]	25	400
Max. pulse length [μs]	400	400
Source level [dB]	242	210
Max. range [m]	600	35
Min. beam width [deg]	0.21	0.3
Max. beam width [deg]	0.7	0.3
Min. vertical beam width [deg]	40	60
Max. vertical beam width [deg]	40	60

Figure 4. Comparison of System 3000 Towfish and Starfish 990F side-scanning sonars (source: [6; 7]).

Despite their obvious advantages, side-scanning sonars still have one significant disadvantage – a high cost. This limits smaller and underfinanced institutions and projects from applying this device for underwater research.

4. Results of the 2016 Yenisei archeological survey

Due to insignificant information on the location of shipwrecks on the Yenisei and their current condition, it was decided to employ a less expensive method of detecting and positioning the wrecks on the river bottom. This included preparatory archival research, the examination of historical records, maps, navigation charts and satellite imagery, employing echo sounding devices and traditional bathymetric instruments, such as the sounding line and grappling hook, submergible cable video cameras and short-term underwater diving. Since the selected wreck was supposedly located at a depth of less than 6 m, the absence of a side-scanner was not considered a major obstacle by the team.

Relevant to this article is the usage of echo sounding for detecting the location and position of the wreck of *Thames*, a nineteenth century British screw schooner, which sank in the mouth of the Sal'naia Kur'ia river – a tributary of the Yenisei – in 1878. Echo sounding, along with side-scan sonar is one of two hydroacoustic methods, employed for underwater archeology [3]. A Lowrance Elite-3x^{TM} fishfinder was used during the survey; the frequency range of the device is between 83 and 200 kHz, the maximum depth of application in 244 m.

The riverbed topography displayed by the device is rather uninformative so it was decided to use the results of sounding by both sonar and sounding line to build a three-dimensional image in Surfer 13^{TM} countering and surface modeling software (Figs. 5, 6).

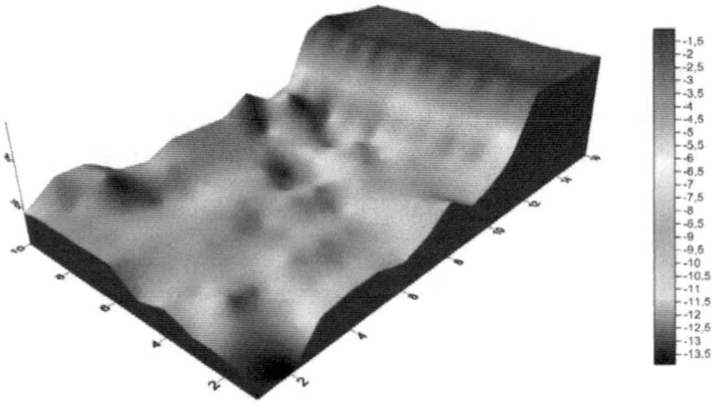

Figure 5. Three-dimensional model of Sal'naia Kur'ia riverbed built in Surfer 13.

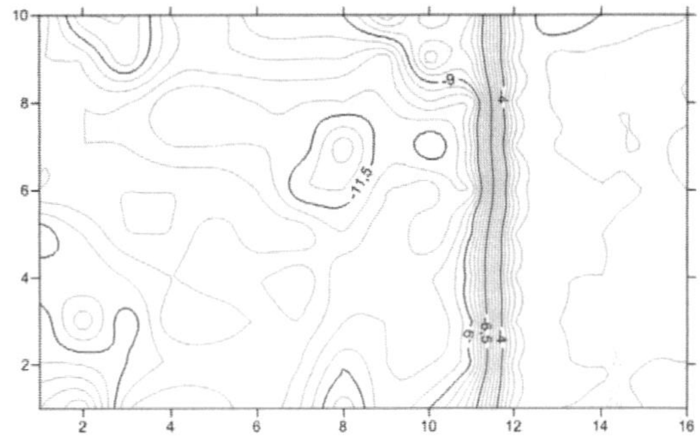

Figure 6. Two-dimensional counter graph of Sal'naia Kur'ia riverbed built in Surfer 13.

The survey revealed a surprisingly deep sector at the confluence of the Sal'naia Kur'ia and the Yenisei (10–13 m) on the south banks of the widening; the north bank has an average 4.5 m depth along the meander. This section stretches for 300 m in an east to west direction; the average depth along the thalweg is 10 m with drops to 13 m and rises to 8 m. As the river narrows there is a sudden rise of the riverbed to 2–2.5 m, which covers an area of 50×30 m at 66°22.762N, 87°34.713E (NW diagonal) and 66°22.722N, 87°35.00E (SE diagonal).

Trawling performed from west to east had the grappling hook repeatedly being snagged in the shallow area. It was decided that the wreck was buried here under a layer of silt and sand. As it had been mentioned, the underwater camera was unable to focus due to floating silt particles. Placing the camera on the riverbed caused silt outs that rendered it useless. Short term underwater diving (snorkeling) revealed part of what is thought to be the stern section of the schooner. Besides scarce historical records which mention the location of the wreck of *Thames* in the Sal'naia Kur'ia [13, p. 19], there is only one pilot chart of the Yenisei where the schooner's position had been marked [14, p. 28]. As there is no record of vessels with the same specifications as *Thames* sailing into the Sal'naia Kur'ia, it can be reported with a great degree of certainty that the wreck is *Thames*. Further archeological work would consist in the removal of over 7,000 m^3 of sand and silt from the vessel, which had been deposited onto it during the last 140 years (this massive amount of sediment suggests that most of it had been on the bottom of the Sal'naia Kur'ia when the ship foundered in 1878, sinking into the silt). This operation should be regarded as futile since most of the vessel's equipment had been removed before she sank. However, it should be noted that the hull might be well preserved as it is buried and protected from the water.

Other practical results include the first sounding of the mouth of the Sal'naia Kur'ia river.

The bathometry showed that the river was not navigable beyond its mouth, which could serve as a harbor for various vessels.

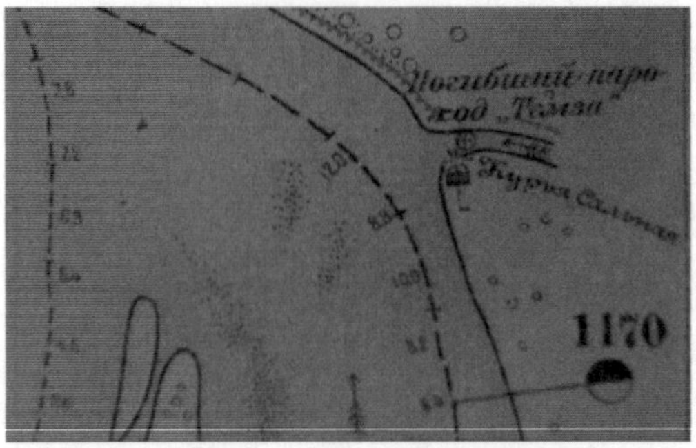

Figure 7. Wreck of *Thames* marked as an obstacle on 1937 pilot's chart.

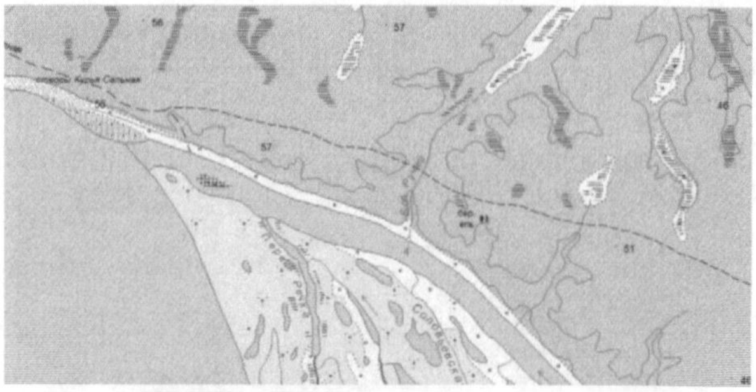

Figure 8. Recent Russian topographical map providing inaccurate information on Sal'naia Kur'ia river.

4. Conclusion

In conclusion it is necessary to say that less sophisticated methods employing simpler active sonars such as fishfinders may find limited application for underwater archeology. Naturally, more advanced equipment such the side-scanning sonar is preferable to use for detecting submerged objects of heritage and performing early research. Despite the vast assortment of side-scanning sonars available on the market, it is preferable to use less energy-consuming and higher frequency devices as they provide better visibility on small depths, which are predominant on the Yenisei; besides they are more mobile and allow investigating lesser uncharted river channels. If, due to a shortage of funds, the researcher is reduced to using simpler sonar technology, such as two-dimensional fishfinders it is possible to use the obtained data to construct topographic models of the riverbed using various software, provided that the area of interest has been divided into equal spaces with depths determined for each space. This can be done easier by affixing the data to the geographic coordinates system through GPS/GLONASS.

References

[1] Ruppe C V, Barstad J F (eds.) 2002 International handbook of underwater archeology Springer Science + Business Media (New York).

[2] Rimskii-Korsakov N A 2005 Gidrolokatsionnye sredstva i praktika podvodnykh issledovanii v IO RAN [Sonars and their use for underwater investigations at the Institute of History of the Russian Academy of Sciences (in Russ.)] *9th International Conference on Science and Technology "Modern Methods of Oceanological Research" Proceedings* (Moscow), Vol. 1, P. 33–46.

[3] Taskaev V N 2010 Metodika provedeniia podvodno-arkheologicheskikh rabot [A technique for performing underwater archeological works] *Voprosy podvodnoi arkheologii* No. 1, P. 45–95.

[4] Kaeser A J, Litts T L (2013) An illustrated guide to low-cost, side scan sonar habitat mapping (Accessed at https://www.fws.gov/panamacity/resources/A)%20Introduction.pdf)

[5] Voronin V A, Tarasov S P (2005) Ispol'zovanie gidrosonara bokovogo obzora dlia resheniia zadach prikladnoi gidroakustiki [Using side-scanning sonar for practical

hydroacoustic problems (in Russ.)] *Izvestiia Iuzhnogo Federal'nogo Universiteta. Tekhnicheskie nauki* **2**(46), P. 65–69.

[6] System 3000 side scan sonar. Dual-frequency single beam sonar (Accessed at http://www.geo-matching.com/upload/246-1902-4263.pdf)

[7] Gidrolokatov bokovogo obzora Starfish 990F [Starfish 990F side-scanning sonar (in Russ.)] (Accessed at http://www.tetis-pro.ru/catalog/StarFish%20990F.pdf)

[8] Flemming, B W 1976 Side-scan sonar: a practical guide *International Hydrographic Review* (Monaco) Vol. **53**, no. 1, P. 66–92.

[9] Tarasov S P, Timoshenko V I, Voronin A V, Kirichenko I A, Pivnev P P, Soldatov G V, Voloshchenko A P, Essi-Ezing A S, Obydennaia, V A Franchuk D A 2012 Measuring phase-frequency characteristics of the reception aerial of the multibeam echo-sounder in hydroacoustic tank (in Russ.) *Inzhenernyi vestnik Dona* No. 4-1, Vol. **22**.

[10] Tyapkin V N, Dmitriev D D, Sokolovskiy A V, Pershin A S 2013 Technique for tuning of adaptive phased antenna array of navigation equipment of GLONASS customer *Vestnik SibGAU* **1(47)** P. 130–134.

[11] Danforth, C W (1998 [2000]) Acoustic applications of phased array technology, John Hopkins University (Accessed at http://casa.colorado.edu/~danforth/science/sonar/sonar1.html)

[12] Chenskii D A, Grigor'ev K A 2014 Schetchik dlinny kabel'-trosa buksiruemogo podvodnogo apparata [A counter for the cable length of an underwater vehicle (in Russ.)] *Inzhenernye i nauchnye prilozheniia na baze tekhnologii NI NIDays – 2014 Conference proceedings (Moscow),* P. 408–409.

[13] Vladimirov, E N (1940) Geroicheskii reis shkhuny "Utreniaia zaria" [The heroic voyage of the schooner *Utreniaia zaria* (in Russ.)] (Moscow–Leningrad).

[14] Lotsmanskaia karta reki Eniseia ot goroda Eniseiska do ust'ia 1927 km [Pilot's chart of the Yenisei from Eniseisk to the estruary, 1927 km (in Russ.)] (1937 [1938]) (Krasnoyarsk [Tomsk]).

Mechanical Activation of Wood for Adhesive-free board Production

V N Ermolin[1], M A Bayandin[2], S N Kazitsin[3]
[1] Professor, Reshetnev Siberian State Aerospace University, Krasnoyarsk, Russian Federation
[2] Associate Professor, Reshetnev Siberian State Aerospace University, Krasnoyarsk, Russian Federation
[3] Assistant, Reshetnev Siberian State Aerospace University, Krasnoyarsk, Russian Federation

E-mail: mihailbayandin@yandex.ru

Abstract. This paper proposes to use hydrodynamic treatment of wood for the manufacture of wood-based panels from sawdust without using adhesive materials. It was found that such a treatment of wood particles (sawdust, dust, wood powder) allows producing panels with high physical-mechanical properties and water resistance. It is proved that the hydrodynamic treatment allows providing maximum energy of autoadhesion interaction in the moulding material due to increase of specific surface with small changes of geometric size of particles in comparison with mechanical methods of milling.
Keywords: wood boards, fraction, hydrodynamic treatment, sawdust, wood dust, wood powder, cavitation, strength, milling.

Introduction

Mechanical conversion of wood produces large quantities of waste. A significant part of the waste is kept in the storehouse or is recycled by burning for heat energy. The problem is particularly acute for reprocessing of milled wood (sawdust) the amount of which is up to 12 % of the volume of raw material [1]. Use of sawdust for the production of boards can be a possible solution of this problem. The most common materials are HDF and MDF boards which are produced in large quantities worldwide and have high mechanical properties [3]. Strength and water resistance of these boards is ensured by addition of synthetic glue, which significantly impairs their environmental specifications. Herewith production of wood boards requires certified raw material, and the use of sawdust as raw material does not allow obtaining the required quality of boards [1].

There are known technologies of board production without using adhesives made from wood particles preliminary chemically, mechanically and biologically treated [2, 4]. This allows to increase the specific surface area and to provide structure formation of the wood boards. Mechanical methods based on the wood milling are the most widespread methods of wood processing. Some authors note [8, 6] that physical-mechanical properties of board materials significantly depend on the fractional composition of the original woodpulp. For example, the paper [7] states that reducing the size of the particles of lignocellulosic raw materials allows to increase physical-mechanical properties of the boards without using adhesive materials. But nevertheless production of material is possible only due to the impact of high pressure and temperature. Therefore material with acceptable properties and a density of at least 1000 kg/m^3 can be produced. These technologies are technically difficult to implement, and such boards have limited field of application.

Being based on the analysis one can come to the conclusion that the use of mechanical disintegrating is ineffective. It is more practical to find methods which allow not only to mill the wood, but also to increase the specific surface area due to fibrillation.

For this purpose the method of wood treatment in a hydrodynamic dispersion machine was chosen. According to the study [5], this method allows to produce homogeneous plastic mass which can be used for production of free-adhesive boards with a wide range of density from 300 kg/m^3 to 1200 kg/m^3. Herewith boards with density comparable to the density of MDF-boards comply with requirements of EN 622: bending strength is 27 MPa, longitudinal strength perpendicular to the sawn face is 0.8 MPa. The amount of through-thickness swelling of these plates does not exceed 8.4 %, which indicates their increased water resistance.

Materials and methods

To confirm the fact that the hydrodynamic treatment allows to increase the specific surface area significantly, the fractional composition of the wood pulp produced by the hydrodynamic machining was studied. At the same time fractional composition of wood powder and wood dust produced by wood grinding was determined. Studies were carried out with the help of cytoanalyzer WU-4. Fractional composition of wood dust and wood powder were determined. Table 1 shows the obtained results.

Table 1 — Fractional composition of wood particles

Fraction size, mm	Sawdust produced by hydrodynamic treatment	Grinding dust	Wood powder
0.5	0	10	7
0.1	70	85	79
less than 0.1	30	5	14

According to the obtained results, size of studied particles does not differ significantly. Special studies were carried out in order to prove that the hydrodynamic treatment increases the specific surface area of the particles due to the fibrillation. Using the same technology there were produced boards from hydrodynamically treated sawdust, dust, hydrodynamically treated dust, wood powder and hydrodynamically treated wood powder. All boards with specified density of 800 kg/m^3 were produced in a single-storeyed laboratory press at temperature of 190 °C and pressure of 3.5 MPa. Specific pressing time is 2.0 min/mm.

Results and discussion

After pressing the boards were weathered in the laboratory for 72 hours and then subjected to physical-mechanical testing. Table 2 shows the obtained results.

Table 2 — Physical-mechanical properties of wood-boards

Physical-mechanical property	Sawdust after hydrodynamic treatment	Grinding dust		Wood powder	
		without treatment	after treatment	without treatment	after treatment
Bending strength, MPa	28.1	7.3	24.6	3.4	16.8
Longitudinal strength perpendicular to the sawn face, MPa	0.82	0.15	0.94	0.11	0.79
Through-thickness swelling over 24 h	7.8	56.7	18.5	55.5	19.9

According to the obtained results, boards produced from hydrodynamically treated sawdust are characterized by maximum strength and minimal swelling. It is noted, that the boards made of wood powder, which is characterized by a comparable particle size, have 8.2 times lower bending strength. Approximately the same difference is found for longitudinal strength perpendicular to the sawn face. Boards of wood dust had a bit higher physical-mechanical properties, but at the same time destruction of the samples at the end of the swelling test was observed. This shows low water resistance of the materials. When wood particles (wood dust and wood powder) were preliminary hydrodynamically treated, a significant improvement of the properties of wood boards was observed. The values of the mechanical properties increased more than 4 times and the swelling amount decreased more than 3 times. Significantly different values of physical-mechanical properties of boards indicate that the hydrodynamic treatment of wood particles with comparable sizes allows to increase the specific surface area and thus to improve energy of autoadhesion interaction. Generally, the obtained results confirm the assumption that in the process of hydrodynamic treatment of the wood with the cavitation effect a more intense fibrillation is possible, which allows to increase specific surface area significantly and to produce free-adhesive materials with different density.

Conclusions

Hydrodynamic treatment allows producing wood particles with a large specific surface. This creates conditions for production of free-adhesive wood boards of different density. Boards made by this method meet the requirements of EN622.

Development of technologies based on hydrodynamic treatment of wood is a promising direction of waste recycling as it allows to produce competitive products.

References

[1] Hakkila P 1989 Utilization of Residual Forest Biomass. p. 581.

[2] Salvado´ J, Vela´squez J A, Ferrando F 2003 Binderless fiberboard from steam exploded *Miscanthus Sinensis: optimization of pressing and pretreatment conditions Wood Science Technology* 37:279–286

IOP Conf. Series: Materials Science and Engineering **155** (2016) 012038 doi:10.1088/1757-899X/155/1/012038

[3] Suchsland O, Woodson G E 1991 Fiberboard – manufacturing practices in the US. *Forest Products Research Society*, USA p 280.
[4] Unbehaun H, Konig S, Spindler D and Kerns G 2008 Enzymatic Modification of Lignocellulosic Substances for the Production of Fiberboards *Moscow University Chemistry Bulletin*, Vol. 63, No. 2, pp. 126-130.
[5] Bayandin M A, Ermolin V N , Kazicin S N 2015 Influence on the formation fine fraction properties of wood-based panels without connections. *Conifers of the boreal area* 33. № 3-4. p. 182-185.
[6] Kawasaki T, Zhang M, Kawai S 1998 Manufacture and properties of ultra-low-density fiberboard *The Japan Wood Research Society* 44:354-360.
[7] Xu J, Widyorini R, Yamauchi H, Kawai S 2004 Development of binderless fiberboard from kenaf core *Journal of wood science* 50:53-61.
[8] Kelvin M Chapman 2006 Wood-based panels: particleboard, fibreboards and oriented strand board *Primary Wood Processing* pp. 425-475.

Model of Synchronous Motion of Links of the Combined Manipulator during Load Lifting

V F Poletaykin[1], P G Kolesnikov[2]

[1] Chairman of the Department of Technologies and machinery of environmental engineering of Reshetnev Siberian State Aerospace University

[2] Chairman of the Department of Vehicles, tractors and forestry machines of Reshetnev Siberian State Aerospace University

E-mail: mfsibgtu@mail.ru

Annotation The combined manipulators composed of telescopic booms and rotary supports with movable in longitudinal vertical plane columns which provide an increased outreach of operative parts and a bigger service space are widely used in different branches of industry (engineering, transport). The article considers development of mathematical model of kinematic links motion of the combined manipulator with a column movable in longitudinal vertical plane. The mode of load lifting in a synchronous movement of boom and column from the position of picking the load up to the transport position was analyzed.

Introduction

Combined hydraulic manipulators are widely used in production of weight-handling, loading and other machinery. Production and improvement of such machines should be based on studying the dynamics of structural elements. Herewith methods of mathematical modeling are considered to be the most effective. On this basis, researches with the use of mathematical models of combined manipulator links motion mode dynamics focused on validation of kinematics parameters and design of the processing equipment, should be considered as relevant.

Foundation of the calculation model

Fig. 1 shows the calculation model of the “processing equipment – load” system. The analyzed mode can be used when the manipulator operates as a processing equipment of forestry, road-building, load-lifting and other machines.

When working element picks the load up, the telescopic boom brings it to the machine by retraction of sections, then the load is set to the transport position by actuation of column rotation hydraulic cylinder (CRHC) and boom lift (BL). Meanwhile the boom rotates around K-axis, and the column rotates around O-axis. Rotation angle of the boom is φ (relative motion), rotation angle of the column is α (transportation motion). Reference point of the angle φ is the lower limit position of the boom; reference point of the angle α is the rightmost position of the column.

Following symbols are used in Fig. 1.

G_1, G_2, G_3: gravitation forces of outer, middle and inner sections of telescopic boom;

$G_{Ц1}$, $G_{Ц2}$, G_0: gravitation forces of hydraulic cylinders while sections’ telescoping and of sections’ telescoping mechanism;

G_3, $G_{ГР}$, G_P: gravitation forces of gripping device, load and rotator;

$G_{Ц3}$, $G_{Ц4}$: gravitation forces of hydraulic cylinders while column rotation and boom lift;

$G_{\Pi P.C}$: total gravitation force of structural elements of the boom and the load brought to point C;

P, P_C: forces on rods of the column rotation and boom lift hydraulic cylinders;

L: boom dimensions at retracted sections;

ℓ_1, ℓ_2, ℓ_3, ℓ_4, ℓ_5: distances from K-axis of boom rotation to the centers of structural elements gravity

ℓ_6, ℓ_7, ℓ_8, ℓ_9, ℓ_{10}, ℓ_{11}: dimensions of manipulator elements.

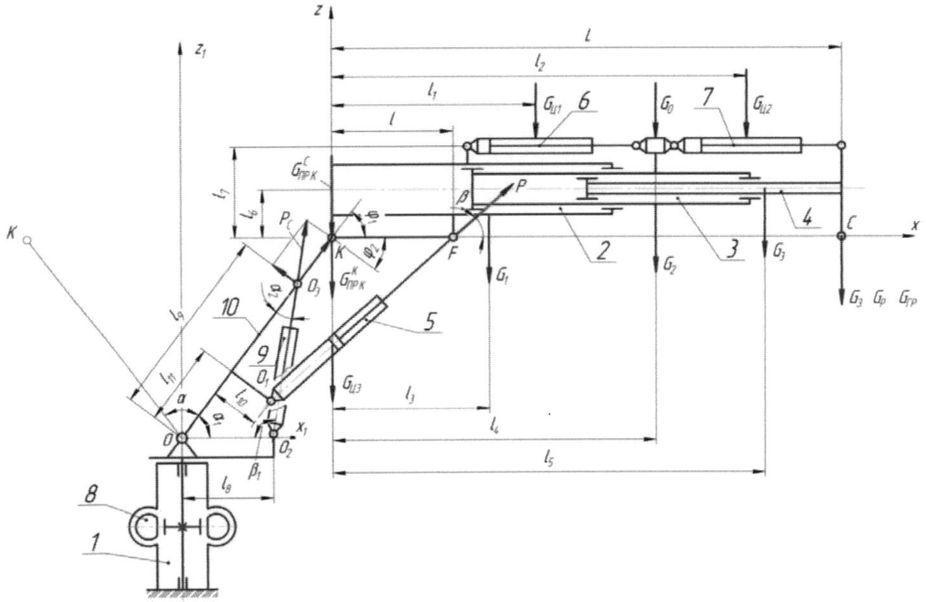

Figure 1. Calculation diagram of the "processing equipment – load" system (manipulator with a tilting column):
1– rotary support; 2, 3, 4 – outer, middle and inner sections of telescopic boom; 5 – boom lift hydraulic cylinder; 6, 7 – hydraulic cylinders of sections' telescoping mechanism; 8 – mechanism of manipulator's rotation in horizontal plane; 9 –hydraulic cylinder of column rotation; 10 – column; $O_1K = C$; $OO_2 = \ell_8 = C_1$

Constructing the equations of motion for the "processing equipment – load" system

The boom rotates in ZKX-plane, the column rotates in Z1OX1-plane. The rotation angles α and φ clearly determine the positions of this elements of the system in the rotation planes. If dimensions of the boom L and the column L_K are known, we can find position of any point by means of specified parameters. Based on this, the system can be considered as a system with two degrees of freedom (K = 2) with generalized coordinates α and φ.

To formulate equations of motion for this mechanical system the Lagrange equations of the 2nd kind was used. According to the number of degrees of freedom of the system two Lagrange equations were formulated

$$\frac{d}{dt}\left(\frac{\partial T}{\partial \dot{\alpha}}\right) - \left(\frac{\partial T}{\partial \alpha}\right) = Q_\alpha,$$
$$\frac{d}{dt}\left(\frac{\partial T}{\partial \dot{\varphi}}\right) - \left(\frac{\partial T}{\partial \varphi}\right) = Q_\varphi, \quad (1)$$

IOP Conf. Series: Materials Science and Engineering **155** (2016) 012039 doi:10.1088/1757-899X/155/1/012039

where Q_{α}, Q_{φ} are generalized forces corresponding to generalized coordinates α and φ respectively.

The motional energy of the system is equal to the sum of motional energies of the column and the boom, i.e. the sum of the motional energies of relative and transportation motions

$$T = T_1 + T_2, \tag{2}$$

where T_1 is motional energy of the given column mass with the given masses of the structural elements mounted on it (column rotation and boom lift hydraulic cylinders and other elements of the hydraulic actuator);

T_2 is motional energy of the given mass of the column and the load.

When the telescopic boom rotates, its sections do not telescope, and the dimensions of the boom L do not change, therefore positions of the mass centers of the structural elements of the boom relatively to the K-axis (radius of mass inertia) remain constant. In this case in order to simplify the determination of the motional energy of the system masses of structural elements of the telescopic boom were brought to point C (point of suspension of the rotator to the boom). Masses of structural elements of the column were brought to K-axis (mounting axis of the boom). Calculation of the mass of structural elements of the boom $m_{ПР.С}$ is based on the condition that motional energy of the reduced mass is equal to the sum of the motional energies of the masses it replaces. Therefore,

$$\frac{G_{ПР.С}\dot{\varphi}^2 L^2}{2g} = \frac{\dot{\varphi}^2}{2g}\cdot\left[\left(G_1\,\ell_3^2 + G_2\,\ell_4^2 + G_3\,\ell_5^2 + G_{Ц1}\,\ell_2^2 + G_0\,\ell_4^2 + G_P + G_З + G_{ГР}\right)\right] \tag{3}$$

On this basis, the boom mass brought to point C is

$$m_{ПР.С} = \frac{G_{ПР.С}}{g} = \frac{G_1}{g}\cdot\frac{\ell_3^2}{L^2} + \frac{G_2}{g}\cdot\frac{\ell_4^2}{L^2} + \frac{G_3}{g}\cdot\frac{\ell_5^2}{L^2} + \frac{G_{Ц1}}{g}\cdot\frac{\ell_1^2}{L^2} + \frac{G_{Ц2}}{g}\cdot\frac{\ell_2^2}{L^2} + + \frac{G_0}{g}\cdot\frac{\ell_4^2}{L^2} + \frac{G_P + G_З + G_{ГР}}{g} \tag{4}$$

or

$$m_{ПР.С} = m_1\,\frac{\ell_3^2}{L^2} + m_2\,\frac{\ell_4^2}{L^2} + m_3\,\frac{\ell_5^2}{L^2} + m_{Ц1}\,\frac{\ell_1^2}{L^2} + m_{Ц2}\,\frac{\ell_2^2}{L^2} + m_0\,\frac{\ell_4^2}{L^2} + m_P + m_З + m_{ГР} \tag{5}$$

where m_1 is mass of the outer section of the boom;

m_2 is mass of the middle section of the boom;

m_3 is mass of the inner section of the boom;

$m_{Ц1}$, $m_{Ц2}$ are masses of hydraulic cylinders of the sections' telescoping mechanism;

m_0 is mass of the sections' telescoping mechanism;

m_P, $m_З$, $m_{ГР}$ are masses of gripping device, load and rotator.

Mass of the column and structural elements mounted on it brought to point K can be calculated with the following equation:

$$\frac{G_{ПР.К}\dot{\alpha}^2 L_K^2}{2g} = \frac{G_K\dot{\alpha}^2\,(0{,}5L_K)^2}{2g} + \frac{0{,}5G_{Ц3}\,\dot{\alpha}^2(0{,}5\ell)^2}{2g} + \frac{0{,}5G_{Ц4}\dot{\alpha}^2(0{,}5\ell_8)^2}{2g} \tag{6}$$

The equation (6) is based on the assumption that gravitation forces of cylinders $G_{Ц3}$ and $G_{Ц4}$ are evenly distributed between the boom and the column as well as between the column and the base of the rotary support; points of their application are $0{,}5\ell$ and $0{,}5\ell_8$ respectively from the mounting axis of the boom (point K).

According to the equation (6) mass of the column and structural elements mounted on it brought to point K is

$$m_{ПР.K} = \frac{G_K}{g} \cdot \frac{(0{,}5L_K)^2}{L_K^2} + \frac{0{,}5G_{Ц3}}{g} \cdot \frac{(0{,}5\ell)^2}{L_K^2} + \frac{0{,}5G_{Ц4}\,(0{,}5L_K)^2}{gL_K^2} = 0{,}25\frac{G_K}{g} + 0{,}125\frac{G_{Ц3}}{g} \cdot \frac{\ell^2}{L_K^2} + +0{,}125m_{Ц4} \cdot \frac{\ell_8^2}{L_K^2} \quad (7)$$

or $m_{ПР.K} = 0{,}25\, m_K + 0{,}125 m_{Ц3} \cdot \frac{\ell^2}{L_K^2} + 0{,}125 m_{Ц4} \cdot \frac{\ell_8^2}{L_K^2}$ (8)

According to (2) the motional energy of the system is

$$T = T_1 + T_2 = \frac{I_0\,\dot{\alpha}^2}{2} + \frac{1}{2}(m_{ПР.C}V_{ac}^2 + I_C\dot{\varphi}^2)\,, \quad (9)$$

where I_0 is moment of inertia of the column relative to O-axis,

I_C is central moment of inertia of the boom assembly;

V_{ac} is an absolute motion velocity of e point C (point of boom mass brining).

The theorem on composition of velocities was applied, according to which absolute velocity of point C is equal to geometric sum of its transportation and relative velocities

$$V_{ac}^2 = V_{ec}^2 + V_{rc}^2 + 2V_{ec} \cdot V_{rc}\cos\gamma,$$

where V_{ec}, V_{rc} are velocities of transportation and relative motion of point C,
γ is an angle between vector directions of transportation and relative motion of point C.

Fig. 2 shows a diagram for determining velocity of an absolute motion of point C - V_{ac}, point of boom and the load mass bringing.

$$V_{ec} = OC \cdot \dot{\alpha}\,;\ V_{rc} = L \cdot \dot{\varphi}\,.$$

According to OKC triangle

$$OC^2 = L_K^2 + L^2 - 2L_K \cdot L\cos(\varphi_H + \varphi).$$

According to the same triangle $L_K^2 = L^2 + OC^2 - 2L \cdot OC \cdot \cos C$.

Fig. 2 shows, that angles γ and C are equal being angles with mutually perpendicular sides. Thus,

$$\cos C = \cos\gamma = \frac{L^2 + OC^2 - L_K^2}{2L * OC}$$

Absolute motion velocity

$$V_{ac}^2 = \dot{\alpha}^2 \cdot OC^2 + \dot{\varphi}^2 \cdot L^2 + 2\dot{\alpha} \cdot \dot{\varphi} \cdot L \cdot OC \cdot \cos\gamma.$$

$\cos\gamma = \cos C$ is expressed by $\cos(\varphi_H + \varphi)$.

$$\cos\gamma = \frac{L^2 + OC^2 - L_K^2}{2L \cdot OC} = \frac{L^2 + L_K^2 + L^2 - 2L_K \cdot L \cdot \cos(\varphi_H + \varphi) - L_K^2}{2L \cdot OC} = \left[\frac{L^2 - L_K \cdot L}{L \cdot OC}\right] \cdot \cos(\varphi_H + \varphi).$$

After substituting

$$V_{ac}^2 = \dot{\alpha}^2 \cdot OC^2 + \dot{\varphi}^2 \cdot L^2 + 2\dot{\alpha} \cdot \dot{\varphi} \cdot L \cdot OC \cdot \left[\frac{L^2 - L_K * L}{L * OC}\right] \cdot \cos(\varphi_H + \varphi) \quad (10)$$

In accordance with (10), equation of motional energy looks like the following

$$T = \frac{m_{\text{ПР.К}} \cdot L_K^2 \cdot \dot{\alpha}^2}{2} + \frac{m_{\text{ПР.С}} \cdot OC^2 \cdot \dot{\alpha}^2}{2} + \frac{m_{\text{ПР.С}} \cdot L^2 \cdot \dot{\varphi}^2}{2} + m_{\text{ПР.С}} \cdot L \cdot OC \cdot \dot{\alpha} \cdot \dot{\varphi} \cdot \left[\frac{L^2 - L_K \cdot L}{L \cdot OC}\right] \cdot cos(\varphi_H + \varphi) + \frac{I_C \cdot \dot{\varphi}^2}{2} \quad (11)$$

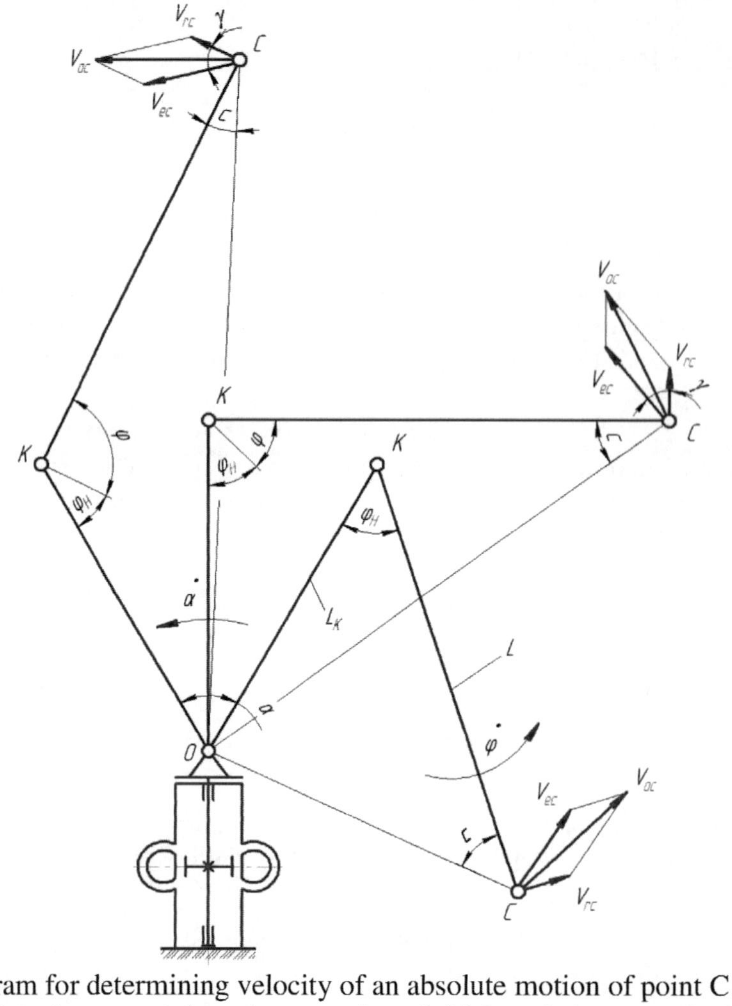

Figure 2. Diagram for determining velocity of an absolute motion of point C in synchronous movement of the boom and the column: φ_H is an initial angle between the boom and the column; φ is rotation angle of the boom (relative motion); α is rotation angle of the column (transportation motion); $\dot{\alpha}$, $\dot{\varphi}$ are angular velocities of the column and the boom, respectively

Equations of motion for the "processing equipment – load" system

By differentiating the equation of the motional energy (11) for the components of the Lagrange equations (1), the following system of equations of motion of the dynamic system of the manipulator is expressed:

$$m_{ПР.К} \cdot L_K^2 \cdot \ddot{\alpha} + m_{ПР.С.} \cdot OC^2 \cdot \ddot{\alpha} + m_{ПР.С} \cdot OC \cdot L \cdot \ddot{\varphi} \left[\frac{L^2 - L_K \cdot L}{L \cdot OC}\right] \cdot \cos(\varphi_н + \varphi) -$$

$$- m_{ПР.С} \cdot OC \cdot L \cdot \dot{\varphi}^2 \cdot \left[\frac{L^2 - L_K \cdot L}{L \cdot OC}\right] \cdot \sin(\varphi_H + \varphi) = Q_\alpha$$

$$(m_{ПР.С}L^2 + I_C)\ddot{\varphi} + m_{ПР.С} \cdot OC \cdot \ddot{\alpha} \cdot L \cdot \left[\frac{L^2 - L_K \cdot L}{L \cdot OC}\right] \cdot \cos(\varphi_H + \varphi) = Q_\varphi \quad (12)$$

Determination of generalized forces Qα and Qφ corresponding to the generalized system coordinates φ and α

To determine generalized forces Q_φ and Q_α the principle of possible motions of the system in direction of the increase of generalized coordinates φ and α - $\Delta\varphi u \Delta\alpha$ was used. For calculation of the generalized force Q_φ *one considers*$\Delta\alpha = 0$, and for calculation of the generalized force $Q_\alpha \Delta\varphi = 0$. Gravitation forces of structural elements and load G_i, and forces on rods of the hydraulic cylinders P_C and P are considered as active forces. Generalized force is considered as a factor while expressing sum of elementary works of active forces in the direction of possible movement:

$$\delta A_\varphi = Q_\varphi \cdot \Delta\varphi; \;\; \delta A_\alpha = Q_\alpha \cdot \Delta\alpha. \quad (13)$$

Let us determine the sum of elementary works of active forces in direction of the generalized coordinate φ. Herewith $\Delta\varphi \neq 0, \Delta\alpha = 0$.

To determine the generalized force Q_φ equations (4), (5) of mass of the boom $m_{ПР.С}$ brought to point C in a horizontal position of the boom are used.

Then $\Sigma\delta A_\varphi = (P\ell \sin\beta - m_{ПР.С}gL)\Delta\varphi$

$$Q_\varphi = P\ell \sin\beta - m_{ПР.С}gL \quad (14)$$

Let us determine the sum of elementary works of active forces and gravitation forces of structural elements in direction of generalized coordinate α. Herewith $\Delta\varphi = 0$; $\Delta\alpha \neq 0$. In the direction of generalized coordinate α active force P_C and gravitation forces of structural elements of the boom and the column perform an operation. In order to simplify the equation of generalized force Q_α gravitation forces of structural elements of boom, load and column to the point K are reduced. In this connection horizontal position of the boom is considered. Bringing the forces to selected points is performed on condition of equality of moments of the reduced force to the sum of moments of reduced forces relative to any point on the plane (Varignon theorem).

In order to determine gravitation force brought to point K, equations of mass of the boom brought to the point C in a horizontal position of the boom (4), (5) $m_{ПР.С}$ are use.

Let us formulate an equation of moments of gravitation force of the boom brought to point C ($G_{ПР.С}$) and gravitation force of the boom brought to point K ($G^C_{ПР.К}$) relative to O-axis.

$$m_{ПР.С}g(L + L_{K}.\cos\alpha_1) = G^C_{ПР.К} \cdot L_{K}.\cos\alpha_1.$$

Therefore,

$$G_{\text{ПР.К}}^{C} = \frac{m_{\text{ПР.С}} g(L + L_K \cdot \cos\alpha_1)}{L_K \cdot \cos\alpha_1} = \frac{m_{\text{ПР.С}} gL + m_{\text{ПР.С}} gL_K \cdot \cos\alpha_1}{L_K \cdot \cos\alpha_1} = \frac{m_{\text{ПР.С}} gL}{L_K \cdot \cos\alpha_1} + m_{\text{ПР.С}} g \qquad (15)$$

Now it is necessary to determine gravitation force of structural elements of the column briught to point K and gravitation force of hydraulic cylinders of G_C, $G_{Ц4}$, $G_{Ц3}$actuators drive. In doing so equations (7), (8) for determination of reduced mass of the said structural elements to the point K are used:

$$G_{\text{ПР.К}}^{K} = m_{\text{ПР.К}} g \qquad (16)$$

Then $\Sigma\delta A_\alpha = \left[P_C \cdot \sin\alpha_2 \cdot \ell_9 - \left(G_{\text{ПР.К}}^{K} + G_{\text{ПР.К}}^{C}\right) L_K \cdot \cos\alpha_1\right]\Delta\alpha$

$$Q_\alpha = P_C \cdot \sin\alpha_2 \cdot \ell_9 - \left(G_{\text{ПР.К}}^{K} + G_{\text{ПР.К}}^{C}\right) L_K \cdot \cos\alpha_1.$$

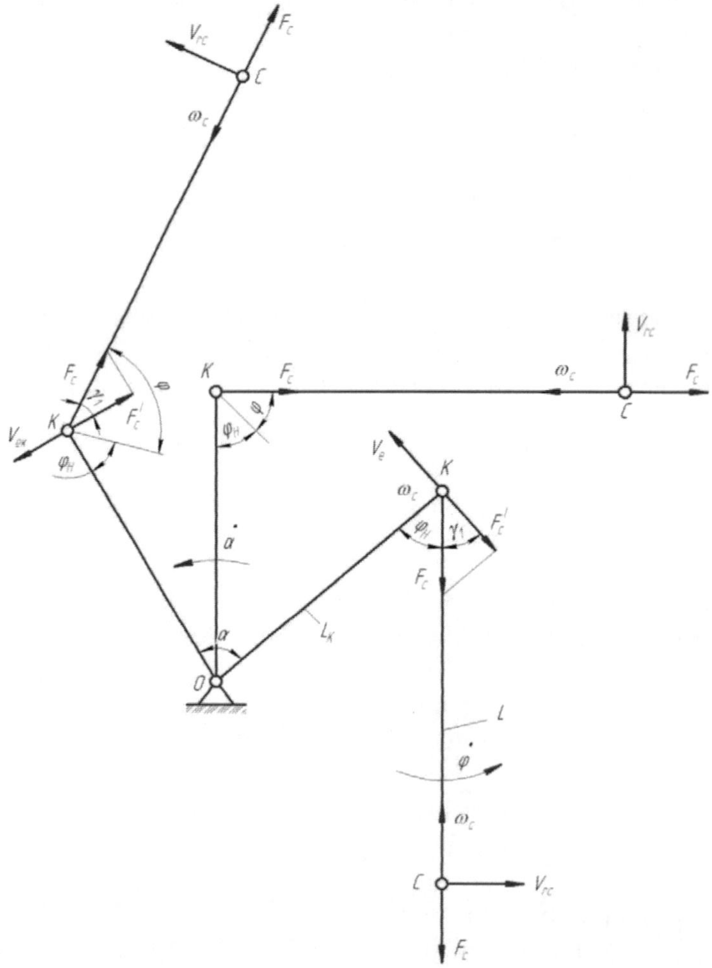

Figure 3. Diagram for determination of Coriolis acceleration and Coriolis forces: ω_C is Coriolis acceleration, F_C is Coriolis force, $V_{\ell c}$, V_{rc} are linear velocities of transportation and relative motion of point C

Since the transportation motion of the system is rotational and while simultaneous rotation of the column and the boom, the distance OC (see Fig. 3) is constantly increasing, so there appears repeated rotating (Coriolis) acceleration ω_C and the Coriolis force F_C.

Coriolis acceleration is calculated by the equation

$$\omega_C = 2\dot{\alpha} \cdot V_{rc} = 2\dot{\alpha} \cdot \dot{\varphi} \cdot L$$

According to the moment of Coriolis force the generalized force Q_α is calculated in the following way:

$$Q_\alpha = P_C \cdot \sin\alpha_2 \cdot \ell_9 - \left(G^K_{\Pi P.K} + G^C_{\Pi P.K}\right)L_K \cdot \cos\alpha_1 - 2m_{\Pi P.C} \cdot \dot{\alpha} \cdot \dot{\varphi} \cdot L \cdot L_K \cdot \cos\gamma_1 . \quad (17)$$

According to the equations (14) and (17) motions of the considered dynamical system are determined as follows:

$$m_{\Pi P.K} \cdot L_K^2 \cdot \ddot{\alpha} + m_{\Pi P.C.} \cdot OC^2 \cdot \ddot{\alpha} + m_{\Pi P.C} \cdot OC \cdot L \cdot \ddot{\varphi}\left[\frac{L^2 - L_K \cdot L}{L \cdot OC}\right] \cdot \cos(\varphi_H + \varphi) -$$

$$- m_{\Pi P.C} \cdot OC \cdot L \cdot \dot{\varphi}^2 \cdot \left[\frac{L^2 - L_K \cdot L}{L \cdot OC}\right] \cdot \sin(\varphi_H + \varphi) =$$

$$= P_C \cdot \sin\alpha_2 \cdot \ell_9 - \left(G^K_{\Pi P.K} + G^C_{\Pi P.K}\right)L_K . \cos\alpha_1 - 2m_{\Pi P.C} \cdot \dot{\alpha} \cdot \dot{\varphi} \cdot L \cdot L_K \cdot \cos\gamma_1 .$$

$$(m_{\Pi P.C}L^2 + I_C)\ddot{\varphi} + m_{\Pi P.C} \cdot OC \cdot \ddot{\alpha} \cdot L \cdot \left[\frac{L^2 - L_K \cdot L}{L \cdot OC}\right] \cdot \cos(\varphi_H + \varphi) = P\ell \sin\beta - m_{\Pi P.C} g L \quad (18)$$

Conclusion

The result of the performed work is a system of inhomogeneous differential equations of secondary order, which are the basis of mathematical models of combined manipulator in the mode of load lifting with synchronous movement of the boom and the column. The obtained system of equations allows analyzing influence of parameters of kinematic scheme on the level of loads on processing equipment in the process of modeling ($\ddot{\varphi}, \ddot{\alpha}$, L, L_K ,$\dot{\varphi}, \varphi, \dot{\alpha}, \alpha$, OC, l_9 , $\alpha_2, \beta, l, \gamma_1$), as well as analyzing weight value of load and processing equipment ($m_{\Pi P.C}$, $m_{\Pi P.K}$). Besides, by changing the rotation rate of shafts of hydraulic pumps motion rate of cylinder rods are changed and, consequently, an absolute velocity of point of reduction of mass of the boom and the load (point C), angular velocity and angular acceleration of the boom and the column are changed as well. The equations allow estimating influence of these factors on amount of load on structural elements of processing equipment.

References

[1] Kanunnik I A, Kilina M I Principals of mechanics of robots. Dynamics of industrial robots. Section II. – Krasnoyarsk: STI, 1992. – p. 64.

[2] Yablonskiy A A Course of theoretical mechanics. Part II. Dynamics. – M.: Higher school, 1966. – p. 411.

IOP Conf. Series: Materials Science and Engineering **155** (2016) 012040 doi:10.1088/1757-899X/155/1/012040

Relation optimization of the starting and main stages of rockets with short-running engine on the external oxidizer

L A Ustinov[1], P V Zelenkov[2], I V Kovalev[2]
[1] TULA STAT UNIVERSINY, Tula, 300012, Russia
[2] Reshetnev Siberian State Aerospace University, Krasnoyarsk, 660037, Russia

E-mail: info@tsu.tula.ru

Abstract. The paper considers an optimal choice of the mass relation of the starting and main stages of the rocket with a short-running engine on the external oxidizer from the point of view of achieving the maximum firing range.

Key words: trajectory optimization

One of the possible ways to increase the firing range of rockets is the application of the engine systems on the basis of the external oxidizer.

The main restricted factors at the development of design-layout scheme of the small caliber rocket of the "earth-earth" class with a low aerodynamic quality are a small size of the cross section of the unit and the constancy of its mass-dimensional characteristics that allow maintaining the ability to fire from standard launchers. Taking it into account and considering that the size and payload of the aircraft are not the factors under consideration and are taken as fixed ones, the total mass of the starting and main engines are considered as constant.

In order to simplify the design as much as possible it is preferable to use uncontrolled air intake units and fuel charge with a constant speed of burning.

In this case, choosing the optimal relation of the masses of the starting and main engines let us take the starting weight of the rocket as constant, the total fuel mass of the starting and main stages, characteristics of fuels, aerodynamic characteristics of the rocket, characteristics of the air intake and let us consider the effects of changes in the mass relation of the charges the starting and main stages, the flow of fuel of the main stage per second, the angle of elevation.

The change in time of the main stage operation for the given mass of its charge is revealed as a change in the fuel flow per second and accordingly as the change of the air excess factor with the corresponding change in thrust and specific impulse of the main stage.

The angle of sight can be significant from the point of view of the output of the main stage into the high rise area, i.e. the time restriction of its operation. The angle of sight has no any practical influence on the velocity of the rocket at the end of the active section of the solid rocket booster (SRB), and therefore the initial conditions of the main stage operation. The distance at which it is accelerated to the speed of the rocket where you will run the main stage should be minimal to maximize the potential of its engine in a dense atmosphere.

The primary rating rational relation of the masses of the starting and main charges and the study of the qualitative nature of the additional factors influence (reduction of the passive mass, change in the flow of fuel per second, the angle of attack change) were carried out using the software package "Trajectory".

This package allows solving the problem of determining the optimal values of the control parameters of the spacecraft where it reaches its maximum flying range.

The results of the numerical and experimental studies including the weight, size and aerodynamic characteristics of the rocket and the traction and power characteristics of the main stage are not given in this paper due to their big volume used as input data.

The minimum charge weight of the starting SRB was chosen to achieve the smallest speed at which the air intake device would provided the designing the required structure of the jump, then taking into account free volume of the dimensions of the rocket, the charge of the main stage which is then decreased in increments of 25% was given and accordingly the charge of solid propellant was increased. Respectively the trajectory parameters of the rocket were changed as the initial conditions for the main stage by the time of the operation of the SRB completion.

The calculations were performed for four cases of the fuel consumption per second: deliberately inflated at the air excess factor $\alpha \leq 1{,}7$ which will be considered as the basic one (100%), two intermediate (50% and 75% of the basic one) with the air excess factor α =2...3 and for the case with a high flow of air when the air excess factor reached 4, and the second fuel consumption was 25% of the basic one.

According to the results of calculations optimization for the rocket with different operation time of the main stage as combinations of parameters, the basic case which traction was significally lower than the traction of other three case showed the worst results.

The height of the main stage completion was approximately 50-25% of the height at the end of the other case of the fuel consumption. The trajectory of the rocket in this case was like a classical ballistic and decreasing the mass of SRB in the considered case (i.e., speed reduction at the beginning of the passive section of the trajectory for the SRB) and the flying range of the rocket decreased. The disposal of the SRB that is, the reducing of the transverse load as it was expected has led to the further reduction of the firing range. It means that the spacecraft flew at the speed less than it was required to start normal operation of the main stage; its combustion was not taken into account in calculation.

Further various cases of fuel consumption (75% and 50% of the first one) were discussed.

The reduction in fuel consumption per second with the corresponding increase in operation time of the main stage effects positively on the flying range. Also the disposals of the SRB and the angle of attack (planning) on the descending path of the trajectory have a positive effect.

There exists a direct correlation: the more operation time of the main stage, the greater the increase the disposal of the SRB corps is. The benefits from reducing the passive mass taking into account the short time of operation of the main stage, is not obvious. In the case of inertial flight, or a low-thrust sustainer stage this mass may be a battery of kinetic energy, increasing the range. Noticeable was the impact on the range of the angle of attack. The significant height of the top of the flight path provides a reduction in high-speed entry into the dense layers of the atmosphere and, consequently, the effect of increasing the lifting force becomes noticeable.

The comparison of the second (75% of basic one) and third (50% of basic one) of the consumed fuel per second shows that for the case when the operation of the main stage is relatively short, the less consumption is preferred, as despite the lower thrust, the active phase ends at a considerably higher altitude, and the speed and angle of flight of the rocket provide the greater flying range. For the third option fuel consumption per seconds (50% of basic one), the increase in range with increasing mass of the fuel decreases is preferred for the combinations when the operation time of the main stage is comparable to the time of entrance into the sparse atmosphere. This can be explained by the fact that with the reduction of the starting SRB the speed of the rockets at the beginning of the operation of the main stage decreases; it means that it creates less thrust at the same time. Ultimately, this leads to a decrease of the inclination angle of the trajectory and a lower speed of the rocket at the beginning of the passive section of the trajectory but it is balanced only partially by a greater flying range at that moment.

In addition, as the charge weight of the main stage (and proportional to its time) of a certain moment is increased, the height of the end of the operation of the main stage is not changed, as it does not create significant thrust higher. That is, the role of the gently sloping section of flight in the upper section of the trajectory is increasing and the decrease in the angle of the trajectory with a corresponding increase in the horizontal component of the velocity while flying in the rarefied atmosphere becomes rational.

This trajectory is a transition to the next designs, which is a rocket with a long-running main engine.

In these cases the fuel supply was 25% from the basic one and the main stage was operating with the air excess factor in the range 3...4. This operating mode provides high values of specific impulse but decreases the thrust created by the engine.

Let us consider the results of the optimization calculations for the rocket with different operating time of the main stages.

In the case with the fuel consumption per second a monotonic increasing range with increasing charge of the main stage is absent. A long time operation of the main stage leads to the trajectory close to horizontal with a height close to the maximum when the engine is able to operation. With a certain ratio of the SRB charges and the main stage range begins to decrease; it can also be explained by the thrust decrease with the speed decreasing at the end of the acceleration at the start, and accordingly, the inability of a shell to resist the horizontal part of the trajectory, preserving horizontal speed, even in the presence of angle of attack. The absence of such a decline for the reset of the body of the first SRB planning on the descending section of the trajectory can be explained by the fact that the reduced mass of the rockets enough traction to keep the rocket with some angle of attack on the horizontal section of the trajectory, and, in addition, at the end of the main stage to increase the angle of the trajectory with access to the even greater heights. This assumption is confirmed by the graph for the case where the corps disposal is performed but planning is not used. The height of the end of the operation of the main stage for all the considered ratios of the masses of the SRB charge and the main stage of this mode changes its fluctuations slightly are related to the average mass of the main stage with the increase of the gently sloping section of the trajectory, as for design cases 3 and 4, and the increase for small and large values of the mass charge of the main stage is associated with the appropriateness of the "jerk" for probably a great height with a simultaneous increase of the angle of inclination of the trajectory. Despite the relatively complex profiles of the trajectories the cases of ricochet trajectories were not implemented [2]. This is due to the lack of the operating time of the main stage which makes it inefficient to apply it again.

From the point of view of maximizing the firing range according to the results of the research about the optimal ratios of the starting and main stages for a rocket with short-running marches in stage with the fixed launch mass and low aerodynamic quality the following conclusions were made:

- is it preferable to accelerate the rocket by the starting engine up to speed necessary to run the main stage;
- for the cases with the engine thrust greater than the sum of the rocket weight and the strength of its drag, a disposal of the waste starting stage is preferred; the more operating time of the main stage, the more evident the increase of the range;
- significant lift height allows the rocket to use the effects of planning by increasing the angle of attack on the descending section of its trajectory;
- time of operation of the main stage exceeds time the dense layers of the atmosphere breaking leads to the decrease of the inclination angle of the trajectory, and then the horizontal phase of flight emerged; there a rocket is moving at the angle of attack "standing" on the jet. At the end of the jet it is reasonable to "charge" up with the increasing angle and the maximum height of the trajectory;
- there exists a ratio of thrust of the main stage and the rockets mass lower when the rocket in the active phase of the main stage is not able to use the angle of attack to keep the trajectories close to the horizontal without losing range.

The research had been conducted with financial support from the Ministry of Education and Science of the Russian Federation (agreement № 14.577.21.0084, unique project identifier: RFMEFI57714X0084)

References

[1] Fomin V M, Aulchenko S M, Zvegintsev V I 2010 Flight of a hypersonic aircraft with a ramjet engine according to ricochet trajectory *PMTF* T. **51.** № 4 P. 85 - 94.
[2] Fomin V M, Aulchenko S M, Zvegintsev V I, Ustinov L A 2014 Analysis of aircraft flight paths with ramjet engine *PMTF* № **6** P. 35 - 42.

IOP Conf. Series: Materials Science and Engineering **155** (2016) 012041 doi:10.1088/1757-899X/155/1/012041

Evaluation of availability of cluster distributed disaster tolerant systems for control and information processing based on a cluster quorum

R Yu Tsarev[1], D V Gruzenkin[1], I V Kovalev[2], A V Prokopenko[1], A N Knyazkov[1]
[1] Siberian Federal University, Russia, 79, Svobodny Prospect, Krasnoyarsk, Russia
[2] Siberian State Aerospace University named after Academician M.F. Reshetnev
31 "Krasnoyarskiy Rabochiy" prospect, Krasnoyarsk, 660037, Russia

E-mail: rtsarev@sfu-kras.ru

Abstract. Control and information processing systems, which often executes critical functions, must satisfy requirements not only of fault tolerance, but also of disaster tolerance. Cluster architecture is reasonable to be applied to provide disaster tolerance of these systems. In this case clusters are separate control and information processing centers united by means of communication channels. Thus, clusters are a single hardware resource interacting with each other to achieve system objectives. Remote cluster positioning allows ensuring system availability and disaster tolerance even in case of some units' failures or a whole cluster crash. A technique for evaluation of availability of cluster distributed systems for control and information processing based on a cluster quorum is presented in the paper. This technique can be applied to different cluster distributed control and information processing systems, claimed to be based on the disaster tolerance principles. In the article we discuss a communications satellite system as an example of a cluster distributed disaster tolerant control and information processing system. Evaluation of availability of the communications satellite system is provided. Possible scenarios of communications satellite system cluster-based components failures were analyzed. The analysis made it possible to choose the best way to implement the cluster structure for a distributed control and information processing system.

Introduction

Developing modern control and information processing systems requires high dependability of both software and hardware. There are a lot of areas of science and industry where system failure can lead to essential economical loss and bring damage to people's health and lives [1]. These areas are banks and finance, space and defense industry, underwater and underground research, nuclear and chemical industries [2].

In this connection one of the main problems is elaboration of such approaches and techniques for control and information processing system design, that could provide system tolerance to software and hardware failures and guarantee fault tolerant solution, that main objective is to preserve data and continue functioning in conditions of mass and, probably, consecutive system failures [3], [4], [5].

The technology of failure handling in this case requires considering structure component interconnection and system ability to react specifically on possible sequence and combination of events leading to system failure with the purpose of providing maximum data security [6].

Wilkins et al. claim in [7], that clustering of computer systems to increase application availability has become a common industry practice. For accurate functioning of cluster distributed control and information processing systems used by big companies and corporations, disaster tolerant solution should be found. By "disaster tolerant solution" is meant the set of software and hardware configurations, settings parameters and organizational arrangements, that in total provide critically important data security and possibility for continuing the work of the system in case of occurring cataclysms leading to partial or complete destruction of the system [8], [9], [10]. As a rule, there are two (or more) datacenters for control and information processing connected by redundant communication channels and distributed into large distances which are enough for the possible catastrophe would not touch both datacenters at once. Configuration and structure of each datacenter's hardware and software can differ.

Design and development of control and information processing systems, the disaster tolerance of which is provided with the distributed cluster architecture, demands considering various parameters such as cost, which depends on configuration and geographical remoteness of datacenters, requirements to integrity and rate of the system recovery, frequency of inspections for compliance with system and procedures requirements [8], [10], [11].

In this work the problem of disaster tolerance assurance of distributed control and information processing system based on cluster system structure is considered. Here is presented the result of analysis of different failure scenarios of cluster distributed control and information processing system' components, which allows evaluating availability of the system and its disaster tolerance.

Concept of cluster quorum

In [12] cluster quorum is mentioned as a dynamic characteristic, the value of which presents the cluster integrity at the current moment. For the cluster structure of distributed control and information processing system let us define cluster quorum as a minimal integrity of the cluster, in conditions of which it remains operable.

Cluster quorum is a percentage characteristic, which implies minimal cluster part capable of coping with tasks given. If the value of cluster quorum is 40%, this means that if 60% of a cluster crush, the cluster is still able to work steadily. In that way, in the ideal case cluster quorum equals 0%.

For calculating cluster quorum value each cluster node is set a weight by an expert. After that according to the weights the "importance equivalent" is calculated for each node. Then the variants of cluster integrity disruption are composed with an appropriate workability value. Minimal operable integrity value is taken for cluster quorum.

Results and discussions

Availability analysis was implemented with the help of program complex for analysis of availability and control of the development of cluster structures for automated control systems designed by the authors [13]. This complex can be implemented for designing new cluster distributed control and information processing systems and for development cluster structures of currently existing systems. Functional purpose of the program complex is an analysis of different cluster structures of a distributed control and information processing system, detection of different system components' failure scenarios, and evaluation of availability of control and information processing systems and their disaster tolerance.

The purpose of the program complex is in designing and visualising of cluster structure of a distributed control and information processing system, analysis of different cluster structures of the system and providing support for decision making when choosing the cluster structure of the system.

Let us consider several cluster structure failure scenarios by the example of communications satellite system [14]. It is supposed that cluster structure of distributed control system for a communications satellite system with four datacenters is needed.

When designing and developing the cluster structure of a communications satellite system the most important parameters are dates and expenses spent on it. Two variants of implementation were considered.

According to the first one, the system includes four datacenters, in the first and the third of which there are three computer nodes, in the second and fourth of which there are two computer nodes and three arbitrators, the dates of realization – 365 days, budget – $1 500 000.

According to the second variant, the system includes four datacenters, in each of which there are two computing nodes, two arbitrators, construction period – 240 days, budget – $1 100 000.

Arbitrator is a full-function system, which is a part of a cluster; it accomplishes unity and synchronizing role for all other nodes [7]. Dynamic characteristic which is calculated every time when the cluster node fails is a cluster quorum defining cluster's integrity.

Table 1 presents some failure scenarios of components of the cluster distributed communications satellite system implemented according to the first variant. In Table 1 the column "Out of order" presents components unavailable for current scenario. Unfortunately, the article format does not allow presenting all the failure scenarios.

Table 1. Cluster distributed communications satellite system failure scenarios.

Scenario number	Out of order	Failed components	Cluster integrity	Components left	Consequences
1	None	None	100%	13 out of 13	
2	None	Arbitrator 1	92%	12 out of 13	No consequences
3	None	Arbitrator 2	92%	12 out of 13	No consequences
4	None	Arbitrator 1, arbitrator 2	84%	11 out of 13	No consequences
5	None	Arbitrator 2, arbitrator 3	84%	11 out of 13	No consequences
...	...	...	...	...	...
959598	Node 2, node 3, node 9, node 10, arbitrator 2	Node 1, node 4, node 6, node 7, node 8, arbitrator 3	25%	2 out of 8	Cluster is stopped
959599	Node 2, node3, node 9, node 10, arbitrator 2	Node 1, node 4, node 6, node 7, node 8, arbitrator 1	25%	2 out of 8	Cluster is stopped
...	...	...	...	...	...
1594322	Node 1, node 2, node 3, node 4, node 5, node 6, node 7, node 8, node 9, node 10, arbitrator 1, arbitrator 2	Arbitrator 3	0%	0 out of 1	Cluster is stopped
1594323	Node 1, node 2, node 3, node 4, node 5, node 6, node 7, node 8, node 9, node 10, arbitrator 1, arbitrator 2, arbitrator 3	None	0%	0 out of 0	

Analysis of two different variants of construction of cluster structure for the communications satellite system indicates that the first variant includes 1 594 323 different failure scenarios, in 239 748 of which a cluster stops working completely. This means that 15,04% of failure scenarios lead to the whole communications satellite system failure. The second variant includes 59 049 scenarios, 17 668 of which lead to cluster stopping, that makes 29,92% of scenarios. In the first variant $1 500 000 were spent, while in the second variant $1 100 000 were spent, which is 36,36% less. Thus, it is reasonable to implement cluster structure of the communications satellite system according to the first variant.

Conclusions

Ensuring disaster tolerance of control and information processing systems has been an important problem since the systems of this class began to be applied in critical areas. The use of cluster structure in control and information processing systems with geographically distantly located datacenters allows implementing its functional objectives even in case of catastrophic occasions leading to the failures of components or clusters of the system.

In this article the analysis of cluster distributed control and information processing system is presented. We considered a communications satellite system as an example of a cluster distributed control and information processing system.

The communications satellite system consisted of different computing nodes and datacenters. The analysis of communications satellite system availability was made with the program complex designed by the authors. On the basis of this analysis the more preferable variant of communications satellite system cluster structure can be selected.

The proposed technique for evaluation of availability of a cluster distributed control and information processing system can be applied in designing and modernizing different distributed systems for control and information processing, where disaster tolerance is required.

References

[1] Knight J C 2002 Safety critical systems: Challenges and directions *Proc. of Int. Conf. on Software Engineering* pp. 547-550
[2] Sommerville I 2011 *Software engineering* (New York: Pearson, Addison-Wesley)
[3] Fekih A 2015 Fault tolerant control design for complex systems: Current advances and open research problems *Proc. of the IEEE Int. Conf. on Industrial Technology* pp. 1007-1012
[4] Rehman S, Kriebel F, Shafique M and Henkel J 2014 Reliability-driven software transformations for unreliable hardware *IEEE Trans. Comput.-Aided Design Integr. Circuits Syst.* **33** 1597-1610
[5] Kulyagin V A, Tsarev R Yu, Prokopenko A V, Nikiforov A Yu and Kovalev I V 2015 N-version design of fault-tolerant control software for communications satellite system *Int. Siberian Conf. on Control and Communications, SIBCON 2015 - Proc.* art. no. 7147116
[6] Tekinerdogan B, Sozer H and Aksit M 2008 Software architecture reliability analysis using failure scenarios," *J. Syst. Software* **81** 558-575
[7] Wilkins R S, Du X, Cochran R A and Popp M 2002 Disaster tolerant Wolfpack geo-clusters *Proc. - IEEE Int. Conf. on Cluster Computing, ICCC* pp 222-227
[8] Nguyen T A, Kim D S and Park J S 2016 Availability modeling and analysis of a data center for disaster tolerance *Future Gener. Comp. Sy.* **56** 27-50
[9] Sen A, Mazumder A, Banerjee S, Das A, Zhou C and Shirazipourazad S 2015 Region-based fault-tolerant distributed file storage system design in networks *Networks* **66** 380-395.
[10] Yao W-B, Zhao L, Wang Z, Yao X and Han S 2015 Research on disaster tolerant information system modeling and simulation *Beijing Youdian Daxue Xuebao/Journal of Beijing University of Posts and Telecommunications* **38** 50-54
[11] Qiu Z and Perez J F 2015 Enhancing reliability and response times via replication in computing clusters *Proc. of IEEE INFOCOM* pp 1355-1363
[12] Özsu M T and Valduriez P 2011 *Principles of Distributed Database Systems* (New York: Springer-Verlag)
[13] Glotov A K, Kovalev I V, Tsarev R Yu, Zavyalova O I, Rusakov M A, Kapulin D V and Kovalev D I 2012 Program complex for analysis of dependability and control of the development of cluster structures for automated control systems *Federal service for intellectual property (Rospatent)* certificate no. 2012614895 31.05.2012 Moscow Russia.
[14] Chernigovskiy A S, Tsarev R Yu and Knyazkov A N 2015 Hu's algorithm application for task scheduling in N-version software for satellite communications control systems *Int. Siberian Conf. on Control and Communications, SIBCON 2015 – Proc.* art. no. 7147270

The scheme of combined application of optimization and simulation models for formation of an optimum structure of an automated control system of space systems

A S Chernigovskiy[1], R Yu Tsarev[1], A Yu Nikiforov[1], P V Zelenkov[2]
[1]Siberian Federal University, 79, Svobodny Prospect, Krasnoyarsk, Russia
[2] Reshetnev Siberian State Aerospace University, 31, Krasnoyarskiy rabochiy ave., Krasnoyarsk, 660037, Russia

E-mail: achernigovskiy@sfu-kras.ru

Abstract. With the development of automated control systems of space systems, there are new classes of spacecraft that requires improvement of their structure and expand their functions. When designing the automated control system of space systems occurs various tasks such as: determining location of elements and subsystems in the space, hardware selection, the distribution of the set of functions performed by the system units, all of this under certain conditions on the quality of control and connectivity of components. The problem of synthesis of structure of automated control system of space systems formalized using discrete variables at various levels of system detalization. A sequence of tasks and stages of the formation of automated control system of space systems structure is developed. The authors have developed and proposed a scheme of the combined implementation of optimization and simulation models to ensure rational distribution of functions between the automated control system complex and the rest of the system units. The proposed approach allows to make reasonable hardware selection, taking into account the different requirements for the operation of automated control systems of space systems.

1. Introduction

The automated control systems of space systems represent hierarchical multi-function and multi-loop complexes with advanced means of communication between the control units. They include a set of ground measuring stations of different types, distributed over large areas, the communication channels providing information transfer between system elements, control units and communication units of different hierarchy levels.

The emergence of new classes of spacecrafts and the expansion of their functions require enhancement of structure of automated control systems of space systems (ACS of SS). Designing such systems include determination of elements and subsystems of ACS of SS in space, the selection of a hardware complex, that provide execution of control functions taking into account spatial placement of the complex [1] and accessibility of its units, distribution of executed functions set on system units taking into account correlation on technology of information processing and control [2]. At the same time, system controlling quality requirements must be provided.

2. Problem statement

In case of the synthesis of structure of ACS of SS there are tasks of rational distribution of functions between the controlling complex and remaining units of the system, the task of a hardware selection in units of system and communication channels in between taking into account different tactic-technical and techno-economic requirements to functioning of ACS of SS [3]. At the same time, emerging mathematical problem formulation of system structure synthesis can be formalized with use of the discrete variables at various levels of system detalization. It is necessary to determine the sequence of tasks and stages of structure creation of ACS of space system and to solve the problem of a selection of control units of ACS of space system due to combined implementation of optimization and simulation models.

3. Solution procedure

The sequence of tasks and stages of structure formation of ACS of SS is shown in Fig. 1. At the initial stages requirements [4] to the developed ACS of SS are formulated, the composition of characteristics and efficiency criteria of system structure options is selected. The possible composition of units, options of their construction and territorial placement, and also the characteristic of these options are defined. Control functions in system, which are detailed to the complex of interdependent control tasks and versions of their solution (or to variants of creation of processing procedures of information and information arrays) that provide execution of the given control functions (Fig. 1) are selected.

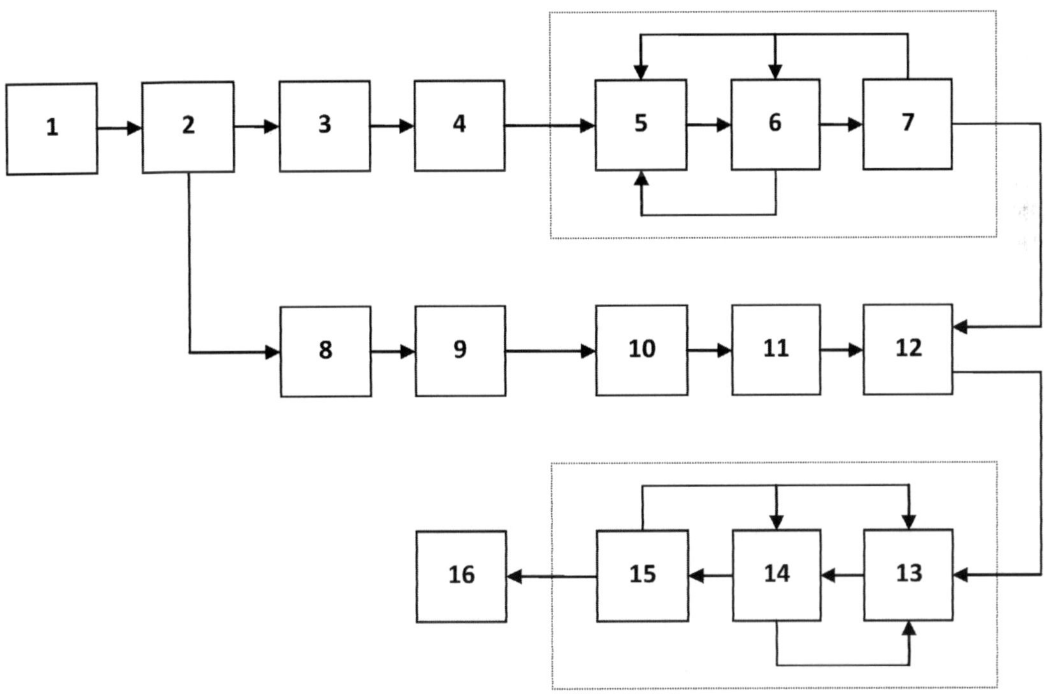

Figure 1. The sequence of tasks and stages of ACS of space system structure formation

IOP Conf. Series: Materials Science and Engineering **155** (2016) 012042 doi:10.1088/1757-899X/155/1/012042

The sequence of tasks and stages of structure formation of ACS of space system shown in fig. 1:

1) formulation of the problem of ACS of space system structure formation;
2) specification and selection of composition of the considered characteristics and efficiency criteria of structure options;
3) determination of possible composition of control units and their correlations;
4) formalization of topological structure construction options of the system (creation of a graph of interdependent control tasks set which are carried out by ACS of SS, G_J);
5) optimization model of topological structure (selection of $G_J^* \subset G_J$);
6) simulation model of topological structure of ACS of space system;
7) analysis and decision correction unit;
8) determination of control tasks functions composition and their correlations;
9) determination of control tasks and functions options of execution;
10) determination of a possible hardware set for control functions and tasks execution;
11) formalization of options of functions and tasks execution (making of a graph of topological structure of ACS of SS reflecting a set of possible options of creation and spatial placement of the controlling units of system and communication links in between, G_I);
12) formalization of creation of the functional structure of ACS of space system (making of map G_I on G_J);
13) optimization model of the system functional structure;
14) simulation model of the system functional structure;
15) analysis and decision correction unit;
16) receiving an optimal alternative of system structure.

Further the selected formation options of interdependent control tasks complex and control units formation options (including hardware controls equipment options) are formalizing in the form of graphs G_I and G_J, which serve as the basis for the formalized stages of the synthesis of the topological and functional structure of the system (highlighted blocks 5-7, 13-15 in Fig. 1).

The considered tasks can be formalized using graph [5] and mapping apparatus. The general diagram of structure formation of ACS of SS includes the following stages:

1) formation of the topological structure graph of ACS of KC G_I , that maps the a set of possible options of formation and spatial placement of the controlling units of the system and communication links in between;

2) formation of a graph of interdependent set of control tasks of G_J which are carried out by KS ACS;

3) formalization of map of the graphs G_I on G_J.

Thus, the task of synthesis of topological structure of system is search of an optimum subgraph on the G_J graph of the ACS of SS units, and the task of synthesis of the functional structure – in search of optimum mapping of the graph G_I to the graph G_J of topological structure of system.

For ACS of SS it is expedient to select the following detalization levels of creation of system:

- a selection of composition and topology of control ground stations (the aggregated variable X);
- a selection of security option of control points from external influences and a hardware complexes selection at units (the aggregated variable Y);

- a selection of option of distribution of functions and control tasks between the selected control units and hardware (the aggregated variable Z).

At the same time the general task of structure formation of ACS of KS can be represented as follows [6]:

$$\text{extr } F(X, Y, Z) = f_1(X) + f_2(X, Y) + f_3(X, Y, Z);$$
$$X \in G_1,\ Y \in G_2(X),\ Z \in G_3(X, Y),$$

where $X = \{x_j\}$, $Y = \{y_{jk}\}$, $Z = \{z_{ijk}\}$ – bool vectors, interconnected as follows: for each $j \in J$, $k \in K$, $i \in I$; $y_{jk} \geq 0$, if $x_j > 0$; $y_{jk} > 0$, if $x_j = 0$; $z_{ijk} \geq 0$, if $y_{jk} > 0$; $z_{ijk} = 0$, if $y_{jk} = 0$.

It is possible to notice, that generally some limitations of the task can be set algorithmically that requires, in turn, the organization of iterative diagrams of the decision of the specified tasks taking into account iterations of the solving of the general task of structure formation of ACS of SS. In practice such difficult procedures are implemented in the form of the human-machine diagrams of synthesis including possibilities of "the designer of system" in the dialog mode to interfere and adjust results of the intermediate decisions and the process of optimization.

The complex of interdependent models and algorithms and iterative diagrams of their interaction in case of synthesis of a topological (spatial) and functional structure of system (fig. 2) is developed for the solution of problems of synthesis of structure of distributed ACS of SS on the basis of methodology of difficult systems structure synthesis.

At the first stage of ACS of SS synthesis (unit 1) the problem of selection of composition and spatial location of control units taking into account costs of their organization, accessibility and globality of control of different classes of the spacecrafts is solving.

It is supposed that in the given territory the set of points included in ACS of SS in which the control units directly interacting with spacecraft can be created is defined. Visibility ranges of different control units break projections of spacecraft motion paths to sections. Set of sections of different points determines accessibility of spacecraft by control units and globality of ACS of SS. The task of structure synthesis of ACS of SS is in the selection of set of control units, minimizing costs of their creation on condition of fulfillment of requirements to globality of control for each class of spacecraft and some other requirements, for example, quantity of the nodes interacting with each class of objects, etc.

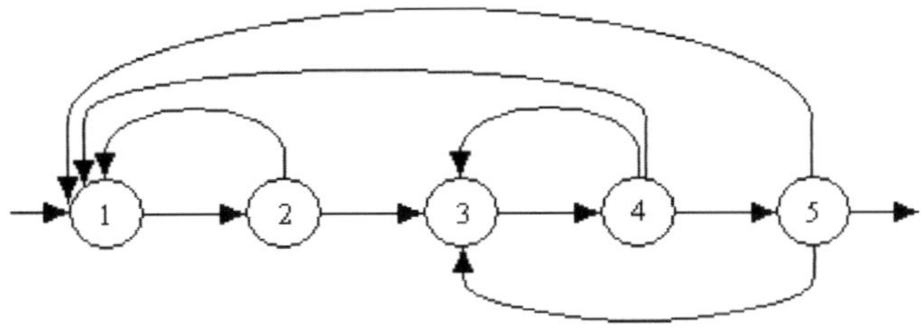

Figure 2. Diagram of combined implementation of optimization and simulation models for a selection of control units

Simulation models [7-8] of interaction of elements of the given topological structure are intended for determination algorithmically given characteristics of functioning of the synthesized option of ACS of SS structure (unit 2). The model allows to analyze functioning of system with different number and layout of ground control stations, different quantity of spacecraft classes and number of spacecrafts in each class, different number of types of the controlling interactions between control centers and spacecrafts. As a result of simulation efficiency characteristics of operation of ground control stations (the general loading, loading by types of interaction and spacecrafts classes) and control quality of spacecrafts of different classes set (number of interactions, their duration, number of unrealized interactions, etc.) are determined.

In units 3, 4 for the selected set of control units and their correlations the options of creation of nodes and communications in between providing increase in survivability of the system determined by probability of execution by system of functions of control taking into account possible adverse external effects are defined.

Different options of creation of the ACS of SS units and communications in between directed to increase in survivability of system are formalized in the form of the graph G_J. For increase in survivability duplicating of nodes and communication links, increase in security of nodes, etc. are used. Each option of creation of nodes or communication links is characterized by costs of their implementation and probability of failure for the considered time interval. For each class of objects the minimum set of nodes which functioning provides execution of the given functions of control is set. It is supposed that such set are the elements connecting an object to command center.

The task of determination of options of creation of ACS of SS consists in minimization of costs of the actions providing execution with system of functions of control in case of the restrictions characterizing survivability of control system for different classes of objects [9].

Such selection occurs as a result of implementation of the analytic-simulative procedure that executes interaction during the synthesis of optimization selection model of a system elements implementation (unit 3) and a simulation model (unit 4).

The complex of simulation models is intended for the detail accounting of probable characteristics of functioning of the generated options of creation and support of survivability of system (unit 4) which allows to analyze functioning of system in case of different values of intensity and for different probability distributions of system elements failure (control units, communication channels, etc.).

According to the general methodology of synthesis of complex systems structure, generation of options for their detail analysis and a selection is carried out by means of optimization models (unit 3). At the same time, the following values are assumed set: $P_{ik}\left(k=\overline{1,K_i}, i=\overline{1,I}\right)$, $P_{jk}\left(k=\overline{1,K_j}, j=\overline{1,J}\right)$, $P_{Rk}\left(k=\overline{1,K_R}\right)$ – probabilities of reliable functioning of *k*-th option of i-th object organization, j-th unit of the ACS of SS, R-th repeater for the considered time interval taking into account adverse external effects; the j = 0 index corresponds to central control unit, and $j=\overline{1,J}$ indexes – sets of ground control units; P_{ijk}, P_{iRk}, P_{Rjk}, P_{j0k} – probabilities of successful functioning of constructing options of communication links between an object and a ground control unit, an object and repeaters, repeaters and a ground control unit, a ground control unit and central unit for the considered time interval taking into account adverse effects; C_{ik}, C_{jk}, C_{Rk}, C_{ijk} – costs of organization options for objects construction options, repeaters, central unit and communication channels, ground control units.

The task of selection of options of the ACS of SS elements composition is in cost minimization:

$$\min\left(\sum_{\alpha}\sum_{k} C_{\alpha k} + \sum_{\beta}\sum_{k} C_{\beta k} x_{\beta k}\right)$$

subject to

$$P_{i0}(x_{ik}, x_{0k})\left[P_{Rk}(x_{Rk}, x_{iRk})\left(1 - P_{ijk}(x_{ijk}, x_{Rjk}, x_{jk})\right) + \right.$$
$$\left. + \left(1 - P_{Rk}(x_{Rk}, x_{iRk})\right)\left(1 - P_{ij}(x_{ijk}, x_{jk})\right)\right] \geq P_i^{\text{доп}}, i = \overline{1, I};$$
$$\sum_{k} x_{\alpha k} = 1, \alpha = i, j, R;$$
$$\sum_{k} x_{\beta k} = 1, \beta = \{iR, ij, jR, j0\},$$

where $x_{ik} = 1$, if k option of i object formation is chosen; $x_{jk} = 1$, if k-th construction option of j-*th* unit is chosen; $x_{Rk} = 1$, if k-th option of repeater construction is chosen; $x_{imk} = 1$, if k-th communication option between i and j, i and R, R and j, j and 0 (central control unit) elements is chosen respectively; $x_{ik} = x_{jk} = x_{Rk} = x_{imk} = 0$ – otherwise.

Values P_{i0}, P_{Rk}, P_{ijk}, P_{ij} calculated by the following expressions:

$$P_{i0}(x_{ik}, x_{0k}) = \left(\sum_{k} P_{ik} x_{ik}\right)\left(\sum_{k} P_{0k} x_{0k}\right);$$
$$P_{Rk}(x_{Rk}, x_{iRk}) = \left(\sum_{k} P_{Rk} x_{Rk}\right)\left(\sum_{k} P_{iRk} x_{iRk}\right);$$
$$P_{ij}(x_{ijk}, x_{jk}) = \prod_{j}\left[1 - \left(\sum_{k} P_{ijk} x_{ijk}\right)\left(\sum_{k} P_{jk} x_{jk}\right)\right];$$
$$P_{ijk}(x_{ijk}, x_{Rjk}, x_{jk}) = \prod_{j}\left[1 - \left(1 - \sum_{k} P_{ijk} x_{ijk}\right)\left(1 - \sum_{k} P_{jk} x_{jk}\right)\right].$$

For the task solving it is offered to use the algorithm based on a "branch-and-bound" method [10]. For the purpose of abbreviation of dimensionality of the task on the graph G_j of options of creation of nodes of system and their correlations separation of sequential sections, determination of the aggregated options of creation of the selected sections and their characteristics is provided. At the same time the aggregated options of sections over which remained dominate are excluded from the graph G_j.

At the subsequent stage of synthesis (unit 5) for the found set of the ACS of SS units that work with objects of different classes, and specified control functions on each class of objects optimum allocation of tasks of interdependent subsystems is determined by levels and units of system and the composition of hardware complex is selected. Control functions are set in the form of a set of the tasks which are carried out by different subsystems (circuits) of control.

The typical control subsystem of ACS of SS is intended for execution of the sequence of interdependent functions (tasks) on measurement, transmission and information processing about set of the spacecraft controlled parameters, framing of the controlling influences and to monitoring over their implementation [11]. The problem is in optimization of distribution of the tasks entering a control circuit on levels and the controlling units of ACS of SS and determination of a hardware set that minimize costs of hardware equipment of units and their maintenance in case of execution of restrictions on efficiency, the hardware reliability of control tasks execution, weight and energy consumption of an onboard equipment, units loading, etc.

Options of distribution of functions and tasks on levels and the ACS of SS units are formalized in the form of set of the interdependent graphs G_0. A specificity of ACS of SS is the way that each individual graph corresponds to a construction option of the corresponding system control circuit. The subgraph, consisting of a set of paths of graphs, sets structure of all system.

4. Conclusion

Thus, within unified approach defines the tasks and stages of the ACS of SS structure formation, including the solutions to interrelated problems of formation of the basic system elements and parts, as well as the selection of a rational option of the structure created by automation space system. The carried-out task analysis of structure formation of ACS of space system showed that it is expedient to select the following detail levels of creation of system:

- a selection of composition and topology of ground control stations;
- a selection of protection from external influences;
- a selection of hardware complexes on the ACS units of space system;
- a selection of option of distribution of functions and control tasks between the selected control units and hardware.

The diagram of combined implementation of optimization and simulation models is developed for the solving of the selection task of control units of ACS of space system. Such diagram allows to use effectively specified models within uniform optimization and simulative approach to structure formation of ACS of space systems that provides rational distribution of functions between the controlling ACS complex and remaining units of the system, and also allows to carry out a reasonable selection of hardware taking into account different requirements to functioning of ACS of space systems.

References

[1] Zelenkov P V, Karaseva M V, Tsareva E A, Tsarev R Y 2015 Definition of the topological structure of the automatic control system of spacecrafts *IOP Conference Series: Materials Science and Engineering* **70** art. no. 012013

[2] Ding K, Zhu K, Chen S, Tian, Z 2015 Research and application of space-ground network technology *ICEIEC 2015 - Proceedings of 2015 IEEE 5th International Conference on Electronics Information and Emergency Communication* (Beijing) pp 444-447

[3] Ting C, Mahfouf H M, Nassef A, Linkens D A, Panoutsos G, Nickel P, Roberts A C, and Hockey G R J 2010 Real-time adaptive automation system based on identification of operator functional state in simulated process control operations *IEEE Transactions on Systems, Man, and Cybernetics Part A:Systems and Humans* **40** 251-262

[4] McGranaghan M and Goodman F 2005 Technical and system requirements for advanced distribution automation *18th International Conference and Exhibition on Electricity Distribution* **5** 93

[5] Zhou G, Feng W, Zhao Q and Zhao H 2015 State tracking and fault diagnosis for dynamic systems using labeled uncertainty graph *Sensors (Switzerland)* **15** 28031-28051

[6] Shipley M F, de Korvin A and Obid R. 1991 A decision making model for multiattribute problems incorporating uncertainty and bias measures *Computational Operations Research* **18** 335–342
[7] Kleijnen J P C 1979 Regression metamodels for generalizing simulation results *IEEE Trans. on Systems, Man and Cybernetics* **2** 93–96.
[8] Kleijnen J P C 1995 Verification and validation of simulation models *European Journal of Operational Research* **82** 145—162.
[9] Van Groenendaal, Kleijnen J P C 1997 On the assessment of economical risk: factorial design versus Monte-Carlo methods *Journal of Reliability Engineering and System Safety* **57** 103–105
[10] Hendy M D and Penny D 1982 Branch and bound algorithms to determine minimal evolutionary trees *Mathematical Biosciences* **59** 277-290
[11] Wang Y, Zhao Y, Shi P and Zheng H 2015 Adaptive control for stabilizing the coupling system with disturbance after capturing spacecraft *Chinese Space Science and Technology* **35** 20-28

IOP Conf. Series: Materials Science and Engineering **155** (2016) 012043 doi:10.1088/1757-899X/155/1/012043

Model and method for optimizing heterogeneous systems

O A Antamoshkin[1,2,4], O A Antamoshkina[1,2,3], P V Zelenkov[1], I V Kovalev[1]
[1] Reshetnev Siberian State Aerospace University, Krasnoyarsk, Russia
[2] Siberian Federal University, Krasnoyarsk, Russia
[3] V.N. Sukachev Institute of Forest SB RAS, Krasnoyarsk, Russia
[4] Krasnoyarsk State Agrarian University, Krasnoyarsk, Russia

E-mail:oleslav24@gmail.com

Abstract. Methodology of distributed computing performance boost by reduction of delays number is proposed. Concept of n-dimentional requirements triangle is introduced. Dynamic mathematical model of resource use in distributed computing systems is described.

Keywords: optimization, resource use, decision support system, distributed computer system.

Introduction. Most modern process automation solutions rely on distributed computer systems (DCS) [1]. These systems consist of thousands of high-performance servers, hundreds of databases and often hundreds of thousands workstations and different communication channels for information exchange both inside the perimeter of network and outside of it with external data sources. Gigabytes of data are being captured and processed by these systems [2]. In most cases information (first of all for monitoring and forecasting) is distributed in global scale, stored and processed in regional data centers of in federal level data centers. Results of data processing are used directly by decision makers [3, 4]. Despite constantly growing processing power of computer systems and networks capacity resources of individual nodes are often insufficient for processing of necessary volume of information. It results in delays in delivery of necessary information that increases time required for decision making and thus decrease the quality of decisions [5].

Model. We will solve the problem of redistribution of load in real time for instant reacting on changing conditions such as increase or decrease of input information flow, servers' breakdown, physical database overload [6].

Variation of provisioning-transfer parameters may be used as one of the variants of redistribution. At the moment when a certain computational node is overloaded decision is made to transfer "raw" data on higher level node. The counter process is possible – when communication channels are overloaded distributed computer systems run necessary computational jobs in place and transfer processed data that has much lower volume.

We further consider dynamic model of network resource distribution for decision support information system that support this task.

Let us denote PBC as N={P,C}, where P is a set of nodes of this network – processing power, data warehouses and communication channels, C – a set of networks edges described by unified transition coefficient that depends on parameters of nodes that are the parts of these edges.

Set of all nodes P may be divided by functionality into three disjoint subsets: P_1 - processing resources (processors), P_2 - data warehouses, P_3 - communication channels.

Is element p_i^1 of P_1 is characterized by the following parameters:

q^i_t- free processing power of p_i node in time point t,

m^i_t- free storage capacity provided for data storage in p_i node in time point t,

w^i_t- p_i node reliability measure in time point t calculated as the possibility of p_i node breakdown.

Data warehouse p_i^2 of P_2 has the following parameters:

f^i_t- disc space allocated for new data,

a^i- database type.

Channels p_i^3 of P_3, are characterized by the following:

v_i- bandwidth of channel p_i^3,

z_i^t- workload of p_i^3 in time point t,

b_i^t-cost of information unit transfer by p_i^3 in time point t,

s^i_t- forecasted delay in p_i^3 in time point t.

To simplify application of optimization algorithms these sets may be joined provided that all parameters of subset elements that relate to other subset are zero. For example for P_2 all parameters except $f^i_t$and a^i are zero.

Considering the above the function that defines each edge may by written as

W_{ji}^t $(q^i_t, m^i_t, w^i_t, f^i_t, a^i, v_i, z_i^t, b_i^t, s^i_t)$, that defines transfer from node j to node in time point t. Function W_{ji}^t is multi-criterion and defines free capacity of edge from i to j in time point t – o_{ji}^t.

As mentioned earlier input data for distributed computer systems are generated mainly by automated data capture systems that capture information from different systems of information gathering. Let us call them sources in terms of network modeling. The main task of distributed computing is to process raw data coming from sources as required (applying predetermined models and methods) and transfer them to destinations for use in decision support systems on workplaces of decision makers [7]. To describe processes applied to data on its way from source to destination we represent it as a flow:

Let us denote q_{ji}^k € Q(set of all flows in the system), k^{th} flow with the source in node p_i, and destination in p_j.

This flow is characterized by the following:

V_{input}^k- input information volume,

V_{output}^k- out put information volume,

c^k- amount of work (number of calculation) to transfer input information into output,

Parametr — requirements to data transfer route: cost of transfer in Rubles, time frame for package transfer, reliability.

If necessary q^k may be divided into N stages for serial or parallel processing, so:

$c^k = \sum c_n^k$,

where c_n^k is the volume of nth stage q^k. Time required for processing and transfer of kth flow is:

$\tau^k = \sum \tau_n{}^k$

where $\tau_n{}^k$ expected processing time on n^{th} stage q^k.

One may easily see that all the flow of input and output data may be divided into separate parts if calculations are divided like this. To denote them we introduce the following:

$V_{input}{}^k{}_n$ input data volume for stage n,

$V_{output}{}^k{}_n$ expected output data volume of stage n.

More generally this task may be represented in the following way: minimizing transfer time of all flows existing in time point t from source in nodes P_i to destinations in p_j for limited resources defined for each edge as $c_{ji}{}^t$ As in each following time point t+1 function may have the other extreme the task has dynamic character.

Formal representation of this statement is like follows:

$\sum \tau^k{}_t$->min

where $\tau^k{}_t$ is expected time for processing of kth flow existing in time point t.

Under limitations:

$\sum c_n{}^k{}_{ij}{}^t \leq o_{ji}{}^t$,

where $\sum c_n{}^k{}_{ij}{}^t$ total of all jobs for running on edge from I to j in time point t.

One of the most important tasks for calculation of this model is calculation of limitations $o_{ji}{}^t$ because they originate not only from internal factors. To calculate $o_{ji}{}^t$ we use parameter $s^i{}_t$ (forecasted delay $p_i{}^3$ in time point t) that depends on external unaccounted effects. In most cases, companies use external communication channels for information transfer with unreliable effectiveness and bandwidth.

The best instrument for forecasting in this case is neuron networks. Statistical methods and time series method may be also useful but configuration of global network is constantly changing and these methods in most cases do not solve the problem [8]. Unlike these methods neural networks are capable not only to run predetermined operations sequence but also analyze inputting information on the fly find patterns in it and make forecasting. Neuron networks are constantly learning basing on previous values.

Method. There are two methods to solve this problem:

1. Static – scaling capacity of computer network and in the first place in bottlenecks or problem nodes.

This approach is not universal for given level of complexity because it does not allow solving problem in real time. It is also ineffective for constantly changing volumes and data flow directions and it is also costly [9].

2. Dynamic – taking off the load from overloaded nodes by idle ones in real time (resources allocation optimization).

This approach allows balancing the load between the nodes of computer system making it more uniform, increase performance and reliability of information processing and decrease total cost of ownership of resource.

The rule of n-dimentional requirement triangle may be used to define optimal set of criteria for equation.

Criteria are being divided into interrelated discrepant triplets. Values of all criteria should be normalized. We assume 1 as the best value of criterion and 0 for the least value. Distance between actual value of Kth criterion and its least value we denote as Δk (segment AO, BO, CO, see Fig. 1). The least requirement value is in point O and the best value - in points A_{max}, B_{max}, C_{max} respectively. In geometrical interpretation it means that in triangle segments from medians intersection (points O) to vertexes of internal triangle are values of analyzed requirements and optimal values of these requirements are in the vertexes of external triangle.

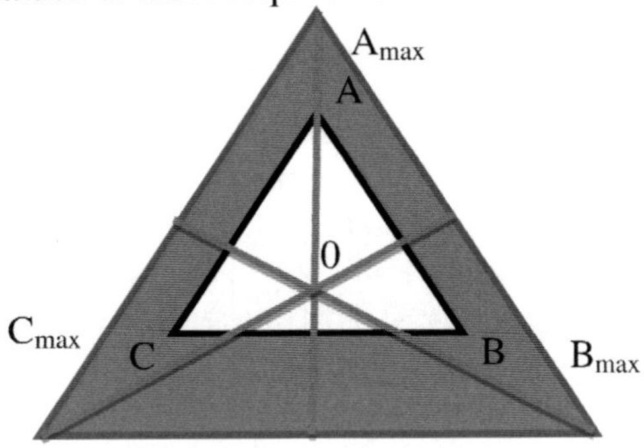

Figure 1. Triangle of requirements' discrepancy.

Basing on discrepancy and interconnection of requirements let us consider each third requirement to be discrepant with two others. It may be interpreted as the system of equations:

$$\begin{cases} AB = \sqrt{AO^2 + BO^2 - 2\,AO\,BO\,\cos\alpha} \\ BC = \sqrt{BO^2 + CO^2 - 2\,BO\,CO\,\cos\beta} \\ AC = \sqrt{AO^2 + CO^2 - 2\,AO\,CO\,\cos\gamma} \\ \quad \alpha + \beta + \gamma = 360 \end{cases}$$

Where [α], [β], [γ] are angles between AOOB, BOOC, AOOC respectively. Segments AB, BC, AC in this case define not only interconnection but also discrepancy.

Let us introduce requirements' balancing function. For 3 requirements it is equal

$$F(x) = \frac{1}{\sqrt{\left(\frac{1}{\Delta_k^1} + \frac{1}{\Delta_k^2} + \frac{1}{\Delta_k^3}\right)\left(\frac{1}{\Delta_k^1} + \frac{1}{\Delta_k^2} - \frac{1}{\Delta_k^3}\right)\left(\frac{1}{\Delta_k^1} + \frac{1}{\Delta_k^3} - \frac{1}{\Delta_k^2}\right)\left(\frac{1}{\Delta_k^3} + \frac{1}{\Delta_k^2} - \frac{1}{\Delta_k^1}\right)}}.$$

Geometric sense of this function is area of ABC triangle.

Equilateral triangle has maximal area. In analyzed case equality of legs is achieved in case of equality or balance of all requirements.

Number of requirements in most real world jobs is much greater that three. To apply the rule of N-dimentional triangle it is necessary to draw triangle in every plane defined by triples of requirements. To simplify task areas of drawn triangles may be multiplied together to go to maximum.

Resume. Proposed methodology allows formalizing the process of distributed computer system resources allocation by dynamic mathematical model. Software implementation of proposed technology experimentally proved the possibility of successful application of theoretical results of proposed research.

References

[1] Ailamaki A, Dash D and Kantere V 2009 Economic Aspects of Cloud Computing. Flash, Informatique, Special HPC: 43-47.
[2] Buyya R, Abramson D and Giddy J 2002 Economic Models for Resource and Scheduling in Grid Computing. *J. Of Concurrency and Computation: Practice and Experience* **14** (5): 1507-1542.
[3] Antamoshkin O A, Antamoshkina O A, Smirnov N A 2016 Multi-agent automation system for monitoring, forecasting and managing emergency situations. *IOP Conf. Series: Materials Science and Engineering* Vol. **122**. - article ID 012003
[4] Antamoshkin O A, Antamoshkina O A, Trofimova N V 2016 Landsat imagery evidences great recent land cover changes induced by wild fires in central Siberia. *IOP Conf. Series: Materials Science and Engineering* Vol. **122** - article ID 012004
[5] Lipaev V V 2006 Program engineering. Methodological base. Moscow: TEIS, pp: 608.
[6] Antamoshkin O A 2011 High-reliable real time systems development. Trudy MAI, 45: 61-72.
[7] Alekseev N A, Bogdanova OV, Kovalev I V and Tsarev R Yu 2010 Plannig of regular jobs in distributed information processing. Data-computing centers and management systems, 8: 11-14.
[8] Antamoshkin O A, Antamoshkina O A and Trofimova N V 2013 Technology of the forecast and assessment of the impact of the negative natural factors on the non-urban lands in agricultural industry with the application of remote sensing data of Earth. Mathematical and Informational Technologies. University of Pristina, Kosovska Mitrovitca, Serbia, pp: 147-148.
[9] Antamoshkin O A 2012 Technology of distributed computer systems resources allocation. Management systems and information technologies, 2(48): 220-224.

The National Forest Heritage of the Altai-Sayan Ecoregion

O A Antamoshkina[1,2,4] , N V Trofimova[3], O A Antamoshkin[1,2,5], P V Zelenkov[1], I V Kovalev[1]
[1] Reshetnev Siberian State Aerospace University, Krasnoyarsk, Russia
[2] Siberian Federal University, Krasnoyarsk, Russia
[3] Altai-Sayan Ecoregional Office WWF, Krasnoyarsk, Russia
[4] V.N. Sukachev Institute of Forest SB RAS, Krasnoyarsk, Russia
[5] Krasnoyarsk State Agrarian University, Krasnoyarsk, Russia

E-mail: sloa@mail.ru

Abstract. The article presents the methodology of the selection and ranking Intact Forest Landscapes of the Altai-Sayan Ecoregion. This method is developed to identify the most valuable areas of forests, which currently does not legally protect. These areas are planned to recommend for inclusion in the National Forest Heritage of Russia.

Keywords: Hierarchy's analysis method, remote sensing, biodiversity, forest protection.

Introduction. In 2012 the Federal Forestry Agency submitted a proposal to establish the National Forest Heritage (hereinafter NFH), including the forest habitats of national-level priority, and it was stated in the Basic Principles of State Policy of use, preservation, protection and reproduction of forests in the Russian Federation for the period up to 2030. To preserve these areas, it is planned to exclude them from economic use.

The most suitable areas for establishing the NFH are Intact Forest Landscapes (hereinafter IFL) - large natural areas of forests with no signs of significant human activity, not less than 50 hectares, not containing settlements, functional transportation communications or any other kinds of anthropogenic influence. Such territories conserve biodiversity at a nationwide scale and maintain viable population of the species including large carnivores and ungulate migration routes. These areas are also crucial for ecological services that forests provide. Besides IFL, the smaller forest areas of outstanding universal value are also to be included in the national forest heritage. Such smaller areas serve the purposes stated above as well as contain objects of historical, scientific or cultural significance [2].

Due to industrial logging, mining operations, infrastructure construction, fires (mostly anthropogenic), the IFL areas in the Altai-Sayan Ecoregion as well as the whole Russia are decreasing.

Nowadays there's an acute problem of preserving the large significant forest habitats not exposed to the anthropogenic impact. They are weakly protected by legally approved mechanisms (various categories of specially protected natural areas, protective and reserve forests), and the categories of high conservation value forests (hereinafter HCVF), providing protection status under the voluntary forest certification under FSC scheme; so other mechanisms for their protection have to be used. The priority now is to preserve them as the National Forest Heritage of Russia [1].

The aim of this publication is to identify the most significant forest habitats in the the Altai-Sayan Ecoregion, which are not protected by law against deforestation for commercial purposes and adjoin fire-exposed areas and logging areas - they must be recommended to be preserved as the National Forest Heritage of Russia.

Identifying the forest habitats suitable to be included in the National Forest Heritage

To identify the legally unprotected IFL areas in the Altai-Sayan Ecoregion, suitable for including in the NFH, it's necessary to exclude protective forests and specially protected natural areas from the list.

The information about the IFL borders of 2013 was supplied by data portal "Valuable forests of Russia" based on the software product by the engineering and technical centre "SCANEX" GeoMixer and is available at http://forest.kosmosnimki.ru The borders of the protective forests and specially protected natural areas in the Altai-Sayan Ecoregion were identified by means of vectorization of index maps on a scale of 1:100 000 – 1:300 000. The protective forests were divided into the following categories: restricted belts of forests next to water bodies; forests alongside spawning ground; forests in water protective zones; forests in pine-nut harvesting areas; anti-erosion forests; forests located in forest, located in the desert, semi-desert, forest-steppe, forest-tundra zones, steppes, mountains (category 2, 4b, 4c, 4d, 4h and 4i according to Art. 102 of the Forestry Code of the Russian Federation).

After specially protected natural areas and protective forests had been excluded from IFL areas, the forests habitats with no legal protection were identified - so called "uncovered territories". These habitats were subsequently considered most suitable to be included in the National forest heritage. The total area of these habitats in the Russian part of the Altai-Sayan Ecoregion is more than 11.5 million hectares. The half of the total area (about 6.5 million hectares) is located in the Republic of Tuva, vast areas (more than 1 million hectares) are located in the Krasnoyarsk Territory and Irkutsk Region.

Allocating the forest habitats suitable for including in the national forest heritage

For allocating the IFL areas to be included in the National Forest Heritage in the Altai-Sayan Ecoregion, the procedure was developed, taking into account specific features of forest ecosystems.

The basic NFH's criteria are:

- areas are the IFL of Russia;
- forests habitats with a high level of biodiversity identified in the framework of international programmes of the biodiversity areas identification;
- forests of scientific or historical significance (protective forests categories).

Identification of the areas to be included in the NFH was based on the analytic hierarchy process. The 4-level incomplete dominance hierarchy of evaluation criteria was used for this. The following determinant were used as a basis for the evaluation criteria:

1) the value of the territory determined by one of two factors: areas with high levels of biodiversity (habitat and migration of rare animal species, important bird and plant areas), or areas with scientific or historical value;

2) the possible anthropogenic threats to the ecological stability of the territory (fires, deforestation, the implementation of large investment projects).

The significance of the aggregate evaluation criteria was calculated using a paired comparison method. While selecting habitats, the methods of spatial analysis for the geographic information system (hereinafter GIS) were used.

The information about the value of areas was obtained from the literature (regional Red Books, the list of Important Bird and Plant Areas (hereinafter IPAs) in the Altai-Sayan Ecoregion [3], etc.), raster and vector maps of habitats of rare species downloaded from the Internet, raster index maps of the forests of scientific or historical significance.

Using the above stated methods within the "uncovered" IFL in the Altai-Sayan Ecoregion, the areas of high conservation value (habitat and migration of rare animal species, important bird and plant areas) as well as the areas of a scientific / historical significance were identified.

The information about the possible anthropogenic threats has been obtained from remote sensing data (satellite Landsat series, Aqua / Terra, SAS Planet program), from the literature, as well as the official websites data on the planned construction of oil and gas pipelines, railways and roads.

The following types of violations threatening conservation of the IFLs in the Altai-Sayan Ecoregion were identified:

– anthropogenic fires;
– logging;
– planned and existing railways;
– planned oil and gas pipelines;
– mining activities;
– recreational impact.

Interpretation of forest areas-victims of forest fires, logging, road construction and mining activities, is based on the spectral response characteristics of the geological substrate [4,5] obtained by the Landsat satellite data, as well as detected by visual interpretation of high-resolution recording (pixel size of 1-5 m) obtained in SASPlanet system.

Another factor threatening the IFLs is areas accessibility (roads which ensure access to forests, population density). The establishment of the 10 km buffer zone around the highways and railways allowed to mitigate their anthropogenic threat. The establishment of the 20 km buffer zone around most-visited tourist places allowed to mitigate recreational impact on the forest ecosystems. The population density was estimated according to statistics.

The GIS analysis of the location of the factors threatening the IFLs' conservation, as well as the areas with high conservation value, made it possible to identify the IFL territories, which are likely to be lost in the near future and can be recommended for inclusion in the NFH. The total area of the habitats recommended to be included in the National Forest Heritage amounts to 1.912 million hectares.

The identified forest habitats were ranked. The threats and values available in every habitat were taken into account, while ranking. The principles of ranking the areas are in the Table 1.

Table 1.

Rank	Values	Threats	Accessibility (roads, population density)
1	The high density of values (habitats for flagship species, important bird areas (hereinafter IBA), habitats for the Red List species)	The high level of threats (the implementation of large infrastructure projects)	High
2	Values available	Threats available	Lack of roads and the high population density
3	Values available	No threats	Lack of roads and the low population density

12 forest habitats recommended to be included in the NFH were identified in the Altai-Sayan Ecoregion. All of them were divided into three levels of conservation significance (equal to the ranks 1-3 in the table 1).

The field research of the forest habitats suitable to be included in the national forest heritage

Field research was conducted to verify that the identified areas are to be included in the NFH. Taking into account transport access and work schedule, 6 areas in the Krasnoyarsk Territory, the Kemerovo Region and the Republic of Tyva were chosen for the field research.

During the field research the temporary sample plots were established according to the statutory requirements of the Industrial Standard 16128-90 "Sample areas for forest management, the methods of establishing". There is a description for every sample plot:

– General characteristics of a plot (angle of slope, exposure, shape, representativeness);

– Forest valuation characteristics of planting (species composition, average diameter, average height, density, the average age of each species, after-growth, shrub layer);

– Significant objects found at the plots (large tree windfall at a different stage of decomposition; individual dry and dead hard- and softwood, high stumps, broken at different heights (stubs); single large old-growth and non-standard trees, trees with nests and hollows; animal shelter; large anthills; natural solonetzic soils).

As a result of the laboratory study of the data obtained from the field research, it was identified that the plots contain the plantations of natural origin where the coniferous trees (fir, cedar, larch) prevail. General condition of the forest stands is satisfactory - within the established sample plots there are no dead coniferous trees, insects on trunks, heart rot in the taken samples. No any business activity is conducted on the researched plots - they are the significant territories with a high level of the natural biodiversity. The plots are the forest ecosystems capable of self-maintenance for indefinitely long time (there were defined age grades from middle-aged to mature).

The Results

Applying the methodological framework of the NFH concept, the regional methodology of evaluation of the intact forest habitats environmental significance was developed, considering conservation of forest biodiversity. The 4-level incomplete dominance hierarchy of evaluation criteria was applied; it allows to rank the habitats for biodiversity conservation.

According to the developed regional methodology, the IFL areas in the Altai-Sayan Ecoregion were identified and ranked for inclusion in the National Forest Heritage. The vector polygon layer of the potential areas to be included in the NFH was prepared with all the necessary attributive information.

The field research allows to prove the applicability of the methodology we developed for the identification the NFH habitats in the Altai-Sayan Ecoregion. It was identified that the plots chosen for research consist of the plantations of natural origin where the coniferous trees prevail. The plots are the forest ecosystems capable of self-maintenance for indefinitely long time (the age grades from middle-aged to mature) and the significant territories with a high level of the natural biodiversity.

The research was financially supported by the Partnership WWF-IKEA project (phase V: 2014-2017).

IOP Conf. Series: Materials Science and Engineering **155** (2016) 012044 doi:10.1088/1757-899X/155/1/012044

References

[1] Russian National Standard FSC - FSC-STD-RUS-V6-1-2012 Russia Natural and Plantations EN.

[2] Aksenov D, Kobyakov K, Shmatkov N, Yaroshenko A Conceptual approaches to the development of the Russian Federation National Forest Heritage *Journal of Sustainable forest management* №**3** (42) Moscow.:2015

[3] Key botanical territories of the Altai-Sayan ecoregion: experience the selection // I.A. Artemov, A.Yu. Koroljuk, N.N. Laschinsky etc .; under the Ed. I.E. Smelyanskiy, G.A. Pronkina - Novosibirsk: Publishing House of the SB RAS, 2009 y., 272 pages.

[4] Kronberg P, remote study of the Earth. - M .: Mir, 1988. - 350 p.

[5] Davis Sh, Landgrebe D A, TL . Phillips and other remote sensing: a quantitative approach. M .: Nedra, 1983, 415 p.